Robust Mixed Model Analysis

Robust Mixed Model Analysis

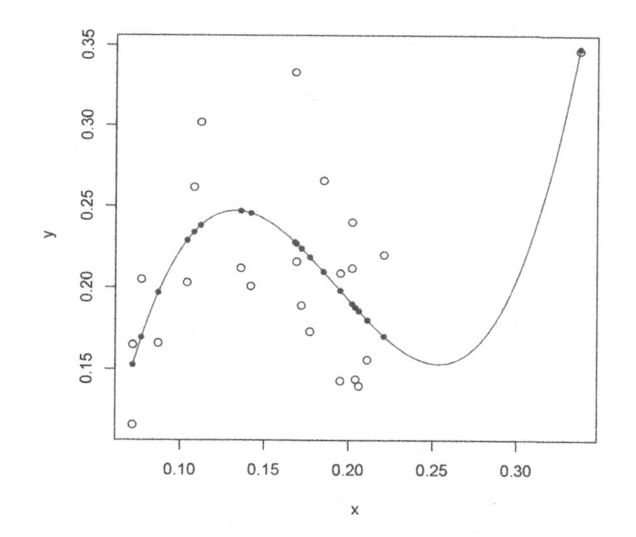

Jiming Jiang
University of California, Davis, USA

World Scientific

EW JERSEY • LONDON • SINGAPORE • BEIJING • SHANGHAI • HONG KONG • TAIPEI • CHENNAI • TOKYO

Published by

World Scientific Publishing Co. Pte. Ltd.

5 Toh Tuck Link, Singapore 596224

USA office: 27 Warren Street, Suite 401-402, Hackensack, NJ 07601

UK office: 57 Shelton Street, Covent Garden, London WC2H 9HE

Library of Congress Cataloging-in-Publication Data
Names: Jiang, Jiming, author.
Title: Robust mixed model analysis / by Jiming Jiang
(University of California, Davis, USA).
Description: New Jersey : World Scientific, 2019. |
Includes bibliographical references and index.
Identifiers: LCCN 2019004891 | ISBN 9789814733830 (hardcover : alk. paper)
Subjects: LCSH: Multilevel models (Statistics)--Problems, exercises, etc. |
Linear models (Statistics)--Problems, exercises, etc. |
Mathematical models--Problems, exercises, etc.
Classification: LCC QA278 .J53 2019 | DDC 519.5/36--dc23
LC record available at https://lccn.loc.gov/2019004891

British Library Cataloguing-in-Publication Data
A catalogue record for this book is available from the British Library.

For any available supplementary material, please visit
https://www.worldscientific.com/worldscibooks/10.1142/9888#t=suppl

Printed in Singapore

To my mother-in-law, Hoàng Thị Vui, and the memory of my father-in-law, Nguyễn Thư, with love

Preface

Over the past decade or so the current author has published a book and a monograph covering topics closely related to mixed effects models. The book, *Linear and Generalized Linear Mixed Models and Their Applications* (Springer 2007), covers two important classes of mixed effects models, namely, linear mixed models (LMMs) and generalized linear mixed models (GLMMs). The monograph, *Asymptotic Analysis of Mixed Effects Models: Theory, Application, and Open Problems* (Chapman & Hall 2017), focuses on asymptotic techniques used in mixed effects models, which include LMMs, GLMMs as well as nonlinear mixed effects models. Having had extensive coverage on mixed effects models in these documents, a question that naturally arises is whether there is necessity for a third one that is, once again, related to the same topics.

The answer to the question is clearly yes. These models are broadly used in practice, especially in situations where the observations are correlated (e.g., medical studies), or where the primary interest is at subject level (e.g., precision medicine), or where variance and covariance are of main interest (e.g., genetics). On the other hand, most mixed effects models are highly parametric relying on specific assumptions not only about the mean functions but also about distributions of the random effects, which are unobservable, adding to the difficulty of checking these assumptions. Furthermore, standard methods of inference about mixed effects models may be sensitive to the impact of unusual data points, sometimes referred to as outliers. It is important to know whether a method of mixed model analysis is robust; if not, what can be done to modify the procedure, or if there are other options, in order for the method to be robust. Here, for the most part, robustness means that a procedure of analysis is not sensitive, or less sensitive, to the impact of something unexpected, if it happens. The current focus is thus on the robustness aspects of the current state-of-the-art mixed model analysis.

The 2007 book is more of a method-application approach; the 2017 monograph is more leaning toward theory. In a way, the current approach is similar to the 2007 book, but with an increased emphasis on applications. Each chapter is supported by a number of real-data applications, or illustrations, some of which would be revisited multiple times in different contexts.

Furthermore, each chapter is supplemented by a number of exercises. The exercises are mostly related to the materials covered in the chapter, and used to help with understanding. In this regard, the current approach is also similar to the 2007 book. Overall, the layout is between a book and a monograph. It may be used as a text for a graduate-level course covering mixed effects models and, in this regard, it has all of the main elements of LMMs and GLMMs. Alternatively, it should also serve as a comprehensive reference on robust mixed model analysis, covering methods, theory and applications.

The author thanks his collaborators, Partha Lahiri, Yihui Luan, Thuan Nguyen, J. Sunil Rao, Hanmei Sun, Mahmoud Torabi, and You-Gan Wang for allowing him to include parts of their research work in this book/monograph. Finally, the author is grateful to his former and current Ph.D. students, Cecilia Dao, Rohosen Bandyopadhyay, and Xiaoyan (Lucia) Liu for their technical supports during the three-year period when the author was working on the book project.

Jiming Jiang
Davis, California
November 2018

Contents

Chapter 1

Introduction

Mixed effects models, or mixed models, have had a wide-ranging impact in modern applied statistics. See, for example, Jiang (2007), McCulloch *et al.* (2008), Demidenko (2013). These models are characterized by random effects that are involved in a certain way. Depending on how the random effects are involved, a mixed model may be classified as a linear mixed model (LMM), generalized linear mixed model (GLMM), non-linear mixed model (NLMM), semi-parametric mixed model (SPMM), or non-parametric mixed model (NAMMA). Due to their broad applications, there has been a growing interest in learning about these models. On the other hand, some of these models, such as LMM and GLMM, are highly parametric, involving distributional assumptions that may not be satisfied in real-life problems. It is important to make sure that a procedure of statistical analysis is robust against violation of the assumptions, or some "bad cases". Here the word *robustness* is formally introduced, for the first time, and we need to make clear what it means. Generally speaking, it means the ability of a procedure to survive some unexpected situations. Life is full of surprises. The unexpected situation could be a violation of an assumption, an outlying observation, or something else. Luckily for the practitioners, there is a rich collection of methods that are currently available and robust, in certain ways. As an introductory example, consider the following.

1.1 Illustrative example

A special class of LMM is the mixed ANOVA model [e.g., Jiang (2007)], which can be expressed as

$$y = X\beta + Z_1\alpha_1 + \cdots + Z_s\alpha_s + \epsilon, \tag{1.1}$$

where X is a known matrix of covariates, β is a vector of unknown parameters (the fixed effects), $Z_1, \dots, Z_s$ are known matrices, $\alpha_1, \dots, \alpha_s$ are vectors of random effects, and ϵ is a vector of errors. The standard assumption assumes that $\alpha_r \sim N(0, \sigma_r^2 I_{m_r})$, where σ_r^2 is an unknown variance, $1 \le r \le s$, and I_n denotes the $n \times n$ identity matrix, and $\epsilon \sim N(0, \tau^2 I_N)$, and that $\alpha_1, \dots, \alpha_s, \epsilon$ are independent. Under such assumptions, restricted maximum likelihood (REML) estimators of the fixed effects, β, and variance components, $\sigma_r^2, 1 \le r \le s$ and τ^2, can be derived. Namely, let A be an $N \times (N-p)$ matrix, where $p = \text{rank}(X)$, such that

$$\text{rank}(A) = N - p \text{ and } A'X = 0. \tag{1.2}$$

The REML estimator of $\psi = (\sigma_1^2, \dots, \sigma_s^2, \tau^2)'$ is the maximum likelihood (ML) estimator of ψ based on $z = A'y$, whose distribution does not depend on β. Once the REML estimator of ψ is obtained, say, $\hat{\psi}$, the REML estimator of β is given by

$$\hat{\beta} = (X'\hat{V}^{-1}X)^{-1}X'\hat{V}^{-1}y, \tag{1.3}$$

where $\hat{V} = \hat{\tau}^2 I_N + \sum_{r=1}^s \hat{\sigma}_r^2 Z_r Z_r'$. It can be shown (Exercise 1.1) that the REML estimator, $\hat{\psi}$, is a solution to the following REML equations:

$$\begin{cases} y'P^2y = \text{tr}(P), \\ y'PZ_rZ_r'Py = \text{tr}(PZ_rZ_r'), \quad 1 \le r \le s, \end{cases} \tag{1.4}$$

where $P = V^{-1} - V^{-1}X(X'V^{-1}X)^{-1}X'V^{-1}$ and V is $\hat{V}$ with $\hat{\sigma}_r^2, 1 \le r \le s$ and $\hat{\tau}^2$ replaced by $\sigma_r^2, 1 \le r \le s$ and τ^2, respectively. Here we assume, for simplicity, that X is full rank.

The point to be made is that the REML equations (1.4), which are derived under the normality assumption, can be derived under a completely different distribution. In fact, exactly the same equations will arise if the distribution of y is assumed to be multivariate-t with the probability density function (pdf)

$$p(y) = \frac{\Gamma\{(n+d)/2\}}{(d\pi)^{n/2}\Gamma(d/2)|V|^{1/2}} \times \left\{1 + \frac{1}{d}(y - X\beta)'V^{-1}(y - X\beta)\right\}^{-(n+d)/2}, \tag{1.5}$$

where d is the degree of freedom of the multivariate t-distribution and $|V|$ denotes the determinant of V. Note that the multivariate normal distribution may be viewed as the limiting distribution of the multivariate-t as $d \to \infty$ (Exercise 1.2). An implication is that the Gaussian REML estimator, that is, REML estimator derived under the normality assumption, is still valid even if the normality assumption is violated in that the actual distribution is multivariate-t. The latter is known to have heavier tails than the multivariate normal distribution.

A further question is how far can one go in violating the normality assumption. For example, what if the actual distribution is unknown? We shall address the issue more systematically later.

1.2 Outline of approaches to robust mixed model analysis

The simplest way of doing something is not to do anything at all.

Not exactly.

Here by not doing anything it merely means that there is no need to make any changes in the existing procedure to make it robust. Still, one has to justify the robustness of the existing procedure. The justification can be by exact derivation, as in Section 1.1 (also Exercise 1.2), by asymptotic arguments, or by empirical studies, such as Monte-Carlo simulations and real-data applications. This is what we call the first approach.

The second approach is to build a procedure of mixed model analysis on weaker assumptions. For example, quasi-likelihood methods avoid full specification of the likelihood function by using estimating functions, or estimating equations. The latter rely only on specification of moments, typically the first two moments. Generally speaking, the more assumptions one makes, the more likely some of these assumptions will be violated. Conversely, a procedure built on weaker assumptions is likely to be more robust than one built on stronger assumptions. Another well-known example is the least squares (LS) method. Under the standard linear regression model, which can be viewed as (1.1) without the term $Z_1\alpha_1 + \cdots + Z_s\alpha_s$, the ML estimator for β is given by $\hat{\beta} = (X'X)^{-1}X'y$, assuming, again, that X is full rank. However, the same estimator can be derived from a seemingly different principle, the least squares (LS), which does not use the normality assumption at all. In fact, the Gaussian ML estimator remains consistent even if the normality fails [e.g., Jiang (2010), sec. 6.7] and, in this sense, the ML estimation is robust. However, the ML, or LS, estimator has another problem: It is not robust to outliers. This brings up the next approach.

The third approach, which is often considered when dealing with outliers, is to robustify an existing method to make it more robust. Here by robustification it means to modify some part of the current procedure with the intention of robustness. For example, generalized estimating equations

(GEE) has been used in the analysis of longitudinal data [e.g., Diggle *et al.* (2002); see Chapter 2 below]. However, there has been concerns that the GEE may not be robust to outlying observations. One way to robustify the GEE is to modify the definition of the residuals so that it has a bounded range, thus reducing the influence of an outlier.

The next approach is to go semi-parametric, or non-parametric. These models are more flexible, or less restrictive, in some ways so that the chance of model misspecification is (greatly) reduced. For example, instead of assuming a linear mixed-effects function, as in LMM, one may assume that the mean function, conditional on the random effects, is an unknown, smooth function. The latter, of course, includes the linear function as a special case so, when the LMM holds, the non-parametric approach would lose some efficiency. However, a trade-off is robustness–the non-parametric mean function would still be valid when the linear function fails. In other words, the mean function would remain correctly specified under the non-parametric approach, even if it is misspecified under the LMM.

Finally, it is always a good practice to carefully choose, and check, the proposed statistical model. This way, the chance, or extent, of model misspecification is likely to be reduced. It is in this sense that mixed model selection and diagnostics are important techniques of robust mixed model analysis. It should be noted that model selection in the context of mixed effects models is not a conventional model selection problem in that the effective sample size is not clear due to correlations among the observations. Thus, for example, a standard information criterion, such as BIC [Schwarz (1978)], would encounter some difficulties due to its dependence on the effective sample size.

1.3 A roadmap

We have outlined a number of main approaches to robust mixed model analysis. However, when it comes to covering these approaches, we have our own strategies. What we are going to do is to first focus on special types of analyses, such as estimation (Chapters 2 and 3), tests (Chapter 4), and mixed model prediction (Chapter 5), where methods of robust mixed model analysis have been developed. In particular, mixed model prediction is one of the characteristics of mixed model analysis, because estimation and testing problems, of course, also appear in other fields of statistics.

After these special topics, the next chapter is devoted to model selection.

Here, following Müller *et al.* (2013), we consider three classes of strategies for mixed model selection, generalized information criteria, the fence methods, and shrinkage mixed model selection.

There are many more topics, methods, or aspects robust mixed model analysis. As a final chapter, we have put together a collection of such topics, including mixed model diagnostics, nonparametric and semiparametric methods, Bayesian analysis, outliers, benchmarking, and further topics on prediction.

It should be noted that there are plenty of intersections of these topics, no matter how one would classify them, that it turns out to be a difficult task to determine which chapter, and section, should be the (main) home of each topic. We have tried our best to make sure that everyone has a home and, hopefully, is happy about its home. Nevertheless, the coverage of relevant topics is not inclusive and there are, unavoidably, some missing topics that are potentially important. The author apologizes for overlooking such work, and would strongly encourage relevant authors to contact him so that the missing work can be included in the next edition.

In each one of these chapters, one or more of the approaches summarized in Section 1.2 will be discussed in detail. It would be helpful to keep this in mind. Of course, the approaches outlined in Section 1.2 only give the "big pictures". Every problem has its specialty, as is often the case.

Note on notation: In probability theory or mathematical statistics, it is customary to use capital letters, such as Y_i, for a random variable, and lower case letters, such as y_i, for an observed, or realized, value of Y_i. Such a distinction is rarely important, however, in applied statistics. We follow the latter convention for notation simplicity, which usually causes no confusion. So, for example, throughout this monograph, y_i represents both a random variable and the observed value of the random variable, depending on the context or occasion.

1.4 Exercises

1.1. Derive the REML equations (1.4). [Hint: First show that the expression for P given below (1.4) is identical to $A(A'VA)^{-1}A'$.] Also show that the REML equations do not depend on the choice of A. In other words, any matrix A satisfying (1.2) results in the same REML equations.

1.2. This exercise is related to the multivariate t-distribution whose pdf is given by (1.5).

a. Show that the multivariate normal distribution is the limit of the multivariate-t as $d \to \infty$.

b. Show that the multivariate t-distribution (1.5) leads to the same REML equation (1.4). More specifically, if one derives the REML estimator following the same procedure as in Exercise 1.1 except with the multivariate normal distribution of y replaced by the multivariate t-distribution, the resulting REML equations are equivalent to (1.4).

Chapter 2

Generalized Estimating Equations

2.1 Method of moments

The method of moments (MM) is thought to be the oldest method of finding point estimators [Casella and Berger (2002), pp. 312]. The method, in its original form, is based on an assumption that the distribution of an observation, y_i, depends on a vector θ of unknown parameters. In order to estimate θ, certain moment conditions are required. Namely, let p be the dimension of θ. It is assumed that $y_1, \dots, y_n$ are independent such that $\mathrm{E}(|y_i|^p) < \infty$. Furthermore, it is assumed that $\mathrm{E}(y_i^k) = \psi_k(\theta), 1 \le i \le n$, where $\psi_k(\cdot)$ is a known function, $1 \le k \le p$. Note that the (pth absolute) moment condition imposed implies that all of the moments involved are finite. Also, the kth moment of y_i ($1 \le k \le p$) does not depend on i. The idea is to equal the up to pth sample moments of the observations to its expected value, that is, $n^{-1}\sum_{i=1}^n y_i^k = \mathrm{E}(n^{-1}\sum_{i=1}^n y_i^k)$, that is,

$$\frac{1}{n}\sum_{i=1}^n y_i^k = \psi_k(\theta), \quad 1 \le k \le p. \tag{2.1}$$

The θ on the right side of (2.1) is supposed to be the true parameter vector. Because θ is unknown, (2.1) is treated as an equation system for solving θ, in which case the latter is treated as a (vector-valued) variable. Note that there are as many equations in (2.1) as the dimension of θ; therefore, one expects to find a solution to (2.1), which is called the MM estimator (MME). Typically, regularity conditions are needed to ensure the existence, and uniqueness, of the MME [e.g., Jiang (1998a)].

The basic idea can be generalized, in several ways. For example, in econometrics, a procedure call generalized method of moments (GMM) is often used [Hansen (1982)]. Let $g(\cdot, \cdot)$ be a vector-valued function satisfying $\mathrm{E}\{g(y_i, \theta)\} = 0$ for every i when θ is the true parameter vector. The generalized sample moments is defined as $m(y;\theta) = n^{-1}\sum_{i=1}^{n} g(y_i, \theta)$. The GMM estimator of θ is obtained by equating $m(y;\theta)$ to its expected value when θ is the true parameter vector, that is, by solving

$$m(y;\theta) = 0. \tag{2.2}$$

It is clear that the MM is a special case of GMM (Exercise 2.1), hence justifying the term GMM.

Another variation of the MM is the method of simulated moments [MSM; e.g., McFadden (1989)]. In some cases, such as under a generalized linear mixed model [GLMM; e.g., Jiang (2007)], the moments have no analytic expressions in terms of θ. We illustrate with an example.

Example 2.1. Suppose that, given the random effects, $\alpha_1, \dots, \alpha_m$, $y_{ij}, 1 \le i \le m, 1 \le j \le k$ are binary outcomes that are (conditionally) independent with $\text{logit}\{\text{P}(y_{ij} = 1|\alpha)\} = \mu + \alpha_i$, where $\alpha = (\alpha_i)_{1 \le i \le m}$, and $\text{logit}(p) = \log\{p/(1-p)\}$. Furthermore, suppose that the random effects are independent and distributed as $N(0, \sigma^2)$. Here, μ, σ are unknown parameters with $\sigma \ge 0$, so let $\theta = (\mu, \sigma)'$. It is more convenient to use the following expression: $\alpha_i = \sigma\xi_i$, $1 \le i \le m$, where $\xi_1, \dots, \xi_m$ are i.i.d. $N(0,1)$ random variables. Consider the following MM equations for estimating θ,

$$\frac{1}{m}\sum_{i=1}^{m} y_{i\cdot} = \text{E}(y_{1\cdot}), \tag{2.3}$$

$$\frac{1}{m}\sum_{i=1}^{m} y_{i\cdot}^2 = \text{E}(y_{1\cdot}^2), \tag{2.4}$$

where $y_{i\cdot} = \sum_{j=1}^{k} y_{ij}$. Note that $y_{i\cdot}, 1 \le i \le m$ are i.i.d. Furthermore, it is easy to show that $\text{E}(y_{1\cdot}) = k\text{E}\{h_\theta(\xi)\}$ and $\text{E}\{y_{1\cdot}^2\} = k\text{E}\{h_\theta(\xi)\} + k(k-1)\text{E}\{h_\theta^2(\xi)\}$, where $h_\theta(x) = \exp(\mu + \sigma x)/\{1 + \exp(\mu + \sigma x)\}$ and $\xi \sim N(0,1)$ (Exercise 2.2).

Situations like Example 2.1 are, actually, typical in GLMM. It is well known that computation is a major issue in inference about GLMMs [e.g., Torabi (2012)]. More specifically, under a GLMM, the likelihood function typically does not have an analytic expression; even more, it may involve high-dimensional integrals, which makes numerical evaluations difficult. For example, consider the following.

Example 2.2. Suppose that, given the random effects $u_1, \dots, u_{m_1}$ and $v_1, \dots, v_{m_2}$, binary responses y_{ij}, $i = 1, \dots, m_1$, $j = 1, \dots, m_2$ are conditionally independent such that, with $p_{ij} = \text{P}(y_{ij} = 1|u, v)$, we have

$\text{logit}(p_{ij}) = \mu + u_i + v_j$, where μ is an unknown parameter, $u = (u_i)_{1 \le i \le m_1}$, and $v = (v_j)_{1 \le j \le m_2}$. Furthermore, the random effects $u_1, \ldots, u_{m_1}$ and $v_1, \ldots, v_{m_2}$ are independent such that $u_i \sim N(0, \sigma_1^2)$, $v_j \sim N(0, \sigma_2^2)$, where the variances σ_1^2 and σ_2^2 are unknown. Thus, the unknown parameters involved are $\theta = (\mu, \sigma_1, \sigma_2)'$ with $\sigma_1, \sigma_2 \ge 0$. It can be shown (Exercise 2.3) that the likelihood function for estimating θ can be expressed as

$$\begin{aligned} & c - \frac{m_1}{2}\log(\sigma_1^2) - \frac{m_2}{2}\log(\sigma_2^2) + \mu y_{\cdot\cdot} \\ & + \log \int \cdots \int \left[\prod_{i=1}^{m_1} \prod_{j=1}^{m_2} \{1 + \exp(\mu + u_i + v_j)\}^{-1} \right] \\ & \times \exp\left(\sum_{i=1}^{m_1} u_i y_{i\cdot} + \sum_{j=1}^{m_2} v_j y_{\cdot j} - \frac{1}{2\sigma_1^2} \sum_{i=1}^{m_1} u_i^2 - \frac{1}{2\sigma_2^2} \sum_{j=1}^{m_2} v_j^2 \right) \\ & du_1 \cdots du_{m_1} dv_1 \cdots dv_{m_2}, \end{aligned} \tag{2.5}$$

where c is a constant, $y_{\cdot\cdot} = \sum_{i=1}^{m_1} \sum_{j=1}^{m_2} y_{ij}$, $y_{i\cdot} = \sum_{j=1}^{m_2} y_{ij}$, and $y_{\cdot j} = \sum_{i=1}^{m_1} y_{ij}$. The multidimensional integral involved in (2.5) has no closed-form expression, and it cannot be further simplified. Furthermore, such an integral is difficult to evaluate even numerically. For example, if $m = n = 40$, the dimension of the integral will be 80. To make it even worse, the integrand involves a product of 1600 terms with each term less than one. This makes it almost impossible to evaluate the integral using a naive Monte Carlo method. To see this, suppose that $u_1, \ldots, u_{40}$ and $v_1, \ldots, v_{40}$ are simulated random effects (from the normal distributions). Then, the product in the integrand (with $m = n = 40$) is numerically zero. Therefore, numerically, the law of large numbers, which is the basis of the (naive) Monte Carlo method, will not yield anything but zero without a huge Monte Carlo sample size. The situation is quite different, however, if one considers, instead of the likelihood, moments of the $y_{i\cdot}$ and $y_{\cdot j}$. For example, it is easy to show that $\text{E}(y_{i\cdot}) = m_2 \text{E}\{h(\mu + \sigma_1 \xi + \sigma_2 \eta)\}$, where $h(x) = e^x/(1 + e^x)$, and ξ, η are independent $N(0, 1)$ (Exercise 2.3). The latter expression only involves a two-dimensional, and the product issue mentioned above does not occur, regardless of the size of m_1 and m_2.

If the moments only involve low-dimensional integrals, such as in Examples 2.1 and 2.2, one can approximate these moments by the corresponding sample moments of generated random variables. The approximation is justified by the law of large numbers [e.g., Jiang (2010), sec. 6.2]. This is the idea of MSM. The method had previously found applications in, for example, econometrics [e.g., McFadden (1989), Lee (1992)].

Table 2.1 **Simulated mean and SE**

		Estimator of μ		Estimator of σ^2	
m	k	Mean	SE	Mean	SE
20	2	0.31	0.52	2.90	3.42
20	6	0.24	0.30	1.12	0.84
80	2	0.18	0.22	1.08	0.83
80	6	0.18	0.14	1.03	0.34

Example 2.1 (continued). To illustrate the MSM, consider, again, Example 2.1. It is more convenient to consider the following equations that are equivalent to (2.3) and (2.4) (verify):

$$\frac{y_{\cdot\cdot}}{mk} = \mathrm{E}\{h_\theta(\xi)\}, \tag{2.6}$$

$$\frac{1}{mk(k-1)}\sum_{i=1}^m (y_{i\cdot}^2 - y_{i\cdot}) = \mathrm{E}\{h_\theta^2(\xi)\}, \tag{2.7}$$

where $\xi \sim N(0,1)$. The next thing we do is to replace the right sides of (2.6) and (2.7) by the corresponding sample moments of random variables generated under the standard normal distribution, resulting

$$\frac{y_{\cdot\cdot}}{mk} = \frac{1}{L}\sum_{l=1}^L h_\theta(\xi_l), \tag{2.8}$$

$$\frac{1}{mk(k-1)}\sum_{i=1}^m (y_{i\cdot}^2 - y_{i\cdot}) = \frac{1}{L}\sum_{l=1}^L h_\theta^2(\xi_l), \tag{2.9}$$

where $\xi_l, 1 \le l \le L$ are independent $N(0,1)$ random variables generated by a computer. The solution to (2.8) and (2.9) are called the MSM estimator. Jiang (1998a) presented results of a simulation study, in which the true μ and σ are 0.2 and 1.0, respectively. The results, based on 1000 simulation runs, are presented in Table 2.1. Here the MSM estimator of σ^2 is simply the square of the MSM estimator of σ. It is seen that the results improve as either m or k increases.

To describe a general procedure for MSM, we assume that the conditional density of y_i given the vector of random effects, α, has the following form of an exponential family,

$$f(y_i|\alpha) = \exp[(w_i/\phi)\{y_i\eta_i - b(\eta_i)\} + c_i(y_i, \phi)], \tag{2.10}$$

where η_i is a linear predictor that can be expressed as

$$\eta_i = x_i'\beta + z_i'\alpha, \tag{2.11}$$

with x_i, z_i being known vector of covariates and design vector, respectively, and β being a vector of unknown parameters (the fixed effects); ϕ is a dispersion parameter, and w_i a known weight. Typically, $w_i = 1$ for ungrouped data; $w_i = k_i$ for grouped data if the response is an average, where k_i is the group size; and $w_i = 1/k_i$ if the response is a group sum. Here $b(\cdot)$ and $c_i(\cdot,\cdot)$ are known functions associated with the generalized linear models [GLM; McCullagh and Nelder (1989); also see Appendix]. Furthermore, we assume that $\alpha = (\alpha_1', \dots, \alpha_q')'$, where α_r is a random vector whose components are independent and distributed as $N(0, \sigma_r^2)$, $1 \le r \le q$. Suppose that $y_i, 1 \le i \le n$ are observed. We assume that $Z = (z_i')_{1\le i\le n} = (Z_1, \dots, Z_q)$ so that $Z\alpha = Z_1\alpha_1 + \cdots + Z_q\alpha_q$. The following expression of α is sometimes more convenient,

$$\alpha = D\xi, \tag{2.12}$$

where D is blockdiagonal with the diagonal blocks $\sigma_r I_{m_r}$, $1 \le r \le q$, and $\xi \sim N(0, I_m)$ with $m = m_1 + \cdots + m_q$. First assume that ϕ is known. Let $\theta = (\beta', \sigma_1, \dots, \sigma_q)'$. Consider an unrestricted parameter space $\theta \in \Theta = R^{p+q}$. This allows computational convenience for using the MSM because, otherwise, there will be constraints on the parameter space. Of course, this raises the issue of identifiability, because both $(\beta', \sigma_1, \dots, \sigma_q)'$ and $(\beta', -\sigma_1, \dots, -\sigma_q)$ correspond to the same model. Nevertheless, it suffices to make sure that β and $\sigma^2 = (\sigma_1^2, \dots, \sigma_q^2)'$ are identifiable. In fact, Jiang (1998a) showed that, under suitable conditions, the MSM estimators of β and σ^2 are consistent; therefore, the latter parameters are, at least, asymptotically identifiable.

The next thing we do is to We first derive a set of sufficient statistics for θ. It can be shown (Exercise 2.4) that the marginal density function of $y = (y_i)_{1\le i\le n}$ can be expressed as

$$L = \int \exp\left\{c + a(y,\phi) + \frac{b(u,\theta)}{\phi} - \frac{|u|^2}{2} + \left(\sum_{i=1}^n w_i x_i y_i\right)'\left(\frac{\beta}{\phi}\right)\right.$$
$$\left. + \left(\sum_{i=1}^n w_i z_i y_i\right)'\left(\frac{D}{\phi}\right)u\right\} du, \tag{2.13}$$

where c is a constant, $a(y,\phi)$ depends only on y and ϕ, and $b(u,\theta)$ depends only on u and θ. It follows that a set of sufficient statistics for θ is given by

$$\begin{aligned}
S_j &= \textstyle\sum_{i=1}^n w_i x_{ij} y_i, \quad 1 \le j \le p,\\
S_{p+l} &= \textstyle\sum_{i=1}^n w_i z_{i1l} y_i, \quad 1 \le l \le m_1,\\
&\vdots\\
S_{p+m_1+\cdots+m_{q-1}+l} &= \textstyle\sum_{i=1}^n w_i z_{iql} y_i, \quad 1 \le l \le m_q,
\end{aligned}$$

where $Z_r = (z_{irl})_{1 \le i \le n, 1 \le l \le m_r}$, $1 \le r \le q$. Thus, a natural set of MM equations can be formulated as

$$\sum_{i=1}^{n} w_i x_{ij} y_i = \sum_{i=1}^{n} w_i x_{ij} \mathrm{E}_\theta(y_i), \quad 1 \le j \le p, \tag{2.14}$$

$$\sum_{l=1}^{m_r} \left(\sum_{i=1}^{n} w_i z_{irl} y_i \right)^2 = \sum_{l=1}^{m_r} \mathrm{E}_\theta \left(\sum_{i=1}^{n} w_i z_{irl} y_i \right)^2, \quad 1 \le r \le q. \tag{2.15}$$

Although the S_js are sufficient statistics for the model parameters only when ϕ is known (which includes the special cases of binomial and Poisson distributions), one may still use Equations (2.14) and (2.15) to estimate θ even if ϕ is unknown, provided that the right-hand sides of these equations do not involve ϕ. Note that the number of equations in (2.14) and (2.15) is identical to the dimension of θ. However, in order for the right sides of (2.14), (2.15) not to depend on ϕ, some changes have to be made.

For simplicity, in the following we assume that Z_r, $1 \le r \le q$ are standard design matrices in the sense that each Z_r consists only of 0s and 1s, and there is exactly one 1 in each row, and at least one 1 in each column. Then, if we denote the ith row of Z_r by $z'_{ir} = (z_{irl})'_{1 \le l \le m_r}$, we have $|z_{ir}|^2 = 1$ and, for $s \ne t$, $z'_{sr} z_{tr} = 0$ or 1. Let $I_r = \{(s,t) : 1 \le s \ne t \le n, z'_{sr} z_{tr} = 1\} = \{(s,t) : 1 \le s \ne t \le n, z_{sr} = z_{tr}\}$. Then, it can be shown (Exercise 2.5) that

$$\sum_{l=1}^{m_r} \mathrm{E}_\theta \left(\sum_{i=1}^{n} w_i z_{irl} y_i \right)^2$$
$$= \sum_{i=1}^{n} w_i^2 \mathrm{E}_\theta(y_i^2) + \sum_{(s,t) \in I_r} w_s w_t \mathrm{E}(y_s y_t). \tag{2.16}$$

It is seen that the first term on the right side of (2.16) depends on ϕ, and the second term does not depend on ϕ (Exercise 2.5). Therefore, a simple modification of the earlier MM equations that will eliminate ϕ would be to replace (2.15) by the following equations,

$$\sum_{(s,t) \in I_r} w_s w_t y_s y_t = \sum_{(s,t) \in I_r} w_s w_t \mathrm{E}_\theta(y_s y_t), \qquad 1 \le r \le q. \tag{2.17}$$

Furthermore, write $\xi = (\xi'_1, \dots, \xi'_q)'$ with $\xi_r = (\xi_{rl})_{1 \le l \le m_r}$. Note that $\xi_r \sim N(0, I_{m_r})$. Then, the right side of (2.14) can be expressed as

$$X'_j W \mathrm{E}\{e(\theta, \xi)\},$$

where X_j is the jth column of X, $W = \text{diag}(w_i, 1 \le i \le n)$, and $e(\theta, \xi) = \{b'(\eta_i)\}_{1 \le i \le n}$ with $\eta_i = \sum_{j=1}^p x_{ij}\beta_j + \sum_{r=1}^q \sigma_r z'_{ir}\xi_r$ (Exercise 2.5). Similarly, the right side of (2.17) can be expressed as

$$\mathrm{E}\{e(\theta, \xi)'WH_rWe(\theta, \xi)\},$$

where H_r is the $n \times n$ symmetric matrix whose (s,t) entry is $1_{\{(s,t) \in I_r\}}$. Thus, the final MM equations that do not involve ϕ are given by

$$\sum_{i=1}^n w_i x_{ij} y_i = X'_j W \mathrm{E}\{e(\theta, \xi)\}, \qquad 1 \le j \le p, \tag{2.18}$$

$$\sum_{(s,t) \in I_r} w_s w_t y_s y_t = \mathrm{E}\{e(\theta, \xi)'WH_rWe(\theta, \xi)\}, \qquad 1 \le r \le q, \tag{2.19}$$

where the expectations on the right sides are with respect to $\xi \sim N(0, I_m)$. The next thing we do is to approximate the right sides of (2.18) and (2.19) by Monte Carlo simulation. Let $\xi^{(1)}, \dots, \xi^{(L)}$ be i.i.d. copies of ξ generated by a computer. Then, the right sides of (2.18), (2.19) are approximated by

$$X'_j W \left[\frac{1}{L} \sum_{l=1}^L e\{\theta, \xi^{(l)}\} \right], \qquad 1 \le j \le p, \tag{2.20}$$

$$\frac{1}{L} \sum_{l=1}^L e\{\theta, \xi^{(l)}\}'WH_rWe\{\theta, \xi^{(l)}\}, \qquad 1 \le r \le q, \tag{2.21}$$

respectively. In conclusion, the MSM equations for estimating θ are (2.18) and (2.19) with the right sides approximated (2.20) and (2.21), respectively.

We have shown how the MSM works in estimation of the parameters under a GLMM, but, so far, we have not discussed the connection between MSM and robustness. Note that, unlike the maximum likelihood (ML) method, the MM only involves assumptions about the moments. For example, in (2.14) and (2.17), only the conditional first moments of the data, y_i, given the random effects, need to be specified (Exercise 2.6). It is possible that the conditional distribution of the data given the random effects does not satisfy (2.10) and (2.11), but the conditional first moments are correctly specified. It is in this sense that the MM (or MSM) method is potentially more robust than the ML method. We consider an example.

Example 2.3. (Beta-binomial) If $Y_1, \dots, Y_k$ are correlated Bernoulli random variables, the distribution of $Y = Y_1 + \cdots + Y_k$ is not binomial; in fact, it may not belong to the exponential family. Here we consider a special case. Let p be a random variable with a beta$(\pi, 1 - \pi)$ distribution, where $0 < \pi < 1$. Suppose that, given p, $Y_1, \dots, Y_k$ are independent Bernoulli(p)

random variables, so that $Y|p \sim$ binomial(k,p). Then, it can be shown (Exercise 2.7) that the marginal distribution of Y is given by

$$P(Y=j) = \frac{\Gamma(j+\pi)\Gamma(k-j+1-\pi)}{j!(k-j)!\Gamma(\pi)\Gamma(1-\pi)}, \qquad 1 \le j \le k. \tag{2.22}$$

This distribution is called beta-binomial(k,π). It follows that $E(Y) = k\pi$ (Exercise 2.7). It is seen that the mean under beta-binomial(k,π) is the same as that under the binomial(k,π) distribution. Now suppose that, given the random effects $\alpha_1,\dots,\alpha_m$, the y_{ij}'s are not conditionally independent; hence, one does not have a GLMM. More specifically, the conditional distribution of $y_{i\cdot}$ is beta-binomial(k,π_i), with logit$(\pi_i) = \mu + \alpha_i$, instead of binomial(k,π_i), as in Example 2.1. It follows that the mean on the right side of (2.3) is still given by $kE\{h_\theta(\xi)\}$, as shown in Exercise 2.2.

As long as the mean functions on the right sides of (2.18) and (2.19) are correctly specified, the MSM estimators are consistent under regularity conditions [Jiang (1998a)], even if a key part of the GLMM assumption, that is, the conditional distribution (2.10), fails. We shall return to this later.

2.2 Generalized estimating equations (GEE)

We have discussed MM and MSM in GLMMs. These methods are robust in the sense that even if part of the distributional assumption fails, the MM or MSM estimators may still be consistent. On the other hand, the MM or MSM estimators may be inefficient in that the variance of the estimator may be large, especially under small or moderate sample size. This can be seen, from example, from Table 2.1. The SE for the MSM estimator of σ^2 is quite large when $m=20, k=2$, although it drops quickly when either m or k increases.

One way to improve the efficiency of MM (MSM) is to explore the estimating equations, say, (2.14) and (2.15). Note that the latter equations are constructed by simply adding some of the sufficient statistics, and the squares of the others, derived below (2.13). This seems a bit arbitrary. To look at the issue more clearly, Suppose that there is a vector of what we call *base statistics*, say, S, which typically is of higher dimension that θ. Let the dimension of S and θ be N and r, respectively. One can construct a set of MM equations by letting

$$BS = Bu(\theta), \tag{2.23}$$

where $u(\theta) = \mathrm{E}_\theta(S)$, the expectation of S when θ is the true parameter vector, and B is any $r \times N$ constant matrix. The dimension of B is chosen so that there are exactly as many equations in (2.23) as the dimension of θ so that, hopefully, there is a unique solution to the MM equation. But there are (infinitely) many choice of B. For example, equations (2.14) and (2.15) correspond to one particular choice of B (Exercise 2.8). The question is why the B has to be chosen this way. Is there a better choice, or, perhaps, the best choice of B?

To explore the issue about the optimality of B, let us first carry out a heuristic derivation of the asymptotic covariance matrix (ACVM) of $\hat{\theta}$, the MM estimator defined as a solution to (2.23). Let $M(\theta)$ denote the difference between the two sides of (2.23), that is, $M(\theta) = B\{S - u(\theta)\}$. By the Taylor series expansion, and recalling that $\hat{\theta}$ satisfies (2.23), we have

$$\begin{aligned} 0 &= M(\hat{\theta}) \\ &\approx M(\theta) + \frac{\partial M}{\partial \theta'}(\hat{\theta} - \theta) \\ &\approx M(\theta) + \mathrm{E}\left(\frac{\partial M}{\partial \theta'}\right)(\hat{\theta} - \theta), \end{aligned} \tag{2.24}$$

where for $M = (m_j)_{1 \le j \le r}$ and $\theta = (\theta_k)_{1 \le k \le r}$, $\partial M/\partial \theta'$ is the matrix whose (j, k) entry is $\partial m_j/\partial \theta_k$, θ is the true parameter vector, at which $\partial M/\partial \theta'$ is evaluated. The first approximation in (2.23) is due to the Taylor expansion; the second approximation holds under regularity conditions [e.g., Jiang (2010), sec. 4.7]. If we denote $M(\theta)$ by M, for notation simplicity, then from (2.24) we arrive at the approximation

$$\hat{\theta} - \theta \approx -\left\{\mathrm{E}\left(\frac{\partial M}{\partial \theta'}\right)\right\}^{-1} M; \tag{2.25}$$

hence, one would expect

$$\mathrm{Var}(\hat{\theta}) \approx \left\{\mathrm{E}\left(\frac{\partial M}{\partial \theta'}\right)\right\}^{-1} \mathrm{Var}(M) \left\{\mathrm{E}\left(\frac{\partial M'}{\partial \theta}\right)\right\}^{-1}, \tag{2.26}$$

where $\partial M'/\partial \theta = (\partial M/\partial \theta')'$, and Var denotes the covariance matrix. It should be noted that the above derivation is not completely rigorous–some regularity conditions are needed for the result to hold, but nevertheless trailblazing for finding the optimal estimating equation. To see this, all we have to do is to specify the terms involved in (2.26), and note that $\partial M/\partial \theta' = -B(\partial u/\partial \theta')$ and $\mathrm{Var}(M) = B\mathrm{Var}(S)B'$. It follows that

$$\mathrm{Var}(\hat{\theta}) \approx \{(BU)^{-1}\}BVB'\{(BU)^{-1}\}', \tag{2.27}$$

where $V = \text{Var}(S)$ and $U = \partial u/\partial\theta'$. It can be shown (Exercise 2.9) that the optimal B in the sense of minimizing the right side of (2.27), according to the partial ordering of symmetric matrices, is $B = U'V^{-1}$. Here for two symmetric matrices, A and B, $A \geq B$ if $A - B$ is nonnegative definite (or positive semidefinite), denoted by $A - B \geq 0$. Thus, at least theoretically, the optimal estimating equation, or the best MM equation, is (2.23) with $B = U'V^{-1}$, that is

$$U'V^{-1}\{S - u(\theta)\} = 0. \tag{2.28}$$

A special form of this optimal estimating equation has been used by Liang and Zeger (1986) for the analysis of longitudinal data. Suppose that data are collected from a number of subjects (e.g., patients) over time. Let y_{it} denote the response collected from the ith subject at time t, $i = 1, \dots, m$, $t \in T_i$, where T_i is the set of observational times for the ith subject, which may be subject-dependent. Let $y_i = (y_{it})_{t \in T_i}$ denote the vector of responses from the ith subject, and $\mu_i = (\mu_{it})_{t \in T_i}$ denote the vector of means, for the ith subject. It is assumed that the mean is associated with a vector of unknown parameters, θ, that is, $\mu_i = \mu_i(\theta)$. It is also assumed that the responses from different subjects, $y_1, \dots, y_m$, are independent. Consider $S = y = (y_i)_{1 \leq i \leq m}$ in (2.28), then $u = \mu$. The optimal estimating equation (2.28) can now be expressed as (Exercise 2.10)

$$\sum_{i=1}^{m} \frac{\partial \mu_i'}{\partial \theta} V_i^{-1}(y_i - \mu_i) = 0. \tag{2.29}$$

(2.29) is widely known as the generalized estimating equation [GEE; e.g., Diggle *et al.* (2002)]. We consider some examples.

Example 2.4. (Best linear unbiased estimation) The simplest model for the mean function would be a linear model, in which case $\mu_i = X_i\beta$, where X_i is a matrix of known covariates associated with the ith subject, and β is a vector of unknown regression coefficients. Note that some of the covariates may be time-varying. Thus, in this case, we have $\theta = \beta$, and $\partial\mu_i/\partial\theta' = X_i$. The GEE (2.29) can be expressed as $(\sum_{i=1}^m X_i'V_i^{-1}X_i)\beta = \sum_{i=1}^m X_i'V_i^{-1}y_i$, which leads to a closed-form solution,

$$\tilde{\beta} = \left(\sum_{i=1}^{m} X_i'V_i^{-1}X_i\right)^{-1} \sum_{i=1}^{m} X_i'V_i^{-1}y_i. \tag{2.30}$$

Like the GEE, (2.30) also has a well-known name, call the best linear unbiased estimator, or BLUE [e.g., Jiang (2007), sec. 2.3.1]. Note that the BLUE is not computable unless the V_i's are known. Also note that, if the distribution of y_i is, actually, normal, that is, $y_i \sim N(\mu_i, V_i)$, then the BLUE is the same as the ML estimator.

Example 2.5. Consider a simple mixed logistic model in a matched-pair setting. Let $y_{it}, i = 1, \dots, m, t = 1, 2$ be binary responses that are (conditionally) independent given the subject-specific random effects, $\alpha_1, \dots, \alpha_m$, such that $\text{logit}(p_{i1}) = \beta + \alpha_i$, $\text{logit}(p_{i2}) = \beta + \delta + \alpha_i$, where $p_{it} = \text{P}(y_{it} = 1|\alpha_i)$, β, δ are unknown parameters corresponding the baseline and increment, respectively. For example, y_{i1} and y_{i2} may correspond to the responses from the ith patient before and after a treatment. Suppose that the random effects are independent and normal, with mean 0 and unknown variance σ^2. It is clear that this is a special case of GLMM. The mean function, μ_i, has components $\mu_{it}, t = 1, 2$, where $\mu_{it} = \text{E}\{h(\beta + \delta 1_{(t=2)} + \sigma\xi)\}$ with $h(u) = e^u/(1 + e^u)$ and $\xi \sim N(0, 1)$. Let $\theta = (\beta, \delta, \sigma)'$. It is easy to see that

$$\frac{\partial \mu_{it}}{\partial \beta} = \text{E}\{h'(\beta + \delta 1_{(t=2)} + \sigma\xi)\},$$
$$\frac{\partial \mu_{it}}{\partial \delta} = \text{E}\{h'(\beta + \delta 1_{(t=2)} + \sigma\xi)\}1_{(t=2)},$$
$$\frac{\partial \mu_{it}}{\partial \sigma} = \text{E}\{h'(\beta + \delta 1_{(t=2)} + \sigma\xi)\xi\},$$

where $h'(u) = e^u(1 + e^u)^{-2}$. Furthermore, we have (Exercise 2.11) $\text{var}(y_{it}) = \mu_{it}(1 - \mu_{it}), t = 1, 2$ and $\text{cov}(y_{i1}, y_{i2}) = \text{E}\{h(\beta + \sigma\xi)h(\beta + \delta + \sigma\xi)\} - \mu_{i1}\mu_{i2}$ with μ_{it} given above. With these expressions, the GEE (2.29) can be specified (Exercise 2.11).

Although (2.29) is, theoretically, the optimal estimating equation, it is not available for computation in most practical situations. The reason is that the covariance matrices, $V_i, 1 \le i \le m$, typically are unknown, or involve some unknown parameters. To deal with this complication, Liang and Zeger (1986) noted that one may replace V_i by a "working covariance matrix", $\tilde{V}_i$, that is available, $1 \le i \le m$. The resulting estimating equation is no longer optimal. However, the corresponding estimator is still consistent, under some mild conditions, even though it may not be as efficient as the GEE estimator obtained from (2.29), if one knows the true V_i's. For example, the working covariance matrix may be chosen as the identity matrix, or a matrix based on the best guess of the true covariance matrix. We consider a simple example.

Example 2.6. In case $\tilde{V}_i = I_{n_i}$, where $n_i = |T_i|$ ($|\cdot|$ denotes cardinality), $1 \le i \le m$, and $\mu_i = X_i\beta$, (2.29) becomes $\sum_{i=1}^m X_i'(y_i - X_i\beta) = 0$, with the solution

$$\hat{\beta} = \left(\sum_{i=1}^{m} X_i' X_i\right)^{-1} \sum_{i=1}^{m} X_i' y_i. \tag{2.31}$$

(2.31) is known as the ordinary least squares (OLS) estimator. It is the same as the LS estimator in linear regression except that now, because the data are correlated with an unknown variance-covariance structure (VCS), the OLS estimator does not have some of the nice properties that the LS estimator has in the case of linear regression. For example, the LS estimator is the same as the BLUE (see Example 2.4) under the linear regression model; however, the OLS estimator is typically not the BLUE when the data are correlated.

Another complication is regarding the variation of the estimator. For simplicity, let us focus on the case of linear models, that is, $\mu_i = X_i\beta, 1 \leq i \leq m$, although the idea can be generalized along the same line to nonlinear models. Note that $\theta = \beta$ in this case. By (2.27), the asymptotic covariance of the BLUE is

$$(U'V^{-1}U)^{-1} = \left(\sum_{i=1}^{m} X_i' V_i^{-1} X_i\right)^{-1}. \tag{2.32}$$

(2.32) can also be derived directly from (2.30), of course. However, when the V_i's are replaced by the working covariance matrices, (2.32) no longer holds. In fact, if we denote the GEE estimator with working covariance matrices, that is, the solution to (2.29) with V_i replaced by $\tilde{V}_i, 1 \leq i \leq m$, by $\hat{\beta}$. Then, once again, it follows from (2.27), with $B = U'\tilde{V}^{-1}$, where $\tilde{V} = \text{diag}(\tilde{V}_i, 1 \leq i \leq m)$, that $\text{Var}(\hat{\beta}) \approx (U'\tilde{V}^{-1}U)^{-1}U'\tilde{V}^{-1}V\tilde{V}^{-1}U(U'\tilde{V}^{-1}U)^{-1} =$

$$\left(\sum_{i=1}^{m} X_i'\tilde{V}_i^{-1}X_i\right)^{-1} \left(\sum_{i=1}^{m} X_i'\tilde{V}_i^{-1}V_i\tilde{V}_i^{-1}X_i\right) \left(\sum_{i=1}^{m} X_i'\tilde{V}_i^{-1}X_i\right)^{-1}. \tag{2.33}$$

Clearly, (2.33) reduces to (2.32) when $\tilde{V}_i = V_i, 1 \leq i \leq m$. We consider another example.

Example 2.6 (continued). In the case of OLS, (2.33) gives

$$\text{Var}(\hat{\beta}) \approx \left(\sum_{i=1}^{m} X_i'X_i\right)^{-1} \left(\sum_{i=1}^{m} X_i'V_iX_i\right) \left(\sum_{i=1}^{m} X_i'X_i\right)^{-1}. \tag{2.34}$$

There is an interesting three-way comparison of the covariance matrices. It is known [e.g., Sen and Srivastava (1990), p. 36] that, under linear regression, the covariance matrix of the LS estimator is

$$\sigma^2(X'X)^{-1} = \sigma^2 \left(\sum_{i=1}^{m} X_i'X_i\right)^{-1}, \tag{2.35}$$

where $X = (X_i)_{1 \le i \le m}$ and σ^2 is the variance of the regression errors, which are assumed to be independent with mean zero. Now we know neither (2.35), the covariance matrix of the LS estimator, nor (2.32), the covariance matrix of the BLUE, are the correct covariance matrix of the OLS estimator–the correct one is (2.34). More specifically, (2.35) is correct if $V_i = \sigma^2 I_{n_i}, 1 \le i \le m$; (2.32) is correct if $\tilde{V}_i = V_i, 1 \le i \le m$; and (2.34) is correct, no matter what.

In practice, the covariance matrix, or ACVM, are used in obtaining measures of uncertainty of the estimator. For example, the standard error (s.e.) of an estimator is defined as an estimated standard deviation (s.d.) of the estimator. However, even though (2.33) is the correct ACVM of $\hat{\beta}$, it is not ready to be used, because the V_i's are unknown. To obtain an estimated ACVM, we use the following useful technique. Note that $V_i = \mathrm{E}\{(y_i - X_i\beta)(y_i - X_i\beta)'\}$, where β is the true parameter vector. If we bring this expression to the middle factor in (2.33), and move the expectation sign to the outside of the summation, the middle factor becomes

$$\mathrm{E}\left\{\sum_{i=1}^{m} X_i'\tilde{V}_i^{-1}(y_i - X_i\beta)(y_i - X_i\beta)'\tilde{V}_i^{-1}X_i\right\}. \tag{2.36}$$

The idea is to estimate (2.36) by the expression inside the expectation, with β replaced by $\hat{\beta}$. This leads to the following estimator of $\mathrm{Var}(\hat{\beta})$:

$$\widehat{\mathrm{Var}}(\hat{\beta}) = \left(\sum_{i=1}^{m} X_i'\tilde{V}_i^{-1}X_i\right)^{-1}\left\{\sum_{i=1}^{m} X_i'\tilde{V}_i^{-1}(y_i - X_i\hat{\beta})(y_i - X_i\hat{\beta})'\tilde{V}_i^{-1}X_i\right\} \times \left(\sum_{i=1}^{m} X_i'\tilde{V}_i^{-1}X_i\right)^{-1}. \tag{2.37}$$

(2.37) is widely known as the "sandwich estimator" for a visually intuitive reason. We illustrate the method with a specific example.

Example 2.7. (Growth curve) For simplicity, suppose that for each of the m individuals, the observations are collected over a common set of times $t_1, \dots, t_k$. Suppose that y_{ij}, the observation collected at time t_j from the ith individual, satisfies $y_{ij} = \xi_i + \eta_i x_{ij} + \zeta_{ij} + \epsilon_{ij}$, where ξ_i and η_i represent, respectively, a random intercept and a random slope; x_{ij} is a known covariate; ζ_{ij} corresponds to a serial correlation; and ϵ_{ij} is an error. For each i, it is assumed that ξ_i and η_i are jointly normally distributed with means β_0, β_1, variances σ_0^2, σ_1^2, respectively, and correlation coefficient ρ; and ϵ_{ij}s are independent and distributed as $N(0, \tau^2)$. As for the ζ_{ij}s, it is assumed that they satisfy the following relation of the first order autoregressive process, or AR(1): $\zeta_{ij} = \phi\zeta_{ij-1} + \omega_{ij}$, where ϕ is a constant such that $0 < \phi < 1$, and ω_{ij}s are independent and distributed as $N\{0, \sigma_2^2(1-\phi^2)\}$. Furthermore, the three random components (ξ, η), ζ, and ϵ are independent, and observations from different individuals are independent. This is sometimes called a growth curve model in the sense that both the intercept and the slope for the mean response, as a function of time, depend on the individual [Diggle *et al.* (2002)]. To express the model in terms of a more standard LMM, write $\xi_i = \beta_0 + v_{0i}$, $\eta_i = \beta_1 + v_{1i}$, and $e_{ij} = \zeta_{ij} + \epsilon_{ij}$. Then, the growth curvemodel can be expressed as

$$\begin{aligned} y_{ij} &= \beta_0 + \beta_1 x_{ij} + v_{0i} + v_{1i}x_{ij} + e_{ij} \\ &= X_{ij}'\beta + X_{ij}'v_i + e_{ij}, \end{aligned}$$

where $X_{ij} = (1, x_{ij})'$, $\beta = (\beta_0, \beta_1)'$, and $v_i = (v_{0i}, v_{1i})'$, or,

$$y_i = X_i\beta + X_i v_i + e_i, \quad i = 1, \dots, m, \tag{2.38}$$

where $y_i = (y_{ij})_{1 \le j \le k}$ and $X_i = (X_{ij}')_{1 \le j \le k}$. (2.38) is a special case of the so-called longitudinal LMM [e.g., Jiang (2007), p. 6].

A feature of the longitudinal LMM is that the y_i's are independent as vectors; however, there may be correlations within the vectors. Furthermore, so far as the current model is concerned, the correlation structure is fairly complicated, involving the (vector-valued) random effects, v_i, and errors, e_i. In particular, the correlation structure within e_i (Exercise 2.12) may not be known, in practice. Of course, one may try to model the correlation structure, but what if the model is wrong? As a more robust strategy, one may consider estimating the β parameters using the GEE with a working covariance matrix. Note that, in most cases, the main interest of longitudinal data analysis is inference about the mean function [e.g., Diggle *et al.* (2002)]. Here, because the β's are the only parameters involved in the mean function, they are of main interest. The simplest working covariance matrix is the identity matrix, I_k, which leads to the OLS estimator (2.31), and the estimated $\text{Var}(\hat{\beta})$ (2.37) with $\tilde{V}_i = I_k, 1 \le i \le m$.

Alternatively, one may consider the so-call equicorrelationworking covariance matrix, that is, the entries of $\tilde{V}_i$ are σ^2 on the diagonal, and $\sigma^2\gamma$ on the off-diagonal, where $|\gamma| < 1$. If σ^2 and γ are known, the resulting GEE estimator for β is given by (2.30) with V_i replaced by $\tilde{V}_i$, $1 \le i \le m$ (verify this). If σ^2 and γ are unknown, but their estimators, $\hat{\sigma}^2$ and $\hat{\gamma}$ are available (see the next section), an extended GEE estimator is obtained by (2.30) with V_i replaced by $\hat{V}_i$, $1 \le i \le m$, where $\hat{V}_i$ is $\tilde{V}_i$ with σ^2, γ replaced by $\hat{\sigma}^2, \hat{\gamma}$, respectively. Similarly, an estimated covariance matrix of the estimator is given by (2.37) with $\tilde{V}_i$ replaced by $\hat{V}_i$, $1 \le i \le m$.

2.3 Iterative estimating equations

The concluding example of the previous section has a part that is a bit vague. Namely, if the working covariance matrix involves some unknown parameters, how would one estimate these parameters? Moreover, as noted, there is potentially a loss of efficiency in GEE with the working covariance matrices, because the latter are not the true covariance matrices, or may not be even close to the true covariance matrices. In this section, we discuss a strategy that allows one to eventually "get it right".

We describe the method under a semiparametric regression model and then discuss its application to longitudinal GLMMs. Consider a follow-up study conducted over a set of prespecified visit times $t_1, \dots, t_b$. Suppose that the responses are collected from subject i at the visit times t_j, $j \in J_i \subset J = \{1, \dots, b\}$. Let $y_i = (y_{ij})_{j \in J_i}$. Here we allow the visit times to be dependent on the subject. This enables us to include some cases with missing responses. Let $X_{ij} = (X_{ijl})_{1 \le l \le p}$ represent a vector of explanatory variables associated with y_{ij} so that $X_{ij1} = 1$. Write $X_i = (X_{ij})_{j \in J_i} = (X_{ijl})_{i \in J_i, 1 \le l \le p}$. Note that X_i may include both time-dependent and independent covariates so that, without loss of generality, it may be expressed as $X_i = (X_{i1}, X_{i2})$, where X_{i1} does not depend on j (i.e., time) whereas X_{i2} does. We assume that (X_i, Y_i), $i = 1, \dots, m$ are independent. Furthermore, it is assumed that

$$\mathrm{E}(Y_{ij}|X_i) = g_j(X_i, \beta), \tag{2.39}$$

where β is a $p \times 1$ vector of unknown regression coefficients and $g_j(\cdot, \cdot)$ are fixed functions. We use the notation $\mu_{ij} = \mathrm{E}(Y_{ij}|X_i)$ and $\mu_i = (\mu_{ij})_{j \in J_i}$. Note that $\mu_i = \mathrm{E}(Y_i|X_i)$. In addition, denote the (conditional) covariance matrix of Y_i given X_i as

$$V_i = \mathrm{Var}(Y_i|X_i), \tag{2.40}$$

whose (j, k)th element is $v_{ijk} = \mathrm{cov}(Y_{ij}, Y_{ik}|X_i) = \mathrm{E}\{(Y_{ij} - \mu_{ij})(Y_{ik} - \mu_{ik})|X_i\}$, $j, k \in J_i$. Note that the dimension of V_i may depend on i. Let $D = \{(j, k) : j, k \in J_i \text{ for some } 1 \le i \le n\}$. Our main interest is to estimate β, the vector of regression coefficients. According to the earlier discussion, if the V_is are known, β may be estimated by the GEE (2.29). On the other hand, if β is known, the covariance matrices V_i can be estimated by the MM method, as follows.

Note that for any $(j, k) \in D$, some of the v_{ijk}s may be the same, either by the nature of the data or by the assumptions. Let L_{jk} denote the number of different v_{ijk}s. Suppose that $v_{ijk} = v(j,k,l)$, $i \in I(j,k,l)$, where $I(j,k,l)$ is a subset of $\{1,\ldots,m\}$, $1 \le l \le L_{jk}$. For any $(j,k) \in D$, $1 \le l \le L_{jk}$, define $\hat{v}(j,k,l) =$

$$\frac{1}{n(j,k,l)} \sum_{i \in I(j,k,l)} \{Y_{ij} - g_j(X_i,\beta)\}\{Y_{ik} - g_k(X_i,\beta)\}, \tag{2.41}$$

where $n(j,k,l) = |I(j,k,l)|$, the cardinality. Then, define $\hat{V}_i = (\hat{v}_{ijk})_{j,k \in J_i}$, where $\hat{v}_{ijk} = \hat{v}(j,k,l)$, if $i \in I(j,k,l)$.

The main points of the above may be summarized as follows. If the V_is were known, one could estimate β by the GEE; on the other hand, if β were known, one could estimate the V_is by the MM. It is clear that there is a cycle here, which motivates the following iterative procedure. Starting with an initial estimator of β, use (3.26), with β replaced by the initial estimator, to obtain the estimators of the V_is; then use (3.19) to update the estimator of β, and repeat the process. We call such a procedure iterative estimating equations, or IEE. If the procedure converges, the limiting estimator is called the IEE estimator, or IEEE. We consider an example.

Example 2.8. A special case of the semiparametric regression model is the linear model with $\mathrm{E}(y_i) = X_i\beta$ and $\mathrm{Var}(y_i) = V_0$, where X_i is a matrix of known covariates, and $V_0 = (v_{qr})_{1 \le q,r \le k}$ is an unknown covariance matrix. Let $y = (y_i)_{1 \le i \le m}$ and assume that $y_i, 1 \le i \le m$ are independent. Then, $V = \mathrm{Var}(y) = \mathrm{diag}(V_0,\ldots,V_0)$. For this special case, the IEE can be formulated as follows. If β were known, a simple consistent estimator of V would be $\hat{V} = \mathrm{diag}(\hat{V}_0,\ldots,\hat{V}_0)$ with

$$\hat{V}_0 = \frac{1}{m}\sum_{i=1}^{m}(y_i - X_i\beta)(y_i - X_i\beta)' \,. \tag{2.42}$$

On the other hand, if V were known, the BLUE of β is given by (2.46). When both β and V are unknown, when iterates between these two steps, starting with $V_0 = I_k$. This procedure is called iterative WLS, or I-WLS (Jiang *et al.* (2007)).

To apply IEE to a longitudinal GLMM, let us denote the responses by y_{ij}, $i = 1,\ldots,m$, $j = 1,\ldots,n_i$, and let $y_i = (y_{ij})_{1 \le j \le n_i}$. We assume that each y_i is associated with a vector of random effects, α_i, that has dimension d such that

$$g(\mu_{ij}) = x_{ij}'\beta + z_{ij}'\alpha_i, \tag{2.43}$$

where $\mu_{ij} = \mathrm{E}(y_{ij}|\alpha_i)$, g is the link function, and x_{ij}, z_{ij} are known vectors. Furthermore, we assume that the responses from different clusters

$y_1, \ldots, y_m$ are independent. Finally, suppose that

$$\alpha_i \sim f(u|\theta), \tag{2.44}$$

where $f(\cdot|\theta)$ is a d-variate pdf known up to a vector of dispersion parameters θ such that $\mathrm{E}_\theta(\alpha_i) = 0$. Let $\psi = (\beta', \theta')'$. Then, we have

$$\begin{aligned}\mathrm{E}(y_{ij}) &= \mathrm{E}\{\mathrm{E}(y_{ij}|\alpha_i)\}\\ &= \mathrm{E}\{h(x_{ij}'\beta + z_{ij}'\alpha_i)\}\\ &= \int h(x_{ij}'\beta + z_{ij}'u)f(u|\theta)du,\end{aligned}$$

where $h = g^{-1}$. Let $W_i = (X_i \ Z_i)$, where $X_i = (x_{ij}')_{1\le j\le n_i}$, $Z_i = (z_{ij}')_{1\le j\le n_i}$. For any vectors $a \in R^p$, $b \in R^d$, define

$$\mu_1(a, b, \psi) = \int h(a'\beta + b'u)f(u|\theta)du.$$

Furthermore, for any $n_i \times p$ matrix A and $n_i \times d$ matrix B, let $C = (A \ B)$, and $g_j(C, \psi) = \mu_1(a_j, b_j, \psi)$, where a_j' and b_j' are the jth rows of A and B, respectively. Then, it is easy to see that

$$\mathrm{E}(y_{ij}) = g_j(W_i, \psi). \tag{2.45}$$

It is clear that (2.45) is simply (2.39) with X_i replaced by W_i, and β replaced by ψ. Note that, because W_i is a fixed matrix of covariates, we have $\mathrm{E}(y_i|W_{ij}) = \mathrm{E}(y_{ij})$. In other words, the longitudinal GLMM satisfies the semiparametric regression model introduced above, hence IEE applies. Again, we consider an example.

Example 2.9. Consider a random-intercept model with binary responses. Let y_{ij} be the response for subject i collected at time t_j. We assume that given a subject-specific random effect α_i, binary responses y_{ij}, $j = 1, \ldots, k$ are conditionally independent with conditional probability $p_{ij} = \mathrm{P}(y_{ij} = 1|\alpha_i)$, which satisfies $\mathrm{logit}(p_{ij}) = \beta_0 + \beta_1 t_j + \alpha_i$, where β_0, β_1 are unknown coefficients. Furthermore, we assume that $\alpha_i \sim N(0, \sigma^2)$, where $\sigma > 0$ and is unknown. Let $y_i = (y_{ij})_{1\le j\le k}$. It is assumed that $y_1, \ldots, y_m$ are independent, where m is the number of subjects. It is easy to show that, under the assumed model, one has

$$\begin{aligned}\mathrm{E}(y_{ij}) &= \int_{-\infty}^{\infty} h(\beta_0 + \beta_1 t_j + \sigma u)f(u)du\\ &\equiv \mu(t_j, \psi), \end{aligned} \tag{2.46}$$

where $h(x) = e^x/(1+e^x)$, $f(u) = (1/\sqrt{2\pi})e^{-u^2/2}$, and $\psi = (\beta_0, \beta_1, \sigma)'$. Write $\mu_j = \mu(t_j, \psi)$, and $\mu = (\mu_j)_{1\le j\le k}$. We have

$$\frac{\partial \mu_j}{\partial \beta_0} = \int_{-\infty}^{\infty} h'(\beta_0 + \beta_1 t_j + \sigma u) f(u) du, \tag{2.47}$$

$$\frac{\partial \mu_j}{\partial \beta_1} = t_j \int_{-\infty}^{\infty} h'(\beta_0 + \beta_1 t_j + \sigma u) f(u) du, \tag{2.48}$$

$$\frac{\partial \mu_j}{\partial \sigma} = \int_{-\infty}^{\infty} h'(\beta_0 + \beta_1 t_j + \sigma u) u f(u) du. \tag{2.49}$$

Also, it is easy to see that the y_is have the same (joint) distribution, hence $V_i = \text{Var}(y_i) = V_0$, an unspecified $k \times k$ covariance matrix, $1 \le i \le m$. Thus, the GEE equation for estimating ψ is given by

$$\sum_{i=1}^{m} \dot{\mu}' V_0^{-1}(y_i - \mu) = 0, \tag{2.50}$$

provided that V_0 is known. On the other hand, if ψ is known, V_0 can be estimated by the method of moments as follows,

$$\hat{V}_0 = \frac{1}{m}\sum_{i=1}^{m}(y_i - \mu)(y_i - \mu)'. \tag{2.51}$$

The IEE procedure then iterates between the two steps when both V_0 and ψ are unknown, starting with $V_0 = I$, the k-dimensional identity matrix.

Note that the mean function, μ_j, in (2.50) and (2.51), is a one-dimensional integral, which can be approximated by a simple Monte Carlo method, as in MSM (see Section 2.1).

Two questions, both of theoretical and practical interest, are: (i) Does the IEE algorithm converge (numerically)? and (ii) if the IEE converges, how does the limit of the convergence behave asymptotically, as an estimator? The short answers to these questions are: (i) Yes, not only the IEE converges, it converges linearly (see below); (ii) Yes, not only the limit of the convergence is a consistent estimator, it is asymptotically as efficient as the solution to the GEE (3.19), as if the true V_i's are known. We discuss these results below, and refer further details to Jiang *et al.* (2007).

We adapt a term from numerical analysis. An iterative algorithm that results in a sequence $x^{(k)}$, $k = 1, 2, \ldots$ converges linearly to a limit x^*, if there is $0 < \rho < 1$ such that $\sup_{k\ge1}\{|x^{(k)} - x^*|/\rho^k\} < \infty$ (e.g., Press *et al.* (1997)). Let

$$L_1 = \max_{1\le i\le m} \max_{j\in J_i} s_{ij},$$

$$s_{ij} = \sup_{|\tilde{\beta}-\beta|\le\epsilon_1} \left|\frac{\partial}{\partial\beta} g_j(X_i, \tilde{\beta})\right|,$$

β be the true parameter vector, ϵ_1 is any positive constant, and

$$\frac{\partial}{\partial\beta}f(\tilde{\beta}) = \left.\frac{\partial f}{\partial\beta}\right|_{\beta=\tilde{\beta}}.$$

Similarly, let $L_2 = \max_{1\le i\le m}\max_{j\in J_i} w_{ij}$, where

$$w_{ij} = \sup_{|\tilde{\beta}-\beta|\le\epsilon_1}\left\|\frac{\partial^2}{\partial\beta\partial\beta'})g_j(X_i,\tilde{\beta})\right\|.$$

Also, let $\mathcal{V} = \{v : \lambda_{\min}(V_i) \ge \lambda_0, \lambda_{\max}(V_i) \le M_0, 1 \le i \le m\}$, where $\lambda_{\min}$ and $\lambda_{\max}$ represent the smallest and largest eigenvalues, respectively, and δ_0 and M_0 are given positive constants. Note that $\mathcal{V}$ is a nonrandom set. An array of nonnegative definite matrices $\{A_{m,i}\}$ is bounded from above if $\|A_{m,i}\| \le c$ for some constant c; the array is bounded from below if $A_{m,i}^{-1}$ exists and $\|A_{m,i}^{-1}\| \le c$ for some constant c. We also refer the notion of O_{P}, including that for random vectors and matrices, to Jiang (2010) (sec. 3.4). Let p and R be the dimensions of β and v, respectively. We assume the following.

A1. For any $(j,k) \in D$, the number of different v_{ijk}s is bounded, that is, for each $(j,k) \in D$, there is a set of numbers $\mathcal{V}_{jk} = \{v(j,k,l), 1 \le l \le L_{jk}\}$, where L_{jk} is bounded, such that $v_{ijk} \in \mathcal{V}_{jk}$ for any $1 \le i \le n$ with $j,k \in J_i$.

A2. The functions $g_j(X_i,\beta)$ are twice continuously differentiable with respect to β; $\mathrm{E}(|Y_i|^4)$, $1 \le i \le m$ are bounded; and L_1, L_2, $\max_{1\le i\le n}(\|V_i\| \vee \|V_i^{-1}\|)$ are $O_{\mathrm{P}}(1)$.

A3 (Consistency of GEE estimator). For any given V_i, $1 \le i \le m$ bounded from above and below, the GEE equation (2.29) has a unique solution $\hat{\beta}$ that is consistent.

A4 (Differentiability of GEE solution). For any v, the solution to (2.29), $\beta(v)$, is continuously differentiable with respect to v, and $\sup_{v\in\mathcal{V}}\|\partial\beta/\partial v\| = O_{\mathrm{P}}(1)$.

A5. $n(j,k,l) \to \infty$ for any $1 \le l \le L_{jk}$, $(j,k) \in D$, as $m \to \infty$.

First consider the numerical convergence of IEE. Let $\hat{\beta}^{(k)}, \hat{v}^{(k)}$ denote the updates of β and v, respectively, at the kth iteration.

Theorem 2.1. Under assumptions *A1–A5*, we have

$$\mathrm{P}(\text{IEE converges}) \to 1$$

as $m \to \infty$. Furthermore, we have

$$\mathrm{P}\left[\sup_{k\ge1}\{|\hat{\beta}^{(k)} - \hat{\beta}^*|/(p\eta)^{k/2}\} < \infty\right] \to 1,$$

$$\mathrm{P}\left[\sup_{k\ge1}\{|\hat{v}^{(k)} - \hat{v}^*|/(R\eta)^{k/2}\} < \infty\right] \to 1$$

as $n \to \infty$ for any $0 < \eta < (p \vee R)^{-1}$, where $(\hat{\beta}^*, \hat{v}^*)$ is the (limiting) IEEE.

Note 1. It is clear that the restriction $\eta < (p \vee R)^{-1}$ is unnecessary (because, for example, $(p\eta_1)^{-k/2} < (p\eta_2)^{-k/2}$ for any $\eta_1 \geq (p \vee R)^{-1} > \eta_2$), but linear convergence would only make sense when $\rho < 1$ (see the definition above).

Note 2. The proof of Theorem 2.1 (see Jiang *et al.* (2007)), in fact, demonstrated that for any $\delta > 0$, there are constants $M_{1,\delta}$, $M_{2,\delta}$, and integer m_δ such that, for all $m \geq m_\delta$,

$$\mathrm{P}\left[\sup_{k\geq 1}\left\{\frac{|\hat{\beta}^{(k)} - \hat{\beta}^*|}{(p\eta)^{k/2}}\right\} \leq M_{1,\delta}\right] > 1 - \delta,$$

$$\mathrm{P}\left[\sup_{k\geq 1}\left\{\frac{|\hat{v}^{(k)} - \hat{v}^*|}{(R\eta)^{k/2}}\right\} \leq M_{2,\delta}\right] > 1 - \delta.$$

Next, we consider asymptotic behavior of the limiting IEEE. For simplicity, the latter is simply called IEEE. The first result is about consistency.

Theorem 2.2. Under the assumptions of Theorem 2.1, the IEEE is consistent.

To establish the asymptotic efficiency of IEEE, we need to strengthen assumptions *A2* and *A5* a little. Define

$$L_{2,0} = \max_{1\leq i\leq m}\max_{j\in J_i}\|\partial^2\mu_{ij}/\partial\beta\partial\beta'\|,$$

and $L_3 = \max_{1\leq i\leq m}\max_{j\in J_i} d_{ij}$, where

$$d_{ij} = \max_{1\leq a,b,c\leq p}\sup_{|\tilde{\beta}-\beta|\leq\epsilon_1}\left|\frac{\partial^3}{\partial\beta_a\partial\beta_b\partial\beta_c}g_j(X_i,\tilde{\beta})\right|.$$

A2′. Same as *A2* except that $g_j(X_i, \beta)$ are three-times continuously differentiable with respect to β, and that $L_2 = O_{\mathrm{P}}(1)$ is replaced by $L_{2,0} \vee L_3 = O_{\mathrm{P}}(1)$.

A5′. There is a positive integer γ such that $m/\{n(j,k,l)\}^\gamma \to 0$ for any $1 \leq l \leq L_{jk}$, $(j,k) \in D$, as $m \to \infty$.

We also need the following additional assumption.

A6. $m^{-1}\sum_{i=1}^m \dot{\mu}_i' V_i^{-1}\dot{\mu}_i$ is bounded away from zero in probability.

Let $\tilde{\beta}$ be the solution to (2.29), where the V_is are the true covariance matrices. Note that $\tilde{\beta}$ is efficient, or optimal, in the sense discussed earlier (see Section 2.2) but is not computable, unless the true V_is are known.

Theorem 2.3. Under assumptions *A1*, *A2′*, *A3*, *A4*, *A5′* and *A6*, we have $\sqrt{m}(\hat{\beta}^* - \tilde{\beta}) \longrightarrow 0$ in probability. Thus, asymptotically, $\hat{\beta}^*$ is as efficient as $\tilde{\beta}$.

Note. The proof of Theorem 2.3 also reveals the following asymptotic expansion,

$$\hat{\beta}^* - \beta = \left(\sum_{i=1}^m \dot{\mu}_i' V_i^{-1} \dot{\mu}_i\right)^{-1} \sum_{i=1}^m \dot{\mu}_i' V_i^{-1}(Y_i - \mu_i) + \frac{o_{\mathrm{P}}(1)}{\sqrt{m}}, \quad (2.52)$$

where $o_{\mathrm{P}}(1)$ represents a term that converges to zero (vector) in probability (e.g., Jiang (2010), sec. 3.4). By Theorem 2.3, (2.52) also holds with $\hat{\beta}^*$ replaced by $\tilde{\beta}$, even though the latter is typically not computable.

Example 2.10 (Real-data example). We use a real-data example to illustrate the convergence of IEE, or I-WLS (see Example 2.8) in this occasion. Jiang *et al.* (2007) analyzed a data set from Hand and Crowder (1996) regarding hip replacements of thirty patients (also see Jiang (2007), sec. 1.7.2). Each patient was measured four times, once before the operation and three times after, for hematocrit, TPP, vitamin E, vitamin A, urinary zinc, plasma zinc, hydroxyprolene (in milligrams), hydroxyprolene (index), ascorbic acid, carotine, calcium, and plasma phosphate (12 variables). An important feature of the data is that there is considerable amount of missing observations. In fact, most of the patients have at least one missing observation for all 12 measured variables. As a result, the observational times are very different for different patients.

Two of the variables are considered: hematocrit and calcium. The first variable was considered by Hand and Crowder (1996), who used the data to assess age, sex, and time differences. The authors assumed an equicorrelated model and obtained Gaussian estimates of regression coefficients and variance components (i.e., MLE under normality). Here we take a robust approach without assuming a parametric covariance structure. The covariates consist of the same variables as suggested by Hand and Crowder (1996). The variables include an intercept, sex, occasion dummy variables (three), sex by occasion interaction dummy variables (three), age, and age by sex interaction. For the hematocrit data, the I-WLS algorithm converged in seven (7) iterations. Here the criterion for the convergence is that the Euclidean distance between consecutive updates of the parameters is less than 10^{-5}. For the calcium data, the I-WLS in thirteen (13) iterations. We shall revisit this example is Subsection 2.6.1.

2.4 Robust estimation in GLMM

So far the main objective of GEE, or IEE, is analysis of longitudinal data, in which estimation of the mean function is of primary interest. In many cases, however, there are interests in estimating the variance components. Furthermore, there are situations of correlated data of which the covariance structure is more complicated than that of the longitudinal data, such as Example 2.2. We need to extend the GEE idea so that the method can be applied to such cases.

In Section 2.1, we considered the MM method which requires minimum distributional assumptions about the data, and produces consistent estimators. However, the MM estimator is inefficient (e.g., Jiang (1998a)). Our goal is therefore two-fold: On the one hand, we need to improve the efficiency of MM; on the other hand, we wish to maintain weak distributional assumptions so that the method is relatively robust to violations of these assumptions. To do so, let us first consider an extension of GLMM. Recall that it is assumed in a GLM [McCullagh and Nelder (1989)] that the distribution of the response is a member of a known exponential family. Thus, for a linear model to fit within the GLM, one needs to assume that the distribution of the response is normal. However, the definition of a linear model does not have to require normality, and many of the methods, such as the least squares, developed in linear models do not require the normality assumption. Thus, in a way, GLM has not fully extended the linear model.

In view of this, we consider a broader class of models than the GLMM, in which the form of the conditional distribution, such as the exponential family, is not required. The method can be described under an even broader framework. Suppose that, given a vector $\alpha = (\alpha_k)_{1 \le k \le m}$ of random effects, responses $y_1, \dots, y_N$ are conditionally independent such that

$$\mathrm{E}(y_i|\alpha) = h(\xi_i), \tag{2.53}$$

$$\mathrm{var}(y_i|\alpha) = a_i(\phi)v(\eta_i), \tag{2.54}$$

where $h(\cdot)$, $v(\cdot)$, and $a_i(\cdot)$ are known functions, ϕ is a dispersion parameter,

$$\xi_i = x_i'\beta + z_i'\alpha, \tag{2.55}$$

where β is a vector of unknown fixed effects, and $x_i = (x_{ij})_{1 \le j \le p}$, $z_i = (z_{ik})_{1 \le k \le m}$ are known vectors. Finally, assume that

$$\alpha \sim F_\theta, \tag{2.56}$$

where F_θ is a multivariate distribution known up to a vector $\theta = (\theta_r)_{1 \le r \le q}$ of dispersion parameters, or variance components. Note that we do not require that the conditional density of y_i given α is a member of the exponential family; instead, only up to second conditional moments are specified, by (2.53) and (2.54). In fact, as will be seen, to obtain consistent estimators, only (3.53) is needed.

We now consider estimation under the extended GLMM by extending the idea described at the beginning of Section 2.2, where the vector of base statistics, S, satisfy the following:

(i) The mean of S is a known function of ψ.

(ii) The covariance matrix of S is a known function of ψ, or at least is consistently estimable.

(iii) Certain smoothness and regularity conditions hold.

Condition (iii) of the above requirement is a bit vague at this point, but it will be specified later when we discuss asymptotic theory. Let the dimension of θ and S be r and N, respectively. If only (i) is assumed, an estimator of θ may be obtained by solving equation (2.23). In fact, this is what we were doing in Section 2.1, where the base statistics are chosen as

$$\begin{aligned} S_j &= \sum_{i=1}^{N} w_i x_{ij} y_i, \quad 1 \le j \le p, \\ S_{p+j} &= \sum_{s \ne t} w_s w_t z_{sk} z_{tk} y_s y_t, \quad 1 \le k \le m. \end{aligned} \tag{2.57}$$

In fact, if $Z = (z_{ik})_{1\le i\le n, 1\le k\le m} = (Z_1 \cdots Z_q)$, where each Z_r is an $N \times m_r$ standard design matrix in that it consists of zeros and ones; there is exactly one 1 in each row, and at least one 1 in each column. Then, by choosing $B = \mathrm{diag}(I_p, 1'_{m_1}, \dots, 1'_{m_q})$, one obtains the MM equations of Section 2.1 that lead to the MSM estimator. Note that, here, B is a constant matrix. In general, we call the solution to (2.23) with a given constant matrix B a first-step estimator.

On the other hand, according to the discussion in the early part of Section 2.2, the optimal B, in the sense of minimizing the asymptotic covariance matrix, (2.27), is $U'V^{-1}$. Unfortunately, the optimal B depends on θ, which is exactly what we wish to estimate. Our approach is to replace the θ involved in the optimal B by $\tilde{\theta}$, a first-step estimator. This leads to the second-step estimator, denoted by $\hat{\theta}$, obtained by solving

$$\tilde{B}S = \tilde{B}u(\theta), \tag{2.58}$$

where $\tilde{B} = U'V^{-1}|_{\theta=\tilde{\theta}}$. It can be shown (Jiang and Zhang (2001)) that, under suitable conditions, the second-step estimator is consistent and asymptotically efficient in the sense that its asymptotic covariance matrix is the

Table 2.2 **Simulation results: mixed logistic model**

Method of	Estimator of μ			Estimator of σ			Overall
Estimation	Mean	Bias	SD	Mean	Bias	SD	MSE
1st-step	.21	.01	.16	.98	$-.02$	.34	.15
2nd-step	.19	$-.01$	.16	.98	$-.02$	.24	.08

same as that of the solution to the optimal estimating equation, that is, (2.28). The following examples show that the second-step estimators can be considerably more efficient than the first-step ones.

Example 2.11. (Mixed logistic model) Consider an extension of Example 2.1 by allowing the number of binary responses to be different for different subjects, that is, replacing k by k_i. The rest of the assumptions remain the same. It is easy to see that (2.57) reduce to $y_{\cdot\cdot}$ and $y_{i\cdot}^2 - y_{i\cdot}$, $1 \leq i \leq m$, where $y_{i\cdot} = \sum_{j=1}^{n_i} y_{ij}$ and $y_{\cdot\cdot} = \sum_{i=1}^{m} y_{i\cdot}$.

If $k_i = k, 1 \leq i \leq m$, that is, we are in the situation of Example 2.1, also known as balanced data, the first-step estimating equations can be shown to be equivalent to the second-step ones (Exercise 2.14). In other words, in the case of balanced data, there is no gain by doing the second-step, and the first-step estimators are already optimal. However, when the data are unbalanced, the first- and second-step estimators are no longer equivalent, and there is a real gain by doing the second-step. To see this, a simulation was carried out (Jiang and Zhang (2001)), in which $m = 100$, $n_i = 2$, $1 \leq i \leq 50$, and $n_i = 6$, $51 \leq i \leq 100$. The true parameters were chosen as $\mu = 0.2$ and $\sigma = 1.0$. The results based on 1000 simulations are summarized in Table 2.2, where SD represents the simulated standard deviation, and the overall MSE is the MSE of the estimator of μ plus that of the estimator of σ. Overall, there is about a 43% reduction of the MSE of the second-step estimators over the first-step ones.

Because the first- and second-step estimators are developed under the assumption of the extended GLMM, the methods apply to some situations beyond (the classical) GLMM. The following is an example.

Example 2.3 (continued). Consider an extension of the beta-binomial example of Example 2.3. It can be shown (Exercise 2.15) that $\mathrm{E}(Y) = k\pi$ and $\mathrm{Var}(Y) = \phi k\pi(1-\pi)$, where $\phi = (k+1)/2$. It is seen that the mean function under beta-binomial(k, π) is the same as that of binomial(k, π), but the variance function is different. In other words, there is an overdispersion.

Now, suppose that, given the random effects $\alpha_i, 1 \leq i \leq m$, which are independent and distributed as $N(0, \sigma^2)$, responses y_{ij}, $1 \leq i \leq m$,

Table 2.3 **Simulation results: beta-binomial mixed model**

Method of	Estimation of μ			Estimation of σ			Overall
Estimation	Mean	Bias	SD	Mean	Bias	SD	MSE
1st-step	.25	.05	.25	1.13	.13	.37	.22
2nd-step	.25	.05	.26	1.09	.09	.25	.14

$1 \leq j \leq n_i$ are independent and distributed as beta-binomial(k, π_i), where $\pi_i = h(\mu + \alpha_i)$ with $h(x) = e^x/(1 + e^x)$. Note that this is not a GLMM under the classical definition, because the conditional distribution of y_{ij} is not a member of the exponential family. However, the model falls within the extended GLMM, because

$$\mathrm{E}(y_{ij}|\alpha) = l\pi_i, \tag{2.59}$$

$$\mathrm{var}(y_{ij}|\alpha) = \phi k\pi_i(1 - \pi_i). \tag{2.60}$$

If only (2.59) is assumed, one may obtain the first-step estimator of (μ, σ), for example, by choosing $B = \mathrm{diag}(1, 1'_m)$. If, in addition, (2.60) is assumed, one may obtain the second-step estimator. To see how much difference there is between the two, a simulation study was carried out with $m = 40$ (Jiang and Zhang (2001)). Again, an unbalanced situation was considered: $n_i = 4$, $1 \leq i \leq 20$ and $n_i = 8$, $21 \leq i \leq 40$. We took $k = 2$, and the true parameters $\mu = 0.2$ and $\sigma = 1.0$. The results based on 1000 simulations are summarized in Table 2.3. Overall, we see about 36% improvement in MSE of the second-step estimators over the first-step ones.

The improvements of the second-step estimators over the first-step ones in the precedent examples are not incidental. We now discuss an asymptotic theory related to this improvement. First, we specify condition (iii) of the requirements for the base statistics stated above (2.57). The results established here are actually more general than extended GLMM.

Let the responses be $y_1, \ldots, y_N$, whose distribution depends on a parameter vector, θ. Let Θ be the parameter space. First, note that B, S, and $u(\theta)$ in (2.58) may depend on N, the sample size; hence in the subsection we use the notation B_N, S_N, and $u_N(\theta)$. Also, the solution to (2.58) is unchanged when B_N is replaced by $C_N^{-1}B_N$, where $C_N = \mathrm{diag}(c_{N,1}, \ldots, c_{N,r})$, $c_{N,j}$ is a sequence of positive constants, $1 \leq j \leq r$, and r is the dimension of θ. Write $M_N = C_N^{-1}B_N S_N$, and $M_N(\theta) = C_N^{-1}B_N u_N(\theta)$. Then the first-step estimator, $\tilde{\theta} = \tilde{\theta}_N$ is the solution to the equation

$$M_N(\theta) = M_N. \tag{2.61}$$

Consider $M_N(\cdot)$ as a map from Θ to a subset of R^r. Let θ denote the true θ everywhere except when defining a function of θ, and $M_N(\Theta)$ be the image of Θ under $M_N(\cdot)$. For $x \in R^r$ and $A \subset R^r$, define $d(x, A) = \inf_{y \in A} |x - y|$. Obviously, $M_N(\theta) \in M_N(\Theta)$. Furthermore, if $M_N(\theta)$ is in the interior of $M_N(\Theta)$, we have $d(M_N(\theta), M_N^c(\Theta)) > 0$. In fact, the latter essentially ensures the existence of the solution to (2.61), as shown by the following theorem. The proof is given in Jiang and Zhang (2001).

Theorem 2.4. Suppose that, as $N \to \infty$,

$$M_N - M_N(\theta) \longrightarrow 0 \tag{2.62}$$

in probability, and

$$\liminf d\{M_N(\theta), M_N^c(\Theta)\} > 0. \tag{2.63}$$

Then, with probability tending to one, the solution to (2.61) exists and is in Θ. If, in addition, there is a sequence $\Theta_N \subset \Theta$ such that

$$\liminf \inf_{\theta_* \notin \Theta_N} |M_N(\theta_*) - M_N(\theta)| > 0, \tag{2.64}$$

$$\liminf \inf_{\theta_* \in \Theta_N, \theta_* \neq \theta} \frac{|M_N(\theta_*) - M_N(\theta)|}{|\theta_* - \theta|} > 0, \tag{2.65}$$

then, any solution $\tilde{\theta}_N$ to (2.61) is consistent.

The lemmas below give sufficient conditions for (2.62)–(2.65). The proofs are fairly straightforward. Let V_N denote the covariance matrix of S_N.

Lemma 2.1. (2.62) holds provided that, as $N \to \infty$,

$$\text{tr}(C_N^{-1} B_N V_N B_N' C_N^{-1}) \longrightarrow 0.$$

Lemma 2.2. Suppose that there is a vector-valued function $M_0(\theta)$ such that $M_N(\theta) \to M_0(\theta)$ as $N \to \infty$. Furthermore, suppose that there exist $\epsilon > 0$ and $N_\epsilon \geq 1$ such that $y \in M_N(\Theta)$ whenever $|y - M_0(\theta)| < \epsilon$ and $N \geq N_\epsilon$. Then (2.63) holds. In particular, if $M_N(\theta)$ does not depend on N, that is, $M_N(\theta) = M(\theta)$, say, then (2.63) holds provided that $M(\theta)$ is in the interior of $M(\Theta)$, the image of $M(\cdot)$.

Lemma 2.3. Suppose that there are continuous functions $f_j(\cdot)$, $g_j(\cdot)$, $1 \leq j \leq r$, such that $f_j\{M_N(\theta)\} \to 0$ if $\theta \in \Theta$ and $\theta_j \to -\infty$, $g_j\{M_N(\theta)\} \to 0$ if $\theta \in \Theta$ and $\theta_j \to \infty$, $1 \leq j \leq r$, uniformly in N. If, as $N \to \infty$,

$$\limsup |M_N(\theta)| < \infty,$$

$$\liminf \min[|f_j\{M_N(\theta)\}|, |g_j\{M_N(\theta)\}|] > 0, \quad 1 \leq j \leq r,$$

then (2.64) holds with $\Theta_N = \Theta_0$, a compact subset of Ψ.

Write $U_N = \partial u_N / \partial \theta'$. Let $H_{N,j}(\theta) = \partial^2 u_{N,j} / \partial\theta\partial\theta'$, where $u_{N,j}$ is the jth component of $u_N(\theta)$, and $H_{N,j,\epsilon} = \sup_{|\theta_* - \theta| \leq \epsilon} \|H_{N,j}(\theta_*)\|$, $1 \leq j \leq L_N$, where L_N is the dimension of u_N.

Lemma 2.4. Suppose that $M_N(\cdot)$ is twice continuously differentiable, and that

$$\liminf \lambda_{\min}(U_N' B_N' C_N^{-2} B_N U_N) > 0,$$

and there is $\epsilon > 0$ such that

$$\limsup \frac{\max_{1\le i\le r} c_{N,i}^{-2}(\sum_{j=1}^{L_N} |b_{N,ij}| H_{N,j,\epsilon})^2}{\lambda_{\min}(U_N' B_N' C_N^{-2} B_N U_N)} < \infty,$$

where $b_{N,ij}$ is the (i,j) element of B_N. Furthermore, suppose, for any compact subset $\Theta_1 \subset \Theta$ such that $d(\theta, \Theta_1) > 0$, we have

$$\liminf \inf_{\theta_* \in \Theta_1} |M_N(\theta_*) - M_N(\theta)| > 0.$$

Then (2.65) holds for $\Theta_N = \Theta_0$, where Θ_0 is any compact subset of Θ that includes θ as an interior point.

Once again, we consider a specific example.

Example 2.1 (continued). As noted (see Example 2.11; also Exercise 2.14), in this case, the first and second-step estimators of $\theta = (\mu, \sigma)'$ are the same, and they both correspond to $B_N = \text{diag}(1, 1_m')$. It can be shown, by choosing $C_N = \text{diag}\{mk, mk(k-1)\}$, that all of the conditions of Lemmas 2.1–2.4 are satisfied.

We now consider the asymptotic normality of the first-step estimator. We say that an estimator $\tilde{\theta}_N$ is asymptotically normal with mean θ and asymptotic covariance matrix $(\Gamma_N' \Gamma_N)^{-1}$ if $\Gamma_N(\tilde{\theta}_N - \theta) \longrightarrow N(0, I_r)$ in distribution, where $r = \dim(\theta)$. Here it is understood that Γ_N is $r \times r$ and non-singular. Let

$$\lambda_{N,1} = \lambda_{\min}(C_N^{-1} B_N V_N B_N' C_N^{-1}),$$
$$\lambda_{N,2} = \lambda_{\min}\{U_N' B_N' (B_N V_N B_N')^{-1} B_N U_N\}.$$

Theorem 2.5. Suppose that (a) the components of $u_N(\theta)$ are twice continuously differentiable; (b) $\tilde{\theta}_N$ satisfies (2.61) with probability tending to one and is consistent; (c) there exists $\epsilon > 0$ such that

$$\frac{|\tilde{\theta}_N - \theta|}{(\lambda_{N,1}\lambda_{N,2})^{1/2}} \max_{1\le i\le r} c_{N,i}^{-1} \left(\sum_{j=1}^{L_N} |b_{N,ij}| H_{N,j,\epsilon} \right) \longrightarrow 0$$

in probability; and (d)

$$\{C_N^{-1} B_N V_N B_N' C_N^{-1}\}^{-1/2} [M_N - M_N(\psi)] \longrightarrow N(0, I_r)$$

in distribution. Then $\tilde{\theta}$ is asymptotically normal with mean θ and asymptotic covariance matrix

$$(B_N U_N)^{-1} B_N V_N B_N' (U_N' B_N')^{-1}. \tag{2.66}$$

Sufficient conditions for existence, consistency, and asymptotic normality of the second-step estimators can be obtained by replacing the conditions of Theorems 2.4 and 2.5 by the corresponding conditions with a probability statement. For example, let ξ_N be a sequence of nonnegative random variables. We say that $\liminf \xi_N > 0$ with probability tending to one if for any $\epsilon > 0$ there is $\delta > 0$ such that $\mathrm{P}(\xi_N > \delta) \geq 1 - \epsilon$ for all sufficiently large N. Note that this is equivalent to $\xi_N^{-1} = O_P(1)$. Then, (2.64) is replaced by (2.64) with probability tending to one. Note that the asymptotic covariance matrix of the second-step estimator is given by (2.66) with $B_N = U_N' V_N^{-1}$, which is $(U_N' V_N U_N)^{-1}$. This is the same as the asymptotic covariance matrix of the solution to (2.58), or (2.61), with the optimal B (B_N). In other words, the second-step estimator is asymptotically optimal. See Jiang and Zhang (2001) for details.

2.5 Robust GEE

So far the focus has been on robustness to violation to distributional assumptions. An issue that has not been addressed is robustness to outliers. As a motivating example, Thall and Vail (1990) analyzed data from a clinical trial involving 59 epileptics. These patients were randomized to a new drug (treatment) or a placebo (control). The number of epileptic seizures was recorded for each patient during an eight-week period, namely, one seizure count during the two-week period before each of four clinic visits. Baseline seizures and the patient's age were available and treated as covariates. As noted by Thall and Vail (1990) [also see Diggle *et al.* (2002), pp. 188-189], patients # 112, 207, 225, and 227 are possible "outliers"; however, there was no clinical basis for excluding these patients from the analysis. It is therefore of interest to carry out the analysis using a method that is robust to potential outliers.

For the ith subject, let y_{ij} denote the response collected at time t_{ij}, $1 \leq j \leq n_i$. In a way, the GEE method is closely related to the LS in that the left side of (2.29) may be viewed as a weighted sum of residuals, $y_i - \mu_i, 1 \leq i \leq m$. However, the residuals are known to be sensitive to heavy-tailed distributions, contaminated data, and outliers. In the case of independent observations, such as linear regression, a standard approach to obtain estimators that are robust to outliers is to use M-estimation that relies on a dispersion function that varies more slowly than the square function, which corresponds to the LS (Huber (1981)). Suppose that

$$y_{ij} = \mu_{ij}(\beta) + \sigma e_{ij}, \tag{2.67}$$

where μ_{ij} is a mean function that depends on a vector β of unknown parameters, σ is a scale parameter, and e_{ij} is a random error. In the context of semiparametric models, He *et al.* (2002) proposed to estimate the parameters by minimizing

$$\sum_{i=1}^{m}\sum_{j=1}^{n_i} \rho(u_{ij}), \tag{2.68}$$

where $u_{ij} = (y_{ij} - \mu_{ij})/\sigma$, which depends on the parameters, and $\rho(\cdot)$ is a convex function such that $\rho(u) \geq \rho(0) = 0$. If $\rho(u) = u^2$, the procedure is the same as LS; if $\rho(u) = |u|$, the procedure corresponds to median regression [e.g., Huber (1981)). The left side of (2.68) may be viewed as a loss function, which is the sum of the dispersion functions for the individuals. This is reasonable if the y_{ij}'s are independent. When the responses are correlated, one would need a multivariate dispersion function to take the correlations into account. For example, with $u_i = (u_{ij})_{1\leq j\leq n_i}$, one may consider $\rho(u_i) = \sqrt{\sum_{j=1}^{n_i} u_{ij}^2}$. The latter is appropriate if the within-subject correlations are the same; otherwise, weights should be considered. Theoretically speaking, the most efficient multivariate dispersion function can be obtained by considering the negative log-likelihood which intrinsically incorporates the possible correlations. But such an approach requires specification of the joint likelihood function, which may be difficult in non-Gaussian situations, and not robust to misspecification of the distribution. Below we take a different approach.

A fundamental idea in the M-estimation is that the central tendency is defined possibly differently than traditionally. Namely, we assume that $\mathrm{E}\{\psi(e_{ij})\} = 0$ instead of $\mathrm{E}(e_{ij}) = 0$, where $\psi(u) = \rho'(u)$, and subderivative will be used when the derivative does not exist. For example, in the case of $\rho(u) = |u|$, we have $\psi(u) = \mathrm{sign}(u)$. The latter corresponds to *median regression* [e.g., Jung (1996), He *et al.* (2002)]. Another example is when e_{ij} has the standard Cauchy distribution. In this case, the loss function $\rho(u) = \log(1 + u^2)$ actually corresponds to the maximum likelihood estimation for independent data (Exercise 2.16). Let $\psi(u_i)$ denote the vector $[\psi(u_{ij})]_{1\leq j\leq n_i}$. A robust version of the GEE equation, (2.29),is

$$\sum_{i=1}^{m} \frac{\partial \mu_i'}{\partial \beta} Q_i^{-1} \psi(u_i), \tag{2.69}$$

where $Q_i = \text{Var}\{\psi(e_i)\}$ and e_i is defined the same way as u_i but with u_{ij} replaced by e_{ij}. Equation (2.69) is motivated by the same optimality theory discussed at the beginning of Section 2.2. As the true Q_i is typically unknown, one may replace it by a working covariance matrix, as in GEE (Liang and Zeger (1986)). It is more convenient to use a working inverse covariance matrix (WICM), W_i, for Q_i^{-1}. In particular, if $Q_i = \tau^2 I_{n_i}, 1 \le i \le m$, for some positive constant τ, (2.69) follows as the usual practice of minimizing (2.68), that is, by differentiation and setting the derivatives equal to zero. The next simplest working covariance matrix is, perhaps, the so-called equicorrelated (EQC) structure, which is closely related to a mixed effects model. Suppose that $\psi(e_{ij})$ can be expressed as

$$\psi(e_{ij}) = \alpha_i + \epsilon_{ij}, \tag{2.70}$$

where α_i is a random effect that has mean 0 and variance σ_α^2, and ϵ_{ij} is an additional error that has mean 0 and variance σ_ϵ^2. Assume that the random effects and errors are uncorrelated. Then, it is easy to show that $\text{var}\{\psi(e_{ij})\} = \sigma_\alpha^2 + \sigma_\epsilon^2$ and $\text{cov}\{\psi(e_{ij}), \psi(e_{ik})\} = \sigma_\alpha^2$, if $j \ne k$. Thus, we have $\text{cor}\{\psi(e_{ij}), \psi(e_{ik})\} = \sigma_\alpha^2/(\sigma_\alpha^2 + \sigma_\epsilon^2)$, which is a constant across all of the subjects and time points. It follows that $Q_i = \sigma_\epsilon^2 I_{n_i} + \sigma_\alpha^2 J_{n_i}$, where I_n and J_n denote the $n \times n$ identity matrix and matrix of 1's. Thus, the WICM has the expression (Exercise 2.17)

$$W_i = Q_i^{-1} = \frac{1}{\sigma_\epsilon^2}\left(I_{n_i} - \frac{\sigma_\alpha^2}{\sigma_\epsilon^2 + n_i\sigma_\alpha^2} J_{n_i}\right). \tag{2.71}$$

Given the WICM, $W_i, 1 \le i \le m$, one can solve

$$\sum_{i=1}^m \frac{\partial \mu_i'}{\partial \beta} W_i \psi(u_i) = 0 \tag{2.72}$$

to obtain an estimator of β. Under suitable conditions, the solution to (2.72) is consistent (Wang *et al.* (2015)). However, the WICM may be very different from the optimal choice which is the true Q_i^{-1}. As a result, the solution to (2.72) may be inefficient. In addition, the scale parameter, σ in (2.67), sometimes can cause a problem that one cannot simply solve (2.72) without knowing σ.

To solve these problems, Wang *et al.* (2015) suggests to supplement (2.72) with a second set of estimating equations to jointly estimate β, and a vector γ of dispersion parameters including those involved in Q_i and possibly σ. More specifically, the authors suggest that the second set of equations has the form

$$U(\gamma; \beta) = \sum_{j=1}^m f_i(y_i, \beta) - g(\gamma) = 0, \tag{2.73}$$

where $f_i(y_i, \beta) = [f_{i,s}]_{1 \le s \le q}$, $q = \dim(\gamma)$, and $g(\gamma) = [g_s(\gamma)]_{1 \le s \le q}$ with

$$g_s(\gamma) = \inf_{\beta} \left[\sum_{i=1}^{m} \mathrm{E}\{f_{i,s}(y_i, \beta)\} \right].$$

The idea is to iterate between (2.72) and (2.73); namely, starting with some initial WICM, one solves (2.72) to update β; then, one replace the β in (2.73) by its update, and solve the latter equation for γ; one then use the latest γ to obtain estimated $Q_i, 1 \le i \le m$, and solve (2.72) again to update β, and so on. Wang *et al.* (2015) showed that, under some regularity conditions, the iterative algorithm has a similar convergence property as the IEE (see Section 2.3). Furthermore, the limiting estimator of β has a similar asymptotic efficiency as IEEE in that its asymptotic covariance matrix is the same as that of the solution to (2.69) with the true Q_i's. We now give a few examples of special cases of (2.73).

Example 2.12. For any i, s, let $a_{i,s} \in R^{n_i}$, and

$$f_{i,s}(y_i, \beta) = a'_{i,s}\{y_i - \mu_i(\beta)\}\{y_i - \mu_i(\beta)\}' a_{i,s}.$$

Then, it is easy to show that

$$\mathrm{E}\{f_{i,s}(y_i, \beta)\} = a'_{i,s}\Sigma_i a_{i,s} + [a'_{i,s}\{\mu_i(\beta) - \mu_i(\beta_0)\}]^2, \tag{2.74}$$

where $\Sigma_i = \mathrm{Var}(y_i)$, the (true) covariance matrix of y_i. Obviously, (2.74) is minimized when $\beta = \beta_0$; hence, in (2.73), one has $g_s(\gamma)$ equal to the first term on the right side of (2.74), where γ is the vector of whatever parameters involved in $\Sigma_i, 1 \le i \le m$.

To consider the next example, let us first introduce a lemma. The proof is left as an exercise (Exercise 2.18). Let $\mathcal{F}$ denote the set of continuous functions f that are nonnegative, even and unimodal, and satisfies $f(x) \to 0$ as $|x| \to \infty$.

Lemma 2.5. For any $f, g \in \mathcal{F}$ and $\mu \in (-\infty, \infty)$, we have

$$\int f(x - \mu) g(x) dx \le \int f(x) g(x) dx.$$

Example 2.13. Suppose that there are constants $c > 0$ such that $c - |\psi(u)| \in \mathcal{F}$. One example of such a ψ is Huber's function, defined as $\psi(u) = u$ if $|u| \le c$; $\psi(u) = -c$ if $u \le -c$, and $\psi(u) = c$ if $u \ge c$. Furthermore, suppose that the pdf of y_{ij}, f_{ij}, is continuous, unimodal, and symmetric about $\mu_{ij} = \mu_{ij}(\beta_0)$. For any positive integer s, let $f(u) = c^s - |\psi(u)|^s$, and $g(u) = f_{ij}(u + \mu_{ij})$. It is easy to show that $f, g \in \mathcal{F}$

(Exercise 2.19). Thus, by Lemma 2.5, we have

$$
\begin{aligned}
c^s - \mathrm{E}\{|\psi(y_{ij} - \mu_{ij}(\beta))|^s\} &= \mathrm{E}\{f(y_{ij} - \mu_{ij}(\beta))\} \\
&= \int f(x - \mu_{ij}(\beta)) f_{ij}(x) dx \\
&= \int f(x - \mu) g(x) dx \quad [\mu = \mu_{ij}(\beta) - \mu_{ij}(\beta_0)] \\
&\leq \int f(x) g(x) dx \\
&= c^s - \mathrm{E}\{|\psi(y_{ij} - \mu_{ij}(\beta_0))|^s\},
\end{aligned}
$$

implying $\mathrm{E}\{|\psi(y_{ij} - \mu_{ij}(\beta))|^s\} \geq \mathrm{E}\{|\psi(y_{ij} - \mu_{ij}(\beta_0))|^s\}$ for any β. Thus, if one defines $f_{i,s}(y_i, \beta) = \sum_{j=1}^{n_i} |\psi(y_{ij} - \mu_{ij}(\beta))|^s$, one has

$$
g(\gamma) = \sum_{i=1}^{m} \sum_{j=1}^{n_i} \mathrm{E}\{|\psi(y_{ij} - \mu_{ij})|^s\}, \tag{2.75}
$$

where γ is the vector of whatever parameters that one needs to know in order to compute the right side of (2.75).

2.6 Real-data examples

We use three real-data examples to illustrate the methods discussed in this chapter. The first example is a continuation of an example discussed earlier on hip replacement. This example is used to further illustrate the IEE method. The second example is regarding a well-known data set on salamander-mating experiment. This example is used to demonstrate the robust estimation method for GLMM. The last example involves longitudinal data from an epileptic seizure study. This data set is used to illustrate the robust GEE method.

2.6.1 *Hip replacement data revisited*

In Example 2.10, we reported the convergence results for IEE in analyzing the hip replacement data. The results of the analyses for the hematocrit and calcium data are presented in Tables 2.4 and 2.5, respectively. The hematocrit data were also analyzed by Hand and Crowder (1996) whose Gaussian estimates are reported for comparison. The parameters correspond to, from left to right, intercept, sex, occasions (three), sex by occasion interaction (three), age, and age by sex interaction; the second row is estimated standard errors corresponding to the IEE, or I-WLS in this case, estimates.

Table 2.4 **Estimates for hematocrit data**

Coef.	β_1	β_2	β_3	β_4	β_5
I-WLS	3.19	0.08	0.65	-0.34	-0.21
s.e.	0.39	0.14	0.06	0.06	0.07
Gaussian	3.28	0.21	0.65	-0.34	-0.21
Coef.	β_6	β_7	β_8	β_9	β_{10}
I-WLS	0.12	-0.051	-0.051	0.033	-0.001
s.e.	0.06	0.061	0.066	0.058	0.021
Gaussian	0.12	-0.050	-0.048	0.019	-0.020

It is seen that the I-WLS estimates are similar to the Gaussian ones, especially for the parameters that are found significant. This is, of course, not surprising, because the Gaussian and I-WLS estimators should both be close to the BLUE, provided that the covariance model suggested by Hand and Crowder is correct (the authors believed that their method was valid in this case). Taking into account the estimated standard errors, we found the coefficients β_1, β_3, β_4, β_5, and β_6 to be significant and the rest of the coefficients insignificant. This suggests that, for example, the recovery of hematocrit improves over time at least for the period of measurement times. The findings are consistent with those of Hand and Crowder with the only exception of β_6. Hand and Crowder considered jointly testing the hypothesis that $\beta_6 = \beta_7 = \beta_8 = 0$ and found an insignificant result. In our case, the coefficients are considered separately, and we found β_7 and β_8 to be insignificant and β_6 to be barely significant at the 5% level. However, because Hand and Crowder did not publish the individual standard errors, this does not necessarily imply a difference. The interpretation of the significance of β_6, which corresponds to the interaction between sex and the first occasion, appears to be less straightforward. As for the calcium data, the covariate variables are the same and listed in the same order. It is seen that, except for β_1, β_3, and β_4, all the coefficients are not significant (at the 5% level). In particular, there seems to be no difference in terms of sex and age. Also, the recovery of calcium after the operation seems to be a little quicker than that of hematocrit, because β_5 is no longer significant. Hand and Crowder (1996) did not analyze this dataset.

2.6.2 *Salamander-mating data*

McCullagh and Nelder (1989, §14.5) presented data from mating experiments regarding two populations of salamanders, Rough Butt and Whiteside. These populations, which are geographically isolated from each other, are found in the southern Appalachian mountains of the eastern United

Table 2.5 **Estimates for calcium data**

Coef.	β_1	β_2	β_3	β_4	β_5
I-WLS	20.1	0.93	1.32	-1.89	-0.13
s.e.	1.3	0.57	0.16	0.13	0.16
Coef.	β_6	β_7	β_8	β_9	β_{10}
I-WLS	0.09	0.17	-0.15	0.19	-0.12
s.e.	0.16	0.13	0.16	0.19	0.09

States. The question whether the geographic isolation had created barriers to the animals' interbreeding was thus of great interest to biologists studying speciation.

Three experiments were conducted during 1986, one in the summer and two in the autumn. In each experiment there were 10 males and 10 females from each population. They were paired according to the design given by Table 14.3 in McCullagh and Nelder (1989). The same 40 salamanders were used for the summer and first autumn experiments. A new set of 40 animals was used in the second autumn experiment. For each pair, it was recorded whether a mating occurred, 1, or not, 0.

The responses are binary and clearly correlated. McCullagh and Nelder (1989) proposed the following mixed logistic model with crossed random effects, which was among one of the earliest, and arguably the most influential example, in the literature of GLMM. For each experiment, let u_i and v_j be the random effects corresponding to the ith female and jth male involved in the experiment. Then, on the logistic scale, the probability of successful mating is modeled in term of fixed effects$+u_i + v_j$. It was further assumed that the random effects are independent and normally distributed with means 0 and variances σ^2 for the females and τ^2 for the males, respectively. Under these assumptions, a GLMM may be formulated as follows. Note that there are 40 different animals of each sex. Suppose that, given the random effects $u_1, \dots, u_{40}$ for the females, and $v_1, \dots, v_{40}$ for the males, the binary responses, y_{ijk}, are conditionally independent such that $\text{logit}\{P(y_{ijk} = 1|u, v)\} = x_{ij}'\beta + u_i + v_j$. Here y_{ijk} represents the kth binary response corresponding to the same pair of ith female and jth male, x_{ij} is a vector of fixed covariates, and β is an unknown vector of regression coefficients. More specifically, x_{ij} consists of an intercept; an indicator of Whiteside female WS_f, an indicator of Whiteside male WS_m, and the product $\text{WS}_f \cdot \text{WS}_m$, representing an interaction.

It should be pointed out that McCullagh and Nelder (1989) made a simplification by assuming that different animals are used in different experiments, and this assumption was followed by subsequent studies [e.g., Karim and Zeger (1992), Lin and Breslow (1996), Jiang (1998a), Booth and Hobert (1999)]. Of course, this is not true in reality, because a group of 40 animals were used twice in a summer and an autumn experiments. However, the situation gets more complicated if this assumption is dropped. This is because there may be serial correlations not explained by the animal-specific random effects. More specifically, the binary responses y_{ijk} may not be conditionally independent given the random effects; as a result, the distribution of y_{ijk} no longer follows a GLMM. Alternatively, one could pool the responses from the two experiments involving the same group of animals, as suggested by McCullagh and Nelder (1989, §4.1), so let $y_{ij\cdot} = y_{ij1} + y_{ij2}$, where y_{ij1} and y_{ij2} represent the responses from the summer and first autumn experiments, respectively, that involved the same group of salamanders. This may avoid the issue of conditional independence; however, a new problem emerges. This is because, given the female and male random effects, the conditional distribution of $y_{ij\cdot}$ is not an exponential family. Note that $y_{ij\cdot}$ is not necessarily binomial given the random effects, because of the potential serial correlation between y_{ij1} and y_{ij2} (see Example 2.3).

Due to such considerations, Jiang and Zhang (2001) considered an extended GLMM for the pooled responses. More specifically, let y_{ij1} be the observed proportion of successful matings between the ith female and jth male in the summer and fall experiments that involved the same group of animals (so $y_{ij1} = 0$, 0.5 or 1), and y_{ij2} be the indicator of successful mating between the ith female and jth male in the last fall experiment that involved a new group of animals. It is assumed that, conditional on the random effects, $u_{k,i}$, $v_{k,j}$, $k = 1, 2$, i, $j = 1, \dots, 20$, which are independent and normally distributed with mean 0 and variances σ_{f}^2 and σ_{m}^2, respectively, y_{ijk}, $(i, j) \in P$, $k = 1, 2$ are conditionally independent, where P represents the set of pairs (i, j) determined by the design, u, and v represent the female and male, respectively; $1, \dots, 10$ correspond to RB, and $11, \dots, 20$ to WS. Furthermore, it was assumed that the conditional mean of the response given the random effects satisfies one of the two models below: (i) (logit model) $\mathrm{E}(y_{ijk}|u, v) = h_1(x_{ij}'\beta + u_{k,i} + v_{k,j})$, $(i, j) \in P$, $k = 1, 2$, where

$$x_{ij}'\beta = \beta_0 + \beta_1 \mathrm{WS_f} + \beta_2 \mathrm{WS_m} + \beta_3 \mathrm{WS_f} \times \mathrm{WS_m}, \tag{2.76}$$

and $h_1(x) = e^x/(1 + e^x)$; (ii) (probit model) same as (i) with $h_1(x)$ replaced by $h_2(x) = \Phi(x)$, where $\Phi(\cdot)$ is the cdf of $N(0, 1)$. Note that it is not assumed that the conditional distribution of y_{ijk} given the random effects is a member of the exponential family. The authors then obtained the first-step estimators (see Section 2.4) of the parameters under both models.

Table 2.6 **First-step estimates with standard errors**

Mean Function	β_0	β_1	β_2	β_3	σ_f	σ_m
Logit	0.95	−2.92	−0.69	3.62	0.99	1.28
	(0.55)	(0.87)	(0.60)	(1.02)	(0.59)	(0.57)
Probit	0.56	−1.70	−0.40	2.11	0.57	0.75
	(0.31)	(0.48)	(0.35)	(0.55)	(0.33)	(0.32)

The results are given in Table 2.5. The numbers in parentheses are the estimated standard errors, obtained from Theorem 2.5 in Section 2.4 under the assumption that the binomial conditional variance is correct. If the latter assumption fails, the standard error estimates are not reliable but the point estimates are still valid.

2.6.3 *Epileptic seizure data*

As an illustration of the robust GEE method, discussed in Section 2.5, we consider the epileptic seizure data from Thall and Vail (1990). Diggle *et al.* (1994, pp. 186–188) suggested Poisson-gamma and Poisson-Gaussian random-effect models. The mean vector for subject i is $\mu_i = \exp(X_i\beta)$, with X_i being the matrix of covariates for the ith patient. We consider five covariates including the intercept, treatment, baseline seizure rate, age of subject, and the interaction between treatment and baseline seizure rate. As in Thall and Vail (1990), the baseline is computed as the logarithm of 1/4 of the 8-week pre-randomization seizure count, and the age is also log-transformed. The treatment variable is a binary indicator for the progabide group. Assuming the log link function, we have $\mu_{ij} = \exp(x'_{ij}\beta)$, where x'_{ij} is the jth row of X_i. Because of the high degree of extra-Poisson variation, it is reasonable to consider a quadratic function $\sigma^2_{ij} = \gamma_1\mu_{ij} + \gamma_2\mu^2_{ij}$ [Bartlett (1936); Morton (1987)]. Note that the negative binomial model restricts γ_1 to be 1.

First consider the Poisson model with overdispersion, which corresponds to fixing γ_2 at 0. Denote $u_{ij} = (y_{ij} - \mu_{ij})/\sigma_{ij}$, where $\sigma^2_{ij} = \gamma\mu_{ij}$. We first ignore possible within-subject correlations and use the independence model. The estimates are $\hat{\beta} = (-2.795, -1.341, 0.897, 0.949, 0.562)'$. The overdispersion parameter γ is estimated as 3.80 using the mean deviance.

As noted earlier, there are some unusual observations. Diggle *et al.* (2002) presented results with and without patient # 207. In fact, the residuals indicate that possible outliers also include patients # 227, 225,112,135. We apply the L_p norm estimation with $\rho(u) = |u|^p$. Also recall that the meaning of μ_{ij} depends on the chosen $\rho(u)$. Recent work on L_p can be found in Lai and Lee (2005). In our case of count data, $\rho(u) = |u|^p$ $(p > 1)$ seems to be a legitimate robust approach for assessing the treatment effect together with other covariates. The case of $p = 2$ corresponds to the GLM or GEE approach depending on whether a correlation structure is incorporated. For illustration purpose, we used two different values for p, namely, 2 and 1.5.

The estimating functions for $(\gamma_1, \gamma_2)'$ are

$$U(\gamma_1; \beta) = \sum_{i=1}^{m} \sum_{j=1}^{n_i} \sigma_{ij}^{-1} \{u_{ij}\psi(u_{ij}) - 1\},$$

$$U(\gamma_2; \beta) = \sum_{i=1}^{m} \sum_{j=1}^{n_i} \mu_{ij}\sigma_{ij}^{-1} \{u_{ij}\psi(u_{ij}) - 1\}.$$

In the case when γ_2 is equal to 0 (Poisson with overdispersion), γ_1 is the overdispersion parameter, which can be estimated by

$$\hat{\gamma}_1 = \left\{ \frac{\sum_i \sum_j \left| \frac{y_{ij} - \mu_{ij}}{\sqrt{\mu_{ij}}} \right|^p}{\sum_i \sum_j \sqrt{\mu}_{ij}} \right\}^{2/p}.$$

As reported by Thall and Vail (1990), there are also strong within-subject correlations. For each value of p and each of the two variance functions, we therefore applied two working models, independence and the unstructured, for the correlation matrix of the M-residuals. Given the fact that the observation times are the same for all of the subjects, the unstructured correlation matrix may be a reasonable choice. For the M-estimation method, we first obtained estimates of γ, including the correlation parameters, using the M-residuals from the independence model, and then obtained estimates of β by solving $U(\beta, \hat{\alpha}) = 0$. We iterated between the correlation matrix, the variance function and estimates of β until there were no noticeable changes in all of the estimates.

The asymptotic covariance matrix of $\hat{\beta}$ is estimated by the modified sandwich estimator. The modification lies in the covariance of the M-residuals for each subject. Instead of using individual subject residuals, we pool all of the subject residual products, $\hat{Q} = \sum_{i=1}^{m} \{\psi(u_i)\psi(u_i)'\}$. This modified sandwich method is very similar to Pan (2001).

Table 2.7 shows the results for the case of $\sigma_{ij}^2 = \gamma_1\mu_{ij} + \gamma_2\mu_{ij}^2$. The parameter estimates and their ratio to the corresponding standard error (z-values) are given. The estimates by the robust methods, especially in terms of the z-values, are quite different from those by the GEE method.

Table 2.7 **Parameter Estimates and z-values for Epileptic Data**

	Intercept	Treatment	log(age)	log(baseline)	Interaction
			L_p norm, $p = 2$		
GLM ($R_w = I$), $\hat{\gamma} = (0.866, 0.481)'$					
$\hat{\beta}$	-1.477	-0.900	0.540	0.892	0.349
z-value	-1.210	-2.114	1.515	6.542	1.645
GEE, $\hat{\gamma} = (1.307, 0.419)'$					
$\hat{\beta}$	-1.988	-1.039	0.684	0.908	0.404
z-value	-1.668	-2.430	1.973	6.917	1.949
			L_p norm, $p = 1.5$		
$R_w = I$, $\hat{\gamma} = (0.997, 0.322)'$					
$\hat{\beta}$	-1.427	-0.833	0.476	0.941	0.291
z-value	-1.295	-2.103	1.483	7.701	1.516
R_w=Unstructured, $\hat{\gamma} = (1.203, 0.289)'$					
$\hat{\beta}$	-1.536	-0.900	0.510	0.939	0.318
z-value	-1.418	-2.281	1.616	7.880	1.683

In all cases, the standard errors are reduced after incorporating correlations among the M-residuals. Given the possible outliers, we believe that the results using $p = 1.5$ are more reliable. In the case of $p = 2$, the M-residuals are known as the Pearson residuals. The plots of M-residuals from the two L_p models (with unstructured correlation matrix and Bartlett variance function) also assure this (see Fig. 2.1).

2.7 Exercises

2.1. Show that the MM is a special case of GMM (see Section 2.1).

2.2. Show that in Example 2.1, $y_{i\cdot}, 1 \leq i \leq m$ are i.i.d., and we have the following expressions: $\mathrm{E}(y_{1\cdot}) = k\mathrm{E}\{h_\theta(\xi)\}$, $\mathrm{E}(y_{1\cdot}^2) = k\mathrm{E}\{h_\theta(\xi)\} + k(k-1)\mathrm{E}\{h_\theta^2(\xi)\}$, where $h_\theta(x) = \exp(\mu+\sigma x)/\{1+\exp(\mu+\sigma x)\}$ and $\xi \sim N(0,1)$.

2.3. This exercise is related to Example 2.2.

a. Show that the log-likelihood function can be expressed as (2.5).

b. Let $\mu = 2.0$, and $m_1 = m_2 = 40$. Generate $u_1, \dots, u_{40}$ and $v_1, \dots, v_{40}$ independently from the $N(0,1)$ distribution, say, using the R software, and compute the product, $\prod_{i=1}^{40} \prod_{j=1}^{40} \{1 + \exp(\mu + u_i + v_j)\}^{-1}$, that is involved in (2.5). What do you get?

c. Show that $\mathrm{E}(y_{i\cdot}) = m_2 \mathrm{E}\{h(\mu+\sigma_1\xi+\sigma_2\eta)\}$, where $h(x) = e^x/(1+e^x)$, and ξ, η are independent $N(0,1)$, which only involves a two-dimensional integration.

2.4. Verify expression (2.13) for the marginal density.

2.5. This exercise is related to some derivations in Section 2.1.

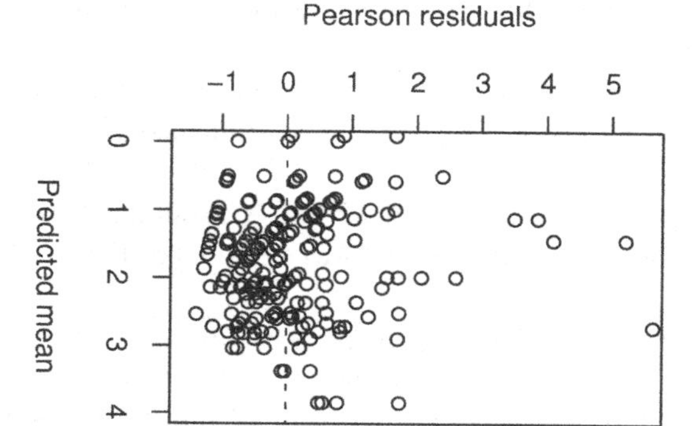

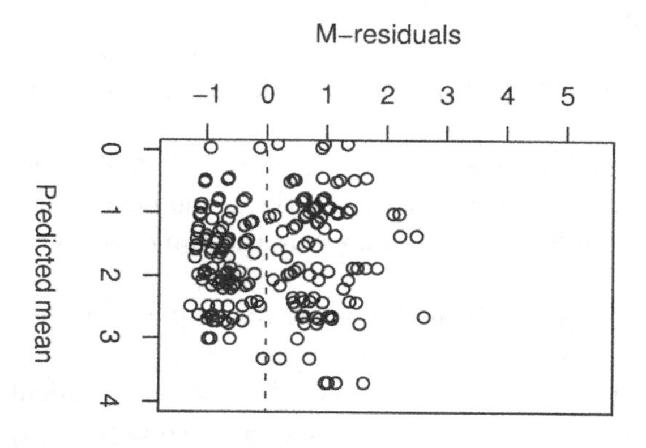

Fig. 2.1 *Residual Plots for Epileptic Seizure Data*

a. Verify (2.16).

b. Show that the first term on the right side of (2.16) depends on ϕ, while the second term does not depend on ϕ.

c. Verify that the right side of (2.14) can be expressed as the right side of (2.18), while the right side of (2.17) can be expressed as the right side of (2.19).

2.6. This exercise has two parts.

a. Show that only the conditional first moments of the y_i's given the random effects are involved in (2.14) and (2.17).

2.7. Show that, under the beta-binomial distribution of Example 2.3, (2.22) holds, with $\mathrm{E}(Y) = k\pi$.

2.8. Show that equations (2.14) and (2.15) are a special case of (2.23). Determine the base statistics S as well as the dimensions N and r in this case.

2.9. Show that the right side of (2.27) is minimized, according to the partial ordering of symmetric matrices [defined below (2.27)], by $B = U'V^{-1}$.

2.10. In (2.28), let $S = y = (y_i)_{1 \le i \le m}$, where $y_i = (y_{it})_{t \in T_i}$, as defined below (2.28). Show that, under the assumption that $y_1, \dots, y_m$ are independent with finite second moments, the optimal estimating equation (2.28) can be expressed as (2.29).

2.11. Show that, in Example 2.5, we have $\mathrm{var}(y_{it}) = \mu_{it}(1 - \mu_{it}), t = 1, 2$ and $\mathrm{cov}(y_{i1}, y_{i2}) = \mathrm{E}\{h(\beta + \sigma\xi)h(\beta + \delta + \sigma\xi)\} - \mu_{i1}\mu_{i2}$. With these expressions, specify the GEE (2.29).

2.12. Derive the variance-covariance structure of the errors, e_i, in (2.38).

2.13. Verify expression (2.46) and also the partial derivatives (2.47)–(2.49).

2.14. Show that, in Example 2.11, if $k_i = k, 1 \le i \le m$, the first-step estimating equation is equivalent to the second-step one. Thus, in this special case, the first-step estimator is the same as the second-step estimator.

2.15. Verify the mean and variance expressions under the beta-binomial distribution in Example 2.3 (continued) in Section 2.4.

2.16. Suppose that $y_{ij}, 1 \le i \le m, 1 \le j \le n_i$ are independent following a Cauchy(μ_{ij}, σ) distribution, where the location parameter, μ_{ij}, depends on a vector β of unknown parameters, i.e., $\mu_{ij} = \mu_{ij}(\beta)$; σ is the scale parameter, and a Cauchy(μ, σ) distribution has the pdf

$$f(x|\mu, \sigma) = \left[\pi\sigma\left\{1 + \left(\frac{x - \mu}{\sigma}\right)^2\right\}\right]^{-1}, \quad -\infty < x < \infty.$$

Show that, in this case, estimating β by minimizing (2.68) with $\rho(u) = \log(1 + u^2)$ is equivalent to maximum likelihood.

2.17. Show that, under (2.70) and the assumptions below it, one has $\mathrm{var}\{\psi(e_{ij})\} = \sigma_\alpha^2 + \sigma_\epsilon^2$ and $\mathrm{cov}\{\psi(e_{ij}), \psi(e_{ik})\} = \sigma_\alpha^2$, if $j \ne k$. Thus, one has $Q_i = \sigma_\epsilon^2 I_{n_i} + \sigma_\alpha^2 J_{n_i}$ and (2.71) holds.

2.18. Prove Lemma 2.5.

2.19. In Example 2.13, show that $f, g \in \mathcal{F}$ for the f, g defined therein.

Chapter 3

Non-Gaussian Linear Mixed Models

The opening illustrative example of Section 1.1 is a special case of inference about a non-Gaussian linear mixed model. Under such a model, the random effects and errors are assumed to be independent, or simply uncorrelated, but their distributions are not assumed to be normal. As a result, the (joint) distribution of the data may not be fully specified (up to a number of parameters). Linear mixed models without the normality assumption is practical because, in practice, the normality assumption rarely holds exactly, or even approximately. It would be useful to develop methods that are robust to violation of such an assumption. Thus, throughout this section, normality is not assumed, unless specially noted.

3.1 Types of models

For the most part, there are two types of non-Gaussian linear mixed models that are not mutually exclusive.

1. Mixed ANOVA model. A non-Gaussian mixed ANOVA model is defined by (1.1), where the components of α_r are i.i.d. with mean 0 and variance σ_r^2, $1 \leq r \leq s$; the components of ϵ are i.i.d. with mean 0 and variance τ^2; and $\alpha_1, \dots, \alpha_s, \epsilon$ are independent. All of the other assumptions are the same as in Section 1.1. Denote the common distribution of the components of α_r by F_r ($1 \leq r \leq s$) and that of the components of ϵ by G. If the parametric forms of $F_1, \dots, F_s, G$ are not assumed, the distribution of y is not full specified, up to a set of parameters. In fact, even if the parametric forms of the F_rs and G are known, as long as they are not normal, the (joint) distribution of y may not have an analytic expression. The vector of variance components, ψ, is defined the same way as in Section 1.1. Alternatively, one may define the Hartley–Rao form of variance components [Hartley and Rao (1967)], $\theta = (\tau^2, \gamma_1, \dots, \gamma_s)'$, where $\gamma_r = \sigma_r^2/\tau^2, 1 \le r \le s$.

A special case of the above model is the so-called balanced mixed ANOVA models. A mixed ANOVA model is balanced if X and Z_r, $1 \le r \le s$ can be expressed as $X = \bigotimes_{l=1}^{w+1} 1_{n_l}^{a_l}$, $Z_r = \bigotimes_{l=1}^{w+1} 1_{n_l}^{b_{r,l}}$, where $(a_1, \dots, a_{w+1}) \in S_{w+1} = \{0,1\}^{w+1}$, $(b_{r,1}, \dots, b_{r,w+1}) \in S \subset S_{w+1}$. In other words, there are w factors in the model; n_l represents the number of levels for factor l $(1 \le l \le w)$; and the $(w+1)$st factor corresponds to "repetition within cells." Thus, we have $a_{s+1} = 1$ and $b_{r,s+1} = 1$ for all r. We consider some examples.

Example 3.1 (One-way random effects model). A mixed model is called a random effects model if the only fixed effect is an unknown mean. Suppose that the observations y_{ij}, $i = 1, \dots, a$, $j = 1, \dots, b_i$ satisfy $y_{ij} = \mu + \alpha_i + \epsilon_{ij}$ for all i and j, where μ is an unknown mean; α_i, $i = 1, \dots, a$ are i.i.d. random effects with mean 0 and variance σ^2; ϵ_{ij}s are i.i.d. errors with mean 0 and variance τ^2; and the random effects and errors are independent. It is easy to show that the model is a special case of the mixed ANOVA model; it is balanced if $b_i = b$ for all i. In the latter case, one has $w = 1$, $n_1 = a$, $n_2 = b$, and $S = \{(0,1)\}$ (Exercise 3.1).

Example 3.2 (Balanced two-way random effects model). For simplicity, let us consider the case of one observation per cell. In this case, the observations y_{ij}, $i = 1, \dots, a$, $j = 1, \dots, b$ satisfy $y_{ij} = \mu + u_i + v_j + e_{ij}$ for all i, j, where μ is as in Example 3.1; u_i, $i = 1, \dots, a$, are i.i.d. random effects with mean 0 and variance σ_u^2; v_j, $j = 1, \dots, b$ are i.i.d. random effects with mean 0 and variance σ_v^2; and e_{ij}s are i.i.d. errors with mean 0 and variance σ_e^2, and u, v, e are independent. Note that this is a special case of the balanced mixed ANOVA model with $w = 2$, $n_1 = a$, $n_2 = b$, $n_3 = 1$, and $S = \{(0,1,1), (1,0,1)\}$ (Exercise 3.1).

2. Longitudinal model. The name "longitudinal" refers to the fact that these models are often used in the analysis of longitudinal data [e.g., Diggle *et al.* (2002)], although such models have been extensively used in other fields as well, such as small area estimation [e.g., Rao and Molina (2015)). A defining feature of these models is that the observations may be divided into independent groups with one random effect (or vector of random effects) associated with each group. In practice, these groups may correspond to different individuals involved in the longitudinal study, or different small areas. Furthermore, there may be serial correlations within each group, which are in addition to the random effect. When the model is used in longitudinal study, there are often time-dependent covariates, which may appear either in X or in Z (see below). Following Datta and Lahiri (2000), a longitudinal model may be expressed as

$$y_i = X_i\beta + Z_i\alpha_i + \epsilon_i, \qquad i = 1, \dots, m, \tag{3.1}$$

where y_i represents the vector of observations from the ith individual; X_i and Z_i are known matrices; β is an unknown vector of regression coefficients; α_i is a vector of random effects; and ϵ_i is a vector of errors. It is assumed that α_i, ϵ_i, $i = 1, \dots, m$ are independent with $\mathrm{E}(\alpha_i) = 0$, $\mathrm{Var}(\alpha_i) = G_i$, $\mathrm{E}(\epsilon_i) = 0$, and $\mathrm{Var}(\epsilon_i) = R_i$), where the covariance matrices G_i and R_i are known up to a vector θ of variance components. We consider an example.

Example 3.3 (Growth curve). Consider, again, Example 2.7, but without the normality assumption. Instead, assume that, for different i's, all of the random variables are independent. Furthermore, for each i, we have $\mathrm{E}(\xi_i) = \mu_1$, $\mathrm{E}(\eta_i) = \mu_2$, $\mathrm{var}(\xi_i) = \sigma_1^2$, $\mathrm{var}(\eta_i) = \sigma_2^2$, and $\mathrm{cor}(\xi_i, \eta_i) = \rho$; also $\mathrm{E}(\epsilon_{ij}) = 0$, $\mathrm{var}(\epsilon_{ij}) = \tau^2$). As for the ζ_{ij}s, it is assumed that they satisfy the following first order autoregressive process, or AR(1): $\zeta_{ij} = \phi\zeta_{\underline{ij-1}} + \omega_{ij}$, where ϕ is a constant such that $0 < \phi < 1$, and ω_{ij}s are independent with mean 0 and variance $\sigma_3^2(1 - \phi^2)$. Finally, three random components (ξ, η), ζ, and ϵ are independent. There is a slight departure of this model from the standard linear mixed model in that the random intercept and slope may have nonzero means. However, by subtracting the means and thus defining new random effects, the model can be expressed in the standard form of a longitudinal model. In particular, the fixed effects are μ_1 and μ_2, and the unknown variance components are σ_j^2, $j = 1, 2, 3$, τ^2, ρ, and ϕ.

3.2 Quasi-likelihood method

In this section we discuss estimation in non-Gaussian linear mixed models. We shall focus on mixed ANOVA models. Some remarks are made at the end of the section on possible extension of the method longitudinal models.

As noted, when normality is not assumed, likelihood-based inference is difficult, or even impossible. To see this, first note that if the distributions of the random effects and errors are not specified, the likelihood function is simply not available. Furthermore, even if the (nonnormal) distributions of the random effects and errors are specified (up to some unknown parameters), the likelihood function is usually complicated. In particular, such a likelihood may not have an analytic expression. Finally, like normality, any other specific distributional assumptions may not hold in practice. These difficulties have led to consideration of methods other than maximum likelihood. One such method is Gaussian-likelihood, or, more generally, quasi-likelihood.

The idea is to use the normality-based estimators, even if the data are not really normal. For the mixed ANOVA models, the REML estimator of the vector of variance components, say, θ, is defined as a root to the (Gaussian) REML equations, provided that the root belongs to the parameter space. Similarly, the ML estimators of β and θ are defined as a root to the (Gaussian) ML equations, provided that they, too, stay in the parameter space. More specifically, under the mixed ANOVA model with the variance components, ψ (see §3.1.1), the REML equations are given by (1.4). With the same model and variance components, the ML equations are

$$\left\{\begin{aligned} X'V^{-1}X\beta &= X'V^{-1}y, \\ y'P^2y &= \mathrm{tr}(V^{-1}), \\ y'PZ_iZ_i'Py &= \mathrm{tr}(Z_i'V^{-1}Z_i), \qquad 1 \le i \le s. \end{aligned}\right. \tag{3.2}$$

Similarly, the REML equations under the ANOVA model with the Hartley–Rao form of variance components, θ (see §3.1.1), are

$$\left\{\begin{aligned} y'Py &= n - p, \\ y'PZ_iZ_i'Py &= \mathrm{tr}(Z_i'PZ_i), \qquad 1 \le i \le s. \end{aligned}\right. \tag{3.3}$$

The ML equations under ANOVA model and the Hartley–Rao form are

$$\left\{\begin{aligned} X'V^{-1}X\beta &= X'V^{-1}y, \\ y'Py &= n, \\ y'PZ_iZ_i'Py &= \mathrm{tr}(Z_i'V^{-1}Z_i), \qquad 1 \le i \le s. \end{aligned}\right. \tag{3.4}$$

At first, it might sound a bit unintuitive: If the normality assumption is wrong, one has the wrong likelihood function; if so, how can the method still work? To answer the question, let us first point out that, although the REML estimator is derived under the normality assumption, Gaussian likelihood is not the only one that can lead to the REML equations. For example, Jiang (1996) noted that exactly the same equations will arise if one starts with a multivariate t-distribution, that is, $y \sim t_n(X\beta, V, d)$, which has a joint pdf

$$p(y) = \frac{\Gamma\{(n+d)/2\}}{(d\pi)^{n/2}\Gamma(d/2)|V|^{1/2}}\left\{1 + \frac{1}{d}(y - X\beta)'V^{-1}(y - X\beta)\right\}^{-(n+d)/2}$$

(Exercise 3.2). Here d is the degree of freedom of the multivariate t-distribution.

More generally, Heyde (1994, 1997) showed that the REML equations can be derived from a quasi-likelihood. Consider a general setting in which y is a vector of responses that is associated with a vector x of explanatory variables. Here we allow x to be random as well. Suppose that the (conditional) mean of y given x is associated with ψ, a vector of unknown parameters. For notational simplicity, write $\mu = \mathrm{E}_\psi(y|x) = \mu(x,\psi)$, and $V = \mathrm{Var}(y|x)$. Here, Var denotes the covariance matrix, and Var or E without subscript ψ means to be taken at the true ψ. Let $\dot{\mu}$ denote the matrix of partial derivatives; that is, $\dot{\mu} = \partial\mu/\partial\psi'$. Consider the following class of vector-valued estimating functions $\mathcal{H} = \{G = B(y-\mu)\}$, where $B = B(x,\psi)$, such that $\mathrm{E}(\dot{G})$ is nonsingular. An estimating equation corresponds to the equation $\mathcal{H} = 0$ to solve for ψ. The following theorem can be derived from Theorem 2.1 of Heyde (1997) [see Jiang (2007), §4.5.1].

Theorem 3.1. Suppose that V is known, and that $\mathrm{E}(\dot{\mu}'V^{-1}\dot{\mu})$ is nonsingular. Then, the optimal estimating function within $\mathcal{H}$ is given by $G^* = \dot{\mu}'V^{-1}(y-\mu)$, that is, with $B = B^* = \dot{\mu}'V^{-1}$.

As it turns out, the REML equations derived under normality and under multivariate-t are equivalent (Exercise 3.2), which is a special case of the quasi-likelihood estimating equation. For such a reason, the (Gaussian) REML estimation may be regarded as a method of quasi-likelihood. Similarly, the (Gaussian) ML estimation may be justified from a quasi-likelihood point of view. For simplicity, the corresponding estimators are still called REML or ML estimators.

Another way to justify the quasi-likelihood method is through asymptotic analysis. It has been shown [Richardson and Welsh (1994), Jiang (1996), Jiang (1997)] that the REML estimator is consistent and asymptotically normal even if normality does not hold, provided that some moment conditions are satisfied. Furthermore, Jiang (1996) showed that the ML estimator has similar asymptotic properties, provided that the number of fixed effects, p, remains bounded or increases at a slower rate than the sample size. Again, the latter result does not require the normality assumption. Therefore, the quasi-likelihood approach is, at least, well-justified from an asymptotic point of view for point estimation.

Nevertheless, to make inference about a model, one needs more than just a point estimator. Standard errors of estimators and interval estimators are often needed. We discuss such problems in the next section.

3.3 Measure of uncertainty

In order to carry out the inference, one needs a measure of uncertainty for the estimator. Under the normality assumption, the uncertainty is typically evaluated using the asymptotic covariance matrix (ACVM) associated with the estimator. The ACM is equal to the inverses of the corresponding Fisher information matrices. The problem is complicated, however, when normality does not hold.

According to the previous section, it is known that even without the normality assumption, the REML estimator is still consistent and asymptotically normal under mild conditions; the same can be said about the MLE, under more restrictive conditions. But, what can we say about their ACVMs? To be specific, let us focus on REML under the mixed ANOVA model with the Hartley–Rao form of variance components, as the idea can be easily extended to other cases. It can be shown that [e.g., Jiang (1996)], without the normality assumption, the ACVM of the REML estimator, $\hat{\theta}$, of the variance components, has the following "sandwich" expression [compare, e.g., with (2.26)]:

$$\Sigma_{\mathrm{R}} = \left\{ \mathrm{E}\left(\frac{\partial^2 l_{\mathrm{R}}}{\partial\theta\partial\theta'} \right) \right\}^{-1} \mathrm{Var}\left(\frac{\partial l_{\mathrm{R}}}{\partial\theta} \right) \left\{ \mathrm{E}\left(\frac{\partial^2 l_{\mathrm{R}}}{\partial\theta\partial\theta'} \right) \right\}^{-1}, \tag{3.5}$$

where l_{R} is the Gaussian restricted log-likelihood given by

$$\begin{aligned} l_{\mathrm{R}}(\theta) = -\frac{1}{2}\Big\{ &(n-p)\log(2\pi) + (n-p)\log(\tau^2) + \log(|A'V_\gamma A|) \\ &+ \frac{1}{\tau^2} y'V(\gamma)y \Big\} \end{aligned} \tag{3.6}$$

under the mixed ANOVA model. When normality indeed holds, one has

$$\mathrm{E}\left(\frac{\partial^2 l_{\mathrm{R}}}{\partial\theta\partial\theta'} \right) = -\mathrm{Var}\left(\frac{\partial l_{\mathrm{R}}}{\partial\theta} \right), \tag{3.7}$$

so (3.5) reduces to the inversed Fisher information matrix:

$$I_{\mathrm{R}}^{-1}(\theta) = \left\{ \mathrm{Var}\left(\frac{\partial l_{\mathrm{R}}}{\partial\theta} \right) \right\}^{-1} = -\left\{ \mathrm{E}\left(\frac{\partial^2 l_{\mathrm{R}}}{\partial\theta\partial\theta'} \right) \right\}^{-1}. \tag{3.8}$$

An attractive feature of (3.8) is that it only depends on the variance components, θ [this is more easily seen using the second expression in (3.8)], of which one already has the (REML) estimator. There may not be such a luxury, however, if (3.7) does not hold.

To see where exactly the problem is, note that $\mathcal{I}_2 = \mathrm{E}(\partial^2 l_{\mathrm{R}}/\partial\theta\partial\theta')$ depends only on θ, even without normality, so this part does not cause any "trouble". What may be troublesome is $\mathcal{I}_1 = \text{Var}(\partial l_{\text{R}}/\partial\theta)$, which may involve higher (than 2nd) moments of the random effects and errors, which are not part of θ. We illustrate with an example.

Example 3.2 (continued). Recall that $I_a, 1_a$ denote the $a \times a$ identity matrix and $a \times 1$ vector of 1's, respectively, and $J_a = 1_a 1_a'$. It can be shown that, in this case, one has

$$\frac{\partial l_{\text{R}}}{\partial \sigma_e^2} = \frac{\xi' H \xi - (ab-1)\sigma_e^2}{2\sigma_e^4}, \tag{3.9}$$

where $H = I_a \otimes I_b + \lambda_1 I_a \otimes J_b + \lambda_2 J_a \otimes I_b + \lambda_3 J_a \otimes J_b$, where $\otimes$ denote the Kronecker product, $\lambda_1 = -b^{-1}\{1-(1+\gamma_1 b)^{-1}\}$, $\lambda_2 = -a^{-1}\{1-(1+\gamma_2 a)^{-1}\}$, and $\lambda_3 = (ab)^{-1}\{1-(1+\gamma_1 b)^{-1} - (1+\gamma_2 a)^{-1}\}$ with $\gamma_1 = \sigma_u^2/\sigma_e^2$ and $\gamma_2 = \sigma_v^2/\sigma_e^2$, and $\xi = y - \mu 1_a \otimes 1_b$. It is clear that, by differentiating (3.9), again, with respective to the variance components, one still ends up with quadratic forms in ξ. Thus, it is easy to see that $\text{E}(\partial^2 l_{\text{R}}/\partial\sigma_e^4)$, etc. are functions of $\theta = (\sigma_e^2, \sigma_u^2, \sigma_v^2)'$ regardless of normality. Similarly, it can be shown that other elements of $\mathcal{I}_2$ are functions of θ. On the other hand, it is easy to see from (3.9) that $\text{var}(\partial l_{\text{R}}/\partial\sigma_e^2)$ involves fourth moments of the random effects and errors (but no third moments; and this has nothing to do with normality, or symmetry; Exercise 3.3). Such quantities are not functions of θ, unless normality holds [in which case one has, for example, $\text{E}(e_{ij}^4) = 3\sigma_e^4$]. In conclusion, in case of non-Gaussian random effects and errors, $\mathcal{I}_1$ involves $\text{E}(e_{ij}^4), \text{E}(u_i^4), \text{E}(v_j^4)$, in addition to θ.

Therefore, In order to make use of the ACVM, one has to find a way to evaluate $\mathcal{I}_1$. Below we consider two approaches.

3.3.1 *Empirical method of moments*

A simple-minded approach would be to estimate the higher moments involved in $\mathcal{I}_1$. Note that estimates of the higher moments are usually not provided in standard packages of mixed model analysis, such as those in SAS or R. Jiang (2003) used an approach, called *empirical method of moments* (EMM) to estimate the higher moments. Let θ be a vector of parameters. Suppose that a consistent estimator of θ, $\hat{\theta}$, is available. Let φ be a vector of additional parameters about which knowledge is needed. Let $\psi = (\theta', \varphi')'$, and $M(\psi, y) = M(\theta, \varphi, y)$ be a vector-valued function of the same dimension as φ that depends on ψ and y, a vector of observations. Suppose that $\text{E}\{M(\psi, y)\} = 0$ when ψ is the true parameter vector. Then, if θ were known, a method of moments estimator of φ would be obtained by solving

$$M(\theta, \varphi, y) = 0 \tag{3.10}$$

for φ. Note that this is more general than the classical method of moments, in which the function M is a vector of sample moments minus their expected values. In econometric literature, this is referred to as generalized method of moments [e.g., Hansen (1982)]. Because θ is unknown, we replace it in (3.10) by $\hat{\theta}$. The result is an EMM estimator of φ, denoted by $\hat{\varphi}$, which is obtained by solving

$$M(\hat{\theta}, \varphi, y) = 0 \ . \tag{3.11}$$

Note that here we use the words "an EMM estimator" instead of "the EMM estimator", because sometimes there may be more than one consistent estimators of θ, and each may result in a different EMM estimator of φ. Jiang (2003) shows that, under suitable conditions, the EMM estimator is consistent for estimating φ. A self-contained consistency result, which is not covered by Jiang (2003), can be found in Section 4.3.

To apply the EMM to non-Gaussian mixed ANOVA model, Jiang (2003) assumed that the third moments of the random effects and errors are zero, that is,

$$\mathrm{E}(\epsilon_1^3) = 0, \ \ \mathrm{E}(\alpha_{r1}^3) = 0, \ \ 1 \le r \le s, \tag{3.12}$$

where ϵ_1 (α_{r1}) is the first component of ϵ (α_r) [recall the components of ϵ (α_r) are i.i.d.] (3.12) is satisfied if, in particular, the distributions of the random effects and errors are symmetric. Under such an assumption, it can be shown that the ACVM of REML estimator of $\psi = (\beta', \psi')'$, where $\psi = (\tau^2, \sigma_1^2, \dots, \sigma_s^2)'$, depends on ψ as well as the kurtoses: $\kappa_0 = \mathrm{E}(\epsilon_1^4) - 3\tau^2$, $\kappa_r = \mathrm{E}(\alpha_{r1}^4) - 3\sigma_r^4, 1 \le r \le s$.

For any matrix $H = (h_{ij})$, define $\|H\|_4 = (\sum_{i,j} h_{ij}^4)^{1/4}$; similarly, if $h = (h_i)$ is a vector, define $\|a\|_4 = (\sum_i h_i^4)^{1/4}$. Let $\mathcal{L}$ be a linear space. Define $L^\perp$ as the linear space $\{v : v'u = 0, \forall u \in \mathcal{L}\}$. If $\mathcal{L}_j, j = 1, 2$ are linear spaces such that $\mathcal{L}_1 \subset \mathcal{L}_2$, then $\mathcal{L}_2 \ominus \mathcal{L}_1$ represents the linear space $\{v : v \in \mathcal{L}_2, v'u = 0, \forall u \in \mathcal{L}_1\}$. If $M_1, \dots, M_k$ are matrices with same number of rows, then $\mathcal{L}(M_1, \dots, M_k)$ represents the linear space spanned by the columns of $M_1, \dots, M_k$. Suppose that the matrices $Z_1, \dots, Z_s$ in (1.1) have been suitably ordered such that

$$\mathcal{L}_r \neq \{0\} \ , \quad 0 \le r \le s \ , \tag{3.13}$$

where $\mathcal{L}_0 = \mathcal{L}(Z_1, \dots, Z_s)^\perp$, $\mathcal{L}_r = \mathcal{L}(Z_r, \dots, Z_s) \ominus \mathcal{L}(Z_{r+1}, \dots, Z_s)$, $1 \le r \le s-1$, and $\mathcal{L}_s = \mathcal{L}(Z_s)$. Let C_r be a matrix whose columns constitute a base of $\mathcal{L}_r$, $0 \le r \le s$. Define $a_{rq} = \|Z_q' C_r\|_4^4$, $0 \le q \le r \le s$. It is easy to see that, under (3.13), one has $a_{rr} > 0$, $0 \le r \le s$ (Exercise 3.4). Let n_r be the number of columns of C_r, and c_{rk} the kth column of C_r, $1 \le k \le n_r$, $0 \le r \le s$. Define

$$b_r(\phi) = 3 \sum_{k=1}^{n_r} \left(\sum_{q=0}^{r} |Z_q' c_{rk}|^2 \phi_v \right)^2 , \quad 0 \le r \le s ,$$

where $\phi = (\phi_r)_{0 \le r \le s}$ with $\phi_0 = \tau^2$ and $\phi_r = \sigma_r^2, 1 \le r \le s$. Let $\kappa = (\kappa_r)_{0 \le r \le s}$. Consider $M(\beta, \phi, \kappa, y) = [M_r(\beta, \phi, \kappa, y)]_{0 \le r \le s}$, where

$$M_r(\beta, \phi, \kappa, y) = \|C_r'(y - X\beta)\|_4^4 - \sum_{q=0}^{r} a_{rq}\kappa_q - b_r(\phi) , \quad 0 \le r \le s .$$

Then, by Lemma 3.1 below and the definition of the C_r's, it is easy to show that $\mathrm{E}\{M(\beta, \phi, \kappa, y)\} = 0$ when β, ϕ, κ are the true parameter vectors (Exercise 3.4). Thus, a set of EMM estimators can be obtained by solving $M(\hat{\beta}, \hat{\phi}, \kappa, y) = 0$, where $\hat{\beta}, \hat{\phi}$ are the REML estimators of β, ϕ, respectively [see (1.3)]. The EMM estimators can be computed recursively, as follows:

$$\begin{aligned} \hat{\kappa}_0 &= a_{00}^{-1} \hat{d}_0 , \\ \hat{\kappa}_r &= a_{rr}^{-1} \hat{d}_r - \sum_{q=0}^{r-1} \left(\frac{a_{rq}}{a_{rr}} \right) \hat{\kappa}_q , \quad 1 \le r \le s , \end{aligned} \tag{3.14}$$

where $\hat{d}_r = \|C_r'(y - X\hat{\beta})\|_4^4 - b_r(\hat{\phi})$, $0 \le r \le s$.

Lemma 3.1. Let $\xi_1, \dots, \xi_n$ be independent random variables such that $\mathrm{E}(\xi_i) = 0$ and $\mathrm{E}(\xi_i^4) < \infty$, and $\lambda_1, \dots, \lambda_n$ be constants. Then, we have

$$\mathrm{E} \left(\sum_{i=1}^{n} \lambda_i \xi_i \right)^4 = 3 \left\{ \sum_{i=1}^{n} \lambda_i^2 \mathrm{var}(\xi_i) \right\}^2 + \sum_{i=1}^{n} \lambda_i^4 \left[\mathrm{E}(\xi_i^4) - 3\{\mathrm{var}(\xi_i)\}^2 \right] .$$

We illustrate the EMM with an example.

Example 3.1 (continued). Consider the balanced case. It is easy to show that $C_0 = I_a \otimes K_b$ and $C_1 = Z_1 = I_a \otimes 1_b$, where

$$K_b = \begin{pmatrix} 1 & \cdots & 1 \\ -1 & \cdots & 0 \\ \vdots & \ddots & \vdots \\ 0 & \cdots & -1 \end{pmatrix}_{b \times (b-1)} .$$

It follows from (3.14) that, in closed-form,

$$\hat{\kappa}_0 = \frac{1}{2a(b-1)} \sum_{i=1}^{a} \sum_{j=2}^{b} (y_{i1} - y_{ij})^4 - 6\hat{\tau}^4 , \tag{3.15}$$

$$\hat{\kappa}_1 = \frac{1}{ab^4} \sum_{i=1}^{a} (y_{i\cdot} - b\hat{\mu})^4 - \frac{1}{2ab^3(b-1)} \sum_{i=1}^{a} \sum_{j=2}^{b} (y_{i1} - y_{ij})^4$$
$$-\frac{3}{b^2}\left(1 - \frac{2}{b}\right)\hat{\tau}^4 - \frac{6}{b}\hat{\tau}^2\hat{\sigma}^2 - 3\hat{\sigma}^4 , \tag{3.16}$$

where $y_{i\cdot} = \sum_{j=1}^{b} y_{ij}$, and $\hat{\mu}$, $\hat{\tau}^2$, and $\hat{\sigma}^2$ are the REML estimators of μ, τ^2, and σ^2, respectively (Exercise 3.5). It is easy to show, either by verifying the conditions of Theorem 1 in Jiang (2003), or by arguing directly, that the EMM estimators are consistent provided only that $a \to \infty$. The only requirement for b is that $b \geq 2$. The latter is reasonable because, otherwise, one cannot separate α and ϵ (Exercise 3.5).

3.3.2 *Partially observed information*

Another idea of estimating the ACVM has to do with the GEE method of estimating the covariance matrix of $\hat{\beta}$, discussed in Subsection 2.2. Recall the derivation of (2.37) from (2.36). The idea is to express a quantity of interest as expectation of sums of random variables. We then dropped the expectation sign and used the expression insided the expectation, with any unknown parameters replaced by their consistent estimator, to approximate the quantity of interest. This is a useful technique; although it should not be "abused". Jiang (2010) (Example 2 in Preface) used the following example to illustrate when the strategy works, and when it does not.

Example 3.4. It might be thought that, with a large sample, one could always approximate the mean of a random quantity by the quantity itself. Before we get into this, let us first be clear on what is meant by a good approximation. An approximation is good if the error of the approximation is of lower order than the approximation itself. In notation, this means that, suppose one wishes to approximate A by B; the approximation is good if $A - B = o(B)$, so that $A = B + A - B = B + o(B) = B\{1 + o(1)\}$. In other words, B is the "main part" of A. Once the concept is clear, it turns out that, in some cases, the approximation technique mentioned above works, while in some other cases, it does not work. For example, suppose that $Y_1, \ldots, Y_n$ are i.i.d. observations such that $\mu = \mathrm{E}(Y_1) \neq 0$. In this case, one can approximate $\mathrm{E}(\sum_{i=1}^{n} Y_i) = n\mu$ by simply removing the expectation sign, that is, by $\sum_{i=1}^n Y_i$. This is because the difference $\sum_{i=1}^n Y_i - n\mu = \sum_{i=1}^n (Y_i - \mu)$ is of the order $O_P(\sqrt{n})$ [provided that $\mathrm{var}(Y_1) < \infty$], which is lower than the order of $\sum_{i=1}^n Y_i$, which is $O_P(n)$. However, the technique completely fails if one considers, for example, approximating $\mathrm{E}(\sum_{i=1}^n Y_i)^2$, where the Y_i's are i.i.d. with mean 0. For simplicity, assume that $Y_i \sim N(0, 1)$. Then, we have $\mathrm{E}(\sum_{i=1}^n Y_i)^2 = n$. On the other hand, $(\sum_{i=1}^n Y_i)^2 = n\chi_1^2$, where χ_1^2 is a random variable with the χ^2 distribution with one degree of freedom. Therefore, $(\sum_{i=1}^n Y_i)^2 - \mathrm{E}(\sum_{i=1}^n Y_i)^2 = n(\chi_1^2 - 1)$, which is of the same order as $(\sum_{i=1}^n Y_i)^2$. Thus, $(\sum_{i=1}^n Y_i)^2$ is not a good approximation to its mean.

It can be seen that a "key-to-success" for this approximation technique is that the random quantity inside the expectation has to be a sum of random variables with "good behavior" (e.g., independence). In the context of likelihood inference, the technique is associated with a term called *observed information*. Consider the case of i.i.d. observations. In this case, the information contained in the entire data, $Y_1, \dots, Y_n$, about the parameter vector, θ, is $\mathcal{I}(\theta) = nI(\theta)$, where, under regularity conditions, one has

$$\mathcal{I}(\theta) = -\sum_{i=1}^n \mathrm{E}\left\{\frac{\partial^2}{\partial\theta\partial\theta'} f(Y_i|\theta)\right\} = -\mathrm{E}\left\{\sum_{i=1}^n \frac{\partial^2}{\partial\theta\partial\theta'} f(Y_i|\theta)\right\}. \quad (3.17)$$

Using the approximation technique, one removes the expectation sign in (3.17) to get an approximation to $\mathcal{I}(\theta)$. The approximation is not yet an estimate, because θ is unknown. Thus, as a last step, one replaced θ in the latest approximation by $\hat{\theta}$, the MLE, to get

$$\widehat{\mathcal{I}(\theta)} = -\sum_{i=1}^n \frac{\partial^2}{\partial\theta\partial\theta'} f(Y_i|\hat{\theta}). \quad (3.18)$$

The estimator (3.18) is called observed information (matrix). Alternatively, especially if $I(\theta)$ has an analytic expression, one can replace θ in this expression by, say, the MLE $\hat{\theta}$ to get

$$\widetilde{\mathcal{I}(\theta)} = nI(\hat{\theta}). \quad (3.19)$$

The estimator (3.19) is called estimated information (matrix). See, for example, Efron and Hinkley (1978), for a discussion and comparison of the two estimators in the i.i.d. case.

However, neither the estimated nor the observed information methods apply to REML estimation. First, as noted, $I_R(\theta)$ is not a function of θ, the vector of variance components, but involves additional parameters, namely the higher moments. Secondly, as it turns out, one cannot express $I_R(\theta)$ as expectation of a sum of random variables that are observable and have good behavior. Note that, under a LMM, the observations are not independent, so the above derivation for the i.i.d. case does not carry through. Nevertheless, it was found that a combination of these two techniques would work just fine. We first use an example to illustrate.

Example 3.2 (continued). It can be shown that the variance of (3.9) can be expressed as $\text{var}(\partial l_{\text{R}}/\partial\sigma_e^2) = S_1 + S_2$, where

$$S_1 = \text{E}\left\{a.\sum_{i,j}\xi_{ij}^4 - a_1\sum_i\left(\sum_j\xi_{ij}\right)^4 - a_2\sum_j\left(\sum_i\xi_{ij}\right)^4\right\}, \quad (3.20)$$

where $\xi_{ij} = y_{ij} - \mu$, $a. = a_0 + a_1 + a_2$, and $a_j, j = 0, 1, 2$, S_2 are functions of θ (Exercise 3.6). Thus, S_2 can be estimated by replacing θ by $\hat{\theta}$, the REML estimator. As for S_1, it can be estimated using the observed information idea, that is, by removing the expectation sign, and replacing μ and θ involved in the expression inside the expectation by their REML estimators.

To describe the method in general, denote the vector of variance components under the mixed ANOVA model by $\theta = (\theta_r)_{0\leq r\leq s}$, where $\theta_0 = \tau^2$ and $\theta_r = \gamma_r, 1 \leq r \leq s$. Then, it can be shown that $\partial l_{\text{R}}/\partial\theta_r = \xi' H_r\xi - h_r, 0 \leq r \leq s$, where $\xi = y - X\beta$, $H_0 = P/2\theta_0$, with P given below (1.4), and $H_r = (\theta_0/2)PZ_rZ_r'P, 1 \leq r \leq s$; $h_0 = (n-p)/2\theta_0$, and $h_r = (\theta_0/2)\text{tr}(PZ_rZ_r'), 1 \leq r \leq s$. Also, with a slight abuse of the notation, let z_{ir}' and z_{rl} be the ith row and lth column of Z_r, respectively, $0 \leq r \leq s$, where $Z_0 = I$ Define $\Gamma(i_1, i_2) = \sum_{r=0}^s \gamma_r(z_{i_1r}\cdot z_{i_2r})$. Here, the dot product of vectors $a_1, \dots, a_k$ of the same dimension is defined as $a_1 \cdot a_2 \cdots a_k = \sum_l a_{1l}a_{2l}\cdots a_{kl}$. Also let $\alpha_0 = \epsilon$ and recall that m_r is the dimension of α_r, $0 \leq r \leq s$ (so, in particular, $m_0 = n$), and that $V = \text{Var}(y) = \theta_0(I_n + \sum_{r=1}^s \theta_r Z_r Z_r')$.

We begin with some expressions for $\text{cov}(\xi_{i_1}\xi_{i_2}, \xi_{i_3}\xi_{i_4})$, where ξ_i is the ith component of ξ, and $\text{cov}(\partial l_{\text{R}}/\partial\theta_q, \partial l_{\text{R}}/\partial\theta_r)$, the (q, r) element of $\mathcal{I}_1$ [see Jiang (2005)].

Lemma 3.2. We have, for any $1 \leq i_j \leq n, j = 1, 2, 3, 4$,

$$\begin{aligned}\text{cov}(\xi_{i_1}\xi_{i_2}, \xi_{i_3}\xi_{i_4}) &= \lambda^2\{\Gamma(i_1, i_3)\Gamma(i_2, i_4) + \Gamma(i_1, i_4)\Gamma(i_2, i_3)\} \\ &\quad + \sum_{r=0}^s \kappa_r z_{i_1r}\cdots z_{i_4r},\end{aligned} \quad (3.21)$$

where $z_{i_1 r} \cdots z_{i_4 r} = z_{i_1 r} \cdot z_{i_2 r} \cdot z_{i_3 r} \cdot z_{i_4 r}$; and, for any $0 \le q, r \le s$,

$$\operatorname{cov}\left(\frac{\partial l_{\mathrm{R}}}{\partial \theta_q}, \frac{\partial l_{\mathrm{R}}}{\partial \theta_r}\right) = 2\operatorname{tr}(H_r V H_r V) + \sum_{t=0}^{s} \kappa_t \sum_{l=1}^{m_t} (z'_{tl} H_q z_{tl})(z'_{tl} H_r z_{tl}). \tag{3.22}$$

Let $f_1, \ldots, f_L$ be the different nonzero functional values of

$$f(i_1, \ldots, i_4) = \sum_{r=0}^{s} \kappa_r z_{i_1 r} \cdots z_{i_4 r}. \tag{3.23}$$

Note that this is the second term on the right side of (3.21). Here by functional value it means $f(i_1, \ldots, i_4)$ as a function of $\kappa = (\kappa_r)_{0 \le r \le s}$. For example, $\kappa_0 + \kappa_1$ and $\kappa_2 + \kappa_3$ are different functions (even if their values may be the same for some κ). Also, let 0 denote the zero function (of κ). Also, let $H_{r,i,j}$ denote the (i, j) element of H_r. Define $\mathcal{A}_l = \{(i_1, ..., i_4) : f(i_1, ..., i_4) = f_l\}, 1 \le l \le L$, and

$$c_{q,r,l} = \frac{1}{|\mathcal{A}_l|} \sum_{(i_1,\ldots,i_4) \in \mathcal{A}_l} H_{q,i_1,i_2} H_{r,i_3,i_4}, \tag{3.24}$$

where $|\cdot|$ denotes cardinality. Note that $c_{q,r,l}$ depends only on θ. Define $c_{q,r}(i_1, \ldots, i_4) = c_{q,r,l}$, if $f(i_1, \ldots, i_4) = f_l$, $1 \le l \le L$. The following result was proved in Jiang (2005). Let $\mathcal{I}_{1,qr}$ denote the (q, r) element of $\mathcal{I}_1$.

Theorem 3.2. For any non-Gaussian mixed ANOVA model, we have, for any $0 \le q, r \le s$, $\mathcal{I}_{1,qr} = \mathcal{I}_{1,1,qr} + \mathcal{I}_{1,2,qr}$, where

$$\mathcal{I}_{1,1,qr} = \mathrm{E}\left\{ \sum_{f(i_1,\ldots,i_4) \ne 0} c_{q,r}(i_1, \ldots, i_4) \xi_{i_1} \cdots \xi_{i_4} \right\}$$

$$\mathcal{I}_{1,2,qr} = 2\operatorname{tr}(H_q V H_r V) - 3\theta_0^2 \sum_{f(i_1,\ldots,i_4) \ne 0} c_{q,r}(i_1, \ldots, i_4) \Gamma(i_1, i_3) \Gamma(i_2, i_4).$$

Theorem 3.2 shows that any element of $\mathcal{I}_1$ can be expressed as the sum of two terms. The first term is expressed as the expectation of a sum with the summands being products of four ξ_i's and a function of θ; the second term is a function of θ only. The first term, $\mathcal{I}_{1,1,qr}$, can be estimated the same way as the observed information, that is, by removing the expectation sign, and replacing β (in the ξ's) and θ by $\hat{\beta}$ and $\hat{\theta}$, the REML estimators, respectively. Denote the estimator by $\hat{\mathcal{I}}_{1,1,qr}$. The second term, $\mathcal{I}_{1,2,qr}$, can be estimated by replacing θ by $\hat{\theta}$. Denote this estimator by $\hat{\mathcal{I}}_{1,2,qr}$. An estimator of $\mathcal{I}_{1,qr}$ is then given by $\hat{\mathcal{I}}_{1,qr} = \hat{\mathcal{I}}_{1,1,qr} + \hat{\mathcal{I}}_{1,2,qr}$. Because the estimator consists partially of an "observed" term and partially an estimated term, it is called *partially observed information*, or POI. Jiang (2005) gives sufficient conditions under which the POI matrix, $\hat{\mathcal{I}}_1 = (\hat{\mathcal{I}}_{1,qr})_{0\le q,r\le s}$, consistently estimate $\mathcal{I}_1$; hence $\hat{\Sigma}_{\mathrm{R}} = \hat{\mathcal{I}}_2^{-1}\hat{\mathcal{I}}_1\hat{\mathcal{I}}_2^{-1}$ consistently estimate Σ_{R} of (3.5), where $\hat{\mathcal{I}}_2$ is $\mathcal{I}_2$ with θ replaced by $\hat{\theta}$, the REML estimator. We consider an example.

Example 3.1 (continued). It is easy to show that $f(i_1j_1,\dots,i_4j_4) = 0$, if not $i_1 = \cdots = i_4$; κ_1, if $i_1 = \cdots = i_4$ but not $j_1 = \cdots = j_4$; and $\kappa_0 + \kappa_1$, if $i_1 = \cdots = i_4$ and $j_1 = \cdots j_4$. Thus, $L = 2$ [note that L is the number of different functional values of $f(i_1j_1,\dots,i_4j_4)$]. Define the following functions of $\theta = (\tau^2,\gamma)'$ and $\gamma = \sigma^2/\tau^2$: $t_0 = 1 - \gamma/(1+\gamma b) - 1/\{(1+\gamma b)ab\}$, $t_1 = (a-1)b/\{a(1+\gamma b)\}$, and $t_3 = \{b(1+\gamma b)^2 - (1+\gamma)^2\}/(b^3-1)$. Then, the POIs are given by $\hat{\mathcal{I}}_{1,qr} = \hat{\mathcal{I}}_{1,1,qr} + \hat{\mathcal{I}}_{1,2,qr}$, $q,r = 0,1$, where

$$\hat{\mathcal{I}}_{1,1,00} = \frac{\hat{t}_1^2 - \hat{t}_0^2 b}{4\hat{\tau}^8 b(b^3-1)}\left\{\sum_i\left(\sum_j \hat{\xi}_{ij}\right)^4 - \sum_{i,j}\hat{\xi}_{ij}^4\right\} + \frac{\hat{t}_0^2}{4\hat{\tau}^8}\sum_{i,j}\hat{\xi}_{ij}^4,$$

$$\hat{\mathcal{I}}_{1,1,01} = \frac{(a-1)(\hat{t}_1 b - \hat{t}_0)}{4\hat{\tau}^6(1+\hat{\gamma}b)^2 a(b^3-1)}\left\{\sum_i\left(\sum_j \hat{\xi}_{ij}\right)^4 - \sum_{i,j}\hat{\xi}_{ij}^4\right\},$$

$$+\frac{(a-1)\hat{t}_0}{4\hat{\tau}^6(1+\hat{\gamma}b)^2 a}\sum_{i,j}\hat{\xi}_{ij}^4,$$

$$\hat{\mathcal{I}}_{1,1,11} = \frac{(a-1)^2}{4\hat{\tau}^4(1+\hat{\gamma}b)^4a^2}\sum_i\left(\sum_j \hat{\xi}_{ij}\right)^4;$$

$$\hat{\mathcal{I}}_{1,2,00} = \frac{1}{2\hat{\tau}^4}\left[ab - 1 - \frac{3}{2}ab\hat{t}_0^2\{(1+\hat{\gamma})^2 - \hat{t}_3\} - \frac{3}{2}a\hat{t}_1^2\hat{t}_3\right],$$

$$\hat{\mathcal{I}}_{1,2,01} = \frac{(a-1)b}{2\hat{\tau}^2(1+\hat{\gamma}b)}\left\{1 - \left(\frac{3}{2}\right)\frac{(\hat{t}_1 b - \hat{t}_0)\hat{t}_3 + (1+\hat{\gamma})^2\hat{t}_0}{1+\hat{\gamma}b}\right\},$$

$$\hat{\mathcal{I}}_{1,2,11} = -\frac{(a-1)(a-3)b^2}{4a(1+\hat{\gamma}b)^2},$$

$\hat{\xi}_{ij} = y_{ij} - \bar{y}_{\cdot\cdot}$, and $\hat{t}_j, j = 0,1,3$ are $t_j, j = 0,1,3$, respectively, with θ replaced by $\hat{\theta}$, the REML estimator.

Note on application: It is seen from Theorem 3.2 that POI or, more specifically, $\mathcal{I}_{1,1}$, is based on first considering $\xi = y - X\beta$ and then replacing the β in ξ by $\hat{\beta}$. On the other hand, $\mathcal{I}_{1,2}$ and $\mathcal{I}_2$ are functions of θ, which are estimated by replacing θ by $\hat{\theta}$. Thus, practically, one can derive the

POI formulae by the following A-B-C steps:

(A) Obtain the REML estimators of θ and β.

(B) Replace y by $\hat{\xi} = y - X\hat{\beta}$, then derive the POI formulae based on the same LMM but without the fixed effects.

(C) Once the formulae in (B) are derived, make sure to use the REML estimator of $\hat{\theta}$ in (A) [not the one based on the LMM model without the fixed effects that is used in (B)].

It can be seen that this procedure was followed in Example 3.1 (continued) discussed above.

3.3.3 *Hypothesis testing: A simulated example*

As an application, we consider using the POI in testing a hypothesis regarding the variance components, or dispersion parameters. The example is regarding Example 3.1 (continued), discussed above. The test is considered robust in the sense that it does not require normality.

Suppose that one wishes to test the hypothesis

$$\mathrm{H}_0: \quad \gamma = 1, \tag{3.25}$$

that is, the variance contribution due to the random effects is the same as that due to the errors. For example, the null hypothesis is equivalent to $\mathrm{H}_0 : h^2 = 2$, where $h^2 = 4\gamma/(1+\gamma)$ is a quantity of genetic interest, called *heritability*. The null hypothesis can be expressed as $\mathrm{H}_0 : K'\theta = 0$ with $K = (0,1)'$. Furthermore, we have $K'\Sigma_{\mathrm{R}}K = \Sigma_{\mathrm{R},11}$, which is the asymptotic variance of $\hat{\gamma}$, the REML of γ. Thus, a test statistic is $\hat{\chi}^2 = (\hat{\gamma}-1)^2/\hat{\Sigma}_{\mathrm{R},11}$, where $\hat{\Sigma}_{\mathrm{R},11}$ is the POI of $\Sigma_{\mathrm{R},11}$. It is easy to show that

$$\hat{\Sigma}_{\mathrm{R},11} = \frac{\hat{\mathcal{I}}_{1,11}\hat{\mathcal{I}}_{2,00}^2 - 2\hat{\mathcal{I}}_{1,01}\hat{\mathcal{I}}_{2,00}\hat{\mathcal{I}}_{2,01} + \hat{\mathcal{I}}_{1,00}\hat{\mathcal{I}}_{2,01}^2}{(\hat{\mathcal{I}}_{2,00}\hat{\mathcal{I}}_{2,11} - \hat{\mathcal{I}}_{2,01}^2)^2}, \tag{3.26}$$

where $\hat{\mathcal{I}}_{1,qr} = \hat{\mathcal{I}}_{1,1,qr} + \hat{\mathcal{I}}_{1,2,qr}$, $q,r = 0,1$, and $\hat{\mathcal{I}}_{1,j,qr}$, $j = 1,2$ are given earlier, but with $\hat{\gamma}$ replaced by 1, its value under H_0. Furthermore, we have $\hat{\mathcal{I}}_{2,00} = -(ab-1)/2\hat{\tau}^4$, $\hat{\mathcal{I}}_{2,01} = -(a-1)b/2\hat{\tau}^2(1+\hat{\gamma}b)$, $\hat{\mathcal{I}}_{2,11} = -(a-1)b^2/2(1+\hat{\gamma}b)^2$, again with $\hat{\gamma}$ replaced by 1, where $\hat{\tau}^2$ is the REML estimator of τ^2 under the null, given by

$$\hat{\tau}^2 = \frac{1}{ab-1}\left(\mathrm{SSE} + \frac{\mathrm{SSA}}{b+1}\right), \tag{3.27}$$

where SSE$= \sum_{i=1}^a \sum_{j=1}^b (y_{ij} - \bar{y}_{i\cdot})^2$, SSA$= b\sum_{i=1}^a (\bar{y}_{i\cdot} - \bar{y}_{\cdot\cdot})^2$ with $\bar{y}_{i\cdot} = b^{-1}\sum_{j=1}^b y_{ij}$ and $\bar{y}_{\cdot\cdot} = (ab)^{-1}\sum_{i=1}^a \sum_{j=1}^b y_{ij}$. The asymptotic null distribution is χ_1^2.

Table 3.1 **RDT versus Jackknife - Size**

Nominal Level	Method	Simulated Size					
		I-i	I-ii	I-iii	II-i	II-ii	II-iii
0.01	RDT	0.022	0.026	0.028	0.011	0.013	0.015
	Jackknife	0.010	0.014	0.020	0.009	0.011	0.013
0.05	RDT	0.070	0.078	0.091	0.054	0.057	0.063
	Jackknife	0.052	0.053	0.068	0.053	0.053	0.060
0.10	RDT	0.123	0.132	0.151	0.106	0.108	0.114
	Jackknife	0.099	0.103	0.122	0.104	0.103	0.109

Jiang (2005) carried out a simulation study on the performance of the dispersion test proposed above, which we call robust dispersion test (RDT), and compared it with an alternative delete-group jackknife method proposed by Arvesen (1969). The latter applies to cases where data can be divided into i.i.d. groups, such as the current situation. We refer to this test as jackknife. More specifically, Arvesen and Schmitz (1970) proposed to use the jackknife estimator with a logarithm transformation, and this is the method that was compared.

We are interested in the situation when a is increasing while b remains fixed. Therefore, the following sample size configurations are considered: (I) $a = 50$, $b = 2$; (II) $a = 400$, $b = 2$. (I) represents a case of moderate sample size while (II), a case of large sample. In addition, we would like to investigate different cases in which normality and symmetry may or may not hold. Therefore, the following combinations of distributions for the random effects and errors are considered: (i) Normal-Normal; (ii) DE-NM$(-2, 2, 0.5)$, where DE represents the double exponential distribution and NM(μ_1, μ_2, ρ) the mixture of two normal distributions with means μ_1, μ_2, variance one, and mixing probability ρ [i.e., the probabilities $1 - \rho$ and ρ correspond to $N(\mu_1, 1)$ and $N(\mu_2, 1)$, respectively]; and (iii) CE-NM$(-4, 1, 0.2)$, where CE represents the centralized exponential distribution, i.e., the distribution of $X - 1$, where $X \sim$ Exponential(1). Note that in case (ii) the distributions are not normal but symmetric, while in case (iii) the distributions are not even symmetric, a further departure from normality. Also note that all these distributions have mean zero. They are standardized so that the distributions of the random effects and errors have variances σ^2 and τ^2, respectively. The true value of μ is set to 1.0. The true value of τ^2 is also chosen as 1.0. Tables 3.1–3.4 are taken from Jiang (2005), in which the simulated size and power, with the nominal level of $\alpha = 0.05$, are reported for testing the null hypothesis against the alternative $\mathrm{H}_1 : \gamma \neq 1$.

Table 3.2 **RDT versus Jackknife - Power (Nominal Level 0.01)**

Alternative	Method	Simulated Power					
		I-i	I-ii	I-iii	II-i	II-ii	II-iii
$\gamma_1 = 0.2$	RDT	0.506	0.616	0.468	1.000	1.000	1.000
	Jackknife	0.487	0.463	0.454	1.000	1.000	1.000
$\gamma_1 = 0.5$	RDT	0.112	0.164	0.122	0.914	0.891	0.793
	Jackknife	0.108	0.121	0.137	0.921	0.866	0.787
$\gamma_1 = 2.0$	RDT	0.354	0.256	0.221	0.995	0.971	0.913
	Jackknife	0.196	0.118	0.072	0.993	0.968	0.887
$\gamma_1 = 5.0$	RDT	0.991	0.954	0.900	1.000	1.000	1.000
	Jackknife	0.954	0.876	0.715	1.000	1.000	1.000

Table 3.3 **RDT versus Jackknife - Power (Nominal Level 0.05)**

Alternative	Method	Simulated Power					
		I-i	I-ii	I-iii	II-i	II-ii	II-iii
$\gamma = 0.2$	RDT	0.747	0.807	0.745	1.000	1.000	1.000
	Jackknife	0.728	0.709	0.668	1.000	1.000	1.000
$\gamma = 0.5$	RDT	0.283	0.336	0.286	0.980	0.966	0.917
	Jackknife	0.277	0.271	0.275	0.981	0.958	0.912
$\gamma = 2.0$	RDT	0.532	0.424	0.369	0.999	0.993	0.973
	Jackknife	0.411	0.317	0.223	0.999	0.993	0.970
$\gamma = 5.0$	RDT	0.997	0.984	0.956	1.000	1.000	1.000
	Jackknife	0.991	0.971	0.903	1.000	1.000	1.000

Table 3.4 **RDT versus Jackknife - Power (Nominal Level 0.10)**

Alternative	Method	Simulated Power					
$\gamma_1 = 0.2$	RDT	0.844	0.875	0.807	1.000	1.000	1.000
	Jackknife	0.829	0.810	0.776	1.000	1.000	1.000
$\gamma_1 = 0.5$	RDT	0.405	0.442	0.396	0.991	0.983	0.954
	Jackknife	0.398	0.382	0.372	0.991	0.979	0.950
$\gamma_1 = 2.0$	RDT	0.633	0.564	0.462	1.000	0.997	0.987
	Jackknife	0.540	0.453	0.350	1.000	0.997	0.986
$\gamma_1 = 5.0$	RDT	0.999	0.992	0.975	1.000	1.000	1.000
	Jackknife	0.998	0.988	0.954	1.000	1.000	1.000

Overall, the jackknife appears to be more accurate in terms of the size, especially when a is relatively small (Case I). On the other hand, the simulated powers for RDT are higher at all alternatives, especially when a is relatively small (Case I). Also note that the jackknife with the logarithmic transformation is specifically designed for this kind of model where the observations are divided into independent groups, while the POI method is for a much richer class of LMM where the observations may not be divided into independent groups. There are examples of the latter kind in which tests about the dispersion parameters are of interest, but we shall leave this to the next chapter.

3.3.4 *Consistency of POI*

In this subsection, we discuss consistency of the estimated ACVM obtained via the POI method. It should be noted that the definition of REML estimator in non-Gaussian LMM differs slightly according to several authors. In Richardson and Welsh (1994), REML estimator is defined as the solution to the (Gaussian) REML equation; in Jiang (1996), REML estimator is defined as the solution to the REML equation plus the requirement that it belongs to the parameter space; in Jiang (1997), REML estimator is defined as the maximizer of the Gaussian restricted likelihood function. In fact, the latter shows that, for a balanced mixed ANOVA model, such a maximizer is a consistent estimator of θ; for a general LMM, it shows that a sieved maximizer is consistent. Note that, from a practical point of view, a sieve puts no restriction on the maximization, because the maximizer is always within a sieve that satisfies the conditions [of Jiang (1997), with a suitable constant]. Therefore, in the following theorem, the REML estimator is understood as the maximizer of the Gaussian restricted likelihood in the sense of Jiang (1997) (with the sieves in the unbalanced case). This eliminates any possible confusion on which solution, or root, to the REML equation to use when there are multiple roots. See, for example, Searle *et al.* (1992, sec. 8.1) for discussion regarding the multiple-root problem. The following theorem is proved in Jiang (2005).

Theorem 3.3. Suppose that (i) $\sigma_r^2 > 0$, $0 < \mathrm{var}(\alpha_{r1}^2) < \infty$, $0 \le r \le s$; (ii)$|x_i|$, $\|z_{ir}\|_1$, $1 \le r \le s$, $1 \le i \le N$ are bounded; (iii) there is a sequence of diagonal matrices $G = \mathrm{diag}(g_0, \dots, g_s)$ with $g_j > 0$, $0 \le j \le s$ such that $G^{-1}\mathcal{G}G^{-1}$ is bounded from above as well as from below; and $\lambda_{\min}(X'V^{-1}X) \to \infty$; (iv) $(g_j g_k)^{-1}\sum_{l=1}^L h_l |c_{j,k,l}|$, $0 \le j,k \le s$ are bounded, and $(g_j g_k)^{-2}\sum_{l_1,l_2=1}^L h_{l_1,l_2}|c_{j,k,l_1}c_{j,k,l_2}| \to 0$, $0 \le j,k \le s$; (v) $(g_j g_k)^{-1} g_{j,k}(\delta) \to 0$, and $(g_j g_k)^{-1}\sum_{l=1}^L h_l d_{j,k,l}(\delta) \to 0$, $0 \le j,k \le s$, uniformly in N as $\delta \to 0$. Then, the POI $\hat{\mathcal{I}}_1$ and POI estimator $\hat{\Sigma}_{\mathrm{R}}$ are both consistent.

Remarks. The first part of condition (iii) (regarding $\mathcal{G}$) is equivalent to the AI^4 condition of Jiang (1996, 1997), which, together with $\sigma_r^2 > 0$, $0 \le r \le s$, guarantees the consistency of the REML estimator $\hat{\theta}$. Furthermore, condition (iii) ensures the consistency of $\hat{\beta} = (X'\hat{V}^{-1}X)^{-1}X'\hat{V}^{-1}y$, where $\hat{V}$ is V with θ replaced by $\hat{\theta}$. Finally, it can be shown (Exercise 3.7) that the first part of condition (v) [regarding $g_{j,k}(\delta)$] is equivalent to that $\tilde{\mathcal{G}}$ is uniformly continuous at θ].

3.4 Real-data example

We use a real-data example to illustrate the potential difference in measure of uncertainty with or without taking into account the non-normality. The data was collected from a Statistics class taught at the University of California, Davis. Partial data is shown in Table 3.5. The first two columns are gender (F for female; M for male and midterm (MD) scores, in percentage (out of a total of 50 points). The rest of the columns are HW scores, also in percentage. There were 112 students enrolled in the class. Each student had a midterm score and six HW scores. The NAs in the data set correspond to HWs that were either missing (e.g., not turned in), or intentionally dropped when computing the overall HW scores. According to the grading policy, the lowest two HW scores were dropped when computing the overall HW scores. The midterm took place near the midpoint of the quarter after the 4th HW (so HW 1, 2, 3, 4 were before the midterm and the rest of the HWs after the midterm). Initially, there were interests in knowing the potential association between the HWs and midterm, and impact of the midterm on how seriously the students completed the HWs. There was also interest in knowing whether there was a difference between male and female in completing the HWs. To address such concerns, the following simple LMM is considered. Let y_{ij} denote the jth HW score (skip the NAs) of the ith student. It is assumed that

$$y_{ij} = \beta_0 + \beta_1 x_{i,1} + \beta_2 x_{i,2} + \beta_3 x_{ij,3} + v_i + e_{ij}, \tag{3.28}$$

$i = 1, \dots, 112$, $j = 1, \dots, 6$, where $x_{i,1}$ is the indicator for gender (1 for F; 0 for M); $x_{i,2}$ is the midterm score; $x_{ij,3}$ is an indicator on whether the HW score is before or after the midterm (0 for before; 1 for after). Note that $x_{i,1}, x_{i,2}$ do not depend on j, but $x_{ij,3}$ depends on both i and j. Furthermore, the β's are known fixed effects; v_i is a student-specific random effect, and e_{ij} is a random error. It is assumed that the random effects are i.i.d. with mean 0 and variance σ_v^2; the errors are i.i.d. with mean 0 and variance σ_e^2; and the random effects and errors are independent.

Note that the data is clearly not normal. For example, the HW scores are percentages out of a total of 100 points for each HW. Such numbers are between 0 and 1, and only up to two decimals. Also, it is seen from Table 3.5 that many numbers are exactly 1, and some exactly 0. Thus, the proposed LMM is a (truly) non-Gaussian LMM.

Table 3.5 **Partial Data**

Gender	MD	HW1	HW2	HW3	HW4	HW5	HW6	HW7	HW8
M	0.66	1.00	0.99	0.96	0.96	NA	1.00	1.00	NA
M	0.68	NA	0.00	0.92	NA	0.78	0.94	0.85	1.00
F	0.82	0.95	0.94	0.98	NA	1.00	1.00	NA	1.00
F	0.82	0.85	NA	0.90	1.00	NA	0.88	0.95	0.96
F	0.72	NA	NA	1.00	1.00	1.00	1.00	1.00	1.00
F	0.80	0.90	0.98	0.96	1.00	0.97	1.00	NA	NA

A question of interest is whether it is necessary to include the student-specific random effects, v_i, in (3.28). The Gaussian REML estimator of the ratio, $\gamma = \sigma_v^2/\sigma_e^2$, is $\hat{\gamma} = 0.193$. On the other hand, the diagonal element of the normality-based ACVM, (3.8), is 2.79×10^{-3}, leading to a standard error (s.e.) for $\hat{\gamma}$ equal to 0.053. Thus, the ratio $\hat{\gamma}/\text{s.e.}(\hat{\gamma}) = 3.65$, which is statistically significant (for testing the hypothesis that γ is zero), say, at 10% level. However, due to the non-normality concern noted above, one would use the POI method to estimate the ACVM. The resulting diagonal element corresponding to γ is 2.60×10^{-2}, leading to a much larger s.e. for $\hat{\gamma}$ equal to 0.161, and the ratio $\hat{\gamma}/\text{s.e.}(\hat{\gamma}) = 1.19$. The result is no longer significant at 10% level. Based on the latter result, one would exclude the random effects from (3.28), and thus simplify the LMM to a linear regression (LR) model.

Note that the analysis results based on the LMM and LR models would lead to quite different conclusions. For example, both analyses find that the coefficient, β_1, corresponding to gender, is insignificant (p-value close to 1 for both). On the other hand, the LMM analysis finds that β_2 is insignificant (p-value ≈ 0.211) while β_3 is significant at 10% level (p-value ≈ 0.085); the LR analysis finds the opposite: β_2 is significant at 10% level (p-value ≈ 0.090) while β_3 is insignificant (p-value ≈ 0.144).

3.5 Exercises

3.1. Show that, in the balanced case of Example 3.1, one has a special case of the balanced mixed ANOVA model defined in §3.1.1. Also show that Example 3.2 is a special case of the balanced mixed ANOVA model.

3.2. Show that the REML equations derived under the multivariate t-distribution (see Section 3.2) are equivalent to those derived under the multivariate normal distribution.

3.3. This exercise has two parts.

a. Show that, in Example 3.2 (continued) in Section 3.3, $\mathrm{E}(\partial^2 l_{\mathrm{R}}/\partial\sigma_e^4)$, $\mathrm{E}(\partial^2 l_{\mathrm{R}}/\partial\sigma_e^2\partial\sigma_u^2)$ and $\mathrm{E}(\partial^2 l_{\mathrm{R}}/\partial\sigma_e^2\partial\sigma_v^2)$ are functions of $\theta = (\sigma_e^2, \sigma_u^2, \sigma_v^2)'$ regardless of normality.

b. Show that $\mathrm{var}(\partial l_{\mathrm{R}}/\partial\sigma_e^2)$ involves the fourth moments of the random effects and errors, but not the third moments.

3.4. This exercise has two parts.

a. Show that, under (3.13), one has $a_{rr} > 0, 0 \leq r \leq s$.

b. Show that, in Subsection 3.3.1, one has $\mathrm{E}\{M(\beta, \phi, \kappa, y)\} = 0$, if β, ϕ, κ are the true parameter vectors. ϕ, κ are the true parameter vectors

3.5. This exercise is associated with Example 3.2 (continued) in Subsection 3.3.1.

a. Verify (3.15) and (3.16).

b. Show that the EMM estimators given by (3.15) and (3.16) are consistent provided that $a \to \infty$ and $b \geq 2$.

c. Explain why the condition $b \geq 2$ in part b above is necessary.

3.6. Consider Example 3.2 (continued) in Subsection 3.3.2. show that the variance of (3.9) can be expressed as $\mathrm{var}(\partial l_{\mathrm{R}}/\partial\sigma_e^2) = S_1 + S_2$, where S_1 is given by (3.20), $\xi_{ij} = y_{ij} - \mu$, $a. = a_0 + a_1 + a_2$, and $a_j, j = 0, 1, 2$, S_2 are functions of θ.

3.7. This exercise has two parts.

a. Prove the following **Lemma**: Let A, G be sequences of positive definite matrices such that $G^{-1}AG^{-1} \to B > 0$. Let $\tilde{A}$ be another sequence of matrices. Then, $A^{-1/2}\tilde{A}A^{-1/2} \to I$, the identity matrix, if and only if $G^{-1}(\tilde{A} - A)G^{-1} \to 0$.

b. Use the above lemma to show that condition (v) of Theorem 3.3 [regarding $g_{j,k}(\delta)$] is equivalent to that $\tilde{\mathcal{G}}$ is uniformly continuous at θ.

Chapter 4

Robust Tests

Part of the previous chapter (see Subsection 3.3.3) has led to the consideration of robust tests in LMM. In this chapter, we continue the coverage on this topic in much greater generality.

4.1 Robust dispersion tests

To be more specific, here in this section, we consider tests regarding only the variance components. Such tests are called *dispersion tests*. Also, we shall focus on tests based on the REML estimators.

Consider the following general hypothesis:

$$\mathrm{H}_0: \quad K'\theta = \varphi, \tag{4.1}$$

where φ is a specified vector, and K is a known $(s+1) \times r$ matrix with $\mathrm{rank}(K) = r$. We assume that the REML estimator $\hat{\theta}$ is asymptotically normal with mean 0 and ACVM Σ_{R}, i.e.,

$$\Sigma_{\mathrm{R}}^{-1/2}(\hat{\theta} - \theta) \longrightarrow N(0, I_{s+1}) \quad \text{in distribution.} \tag{4.2}$$

Sufficient conditions for (4.2) can be found in, e.g., Jiang (1996). It is then easy to show that, under the null hypothesis (4.1),

$$(K'\hat{\theta} - \varphi)'(K'\Sigma_{\mathrm{R}}K)^{-1}(K'\hat{\theta} - \varphi) \longrightarrow \chi_r^2 \quad \text{in distribution.} \tag{4.3}$$

We then replace Σ_{R} by its POI estimator $\hat{\Sigma}_{\mathrm{R}}$, discussed in Subsection 3.3.2, or its EMM estimator, discussed in Subsection 3.3.1, to obtain the test statistic

$$\hat{\chi}^2 = (K'\hat{\theta} - \varphi)'(K'\hat{\Sigma}_{\mathrm{R}}K)^{-1}(K'\hat{\theta} - \varphi). \tag{4.4}$$

The following theorem states that $\hat{\chi}^2$ has the same asymptotic null distribution as (4.3).

Theorem 4.1. Suppose that the conditions of Theorem 3.3 are satisfied. Furthermore, suppose that (4.2) holds. Then, under the null hypothesis, $\hat{\chi}^2 \to \chi_r^2$ in distribution.

In cases that some components of θ are specified under the null hypothesis, it is customary to use these specified values, instead of the estimators, in the POI estimator. Under the null hypothesis, this may improve the accuracy of the POI estimator, although the difference is expected to be small in large sample [because of the consistency of $\hat{\theta}$; e.g., Jiang (1996)]. It can be shown [see Jiang (2005)] that the same conclusion holds after such a modification. We consider some examples.

Example 4.1. Let us revisit the example discussed in Subsection 3.3.3. It is easy to show that, in this case, the test statistic (4.4) reduces to the expression above (3.26), that is, $\hat{\chi}^2 = (\hat{\gamma}_1 - 1)^2/\hat{\Sigma}_{\mathrm{R},11}$, where $\hat{\Sigma}_{\mathrm{R},11}$ is given by (3.26) (Exercise 4.1). According to Theorem 4.1, the asymptotic null distribution is χ_1^2.

Example 4.2. Consider the balanced two-way random effects model of Example 3.2. Suppose that one is interested in testing the following hypothesis H_0: $\sigma_u^2 = \sigma_v^2$, or, equivalently, H_0: $\gamma_1 = \gamma_2$, which means that the two random effect factors contribute equally to the total variation. It is easy to show that the test statistic (4.4) reduces to

$$\hat{\chi}^2 = \frac{(\hat{\gamma}_1 - \hat{\gamma}_2)^2}{\hat{\Sigma}_{\mathrm{R},11} - 2\hat{\Sigma}_{\mathrm{R},12} + \hat{\Sigma}_{\mathrm{R},22}}, \tag{4.5}$$

where $\hat{\gamma}_1$, $\hat{\gamma}_2$ are the REML estimators of γ_1 and γ_2, respectively, and $\hat{\Sigma}_{\mathrm{R},jk}$ is the j,k element of the POI estimator $\hat{\Sigma}_{\mathrm{R}}$ of the ACVM of $\hat{\theta} = (\hat{\sigma}_e^2, \hat{\gamma}_1, \hat{\gamma}_1)'$, the REML estimator. Note that in this case there is no (fully) specified values of the parameters under the null hypothesis, although the latter may still be used in some way (but the difference is expected to be small in large sample; see the remark below Theorem 4.1).

On the other hand, it is interesting to see how the test performs when straight POI estimator is used in the denominator of (4.5). A simulation study was carried out in Jiang (2005) to investigate this. More specifically, it considers performance of the test under both moderate and large sample sizes, as well as departures from normality. The following sample size configurations are considered: (I) $a = 40$, $b = 40$; (II) $a = 200$, $b = 200$. Furthermore, the following combinations of distributions for the random effects and errors are considered: (i) u, $v \sim$ Normal; (ii) u, $v \sim$ DE; (iii) $u \sim$ DE, $v \sim$ CE; and (iv) u, $v \sim$ CE. In all cases, $e \sim$ Normal.

Note that the jackknife method discussed in Subsection 3.3.3 does not

Table 4.1 **Simulated Size**

Nominal	I-i	I-ii	I-iii	I-iv	II-i	II-ii	II-iii	II-iv
0.01	0.014	0.011	0.014	0.011	0.011	0.008	0.011	0.008
0.05	0.071	0.061	0.070	0.066	0.053	0.051	0.055	0.048
0.10	0.135	0.126	0.139	0.136	0.108	0.108	0.109	0.102

Table 4.2 **Simulated Power (Nominal Level 0.01)**

Alternative	I-i	I-ii	I-iii	I-iv	II-i	II-ii	II-iii	II-iv
$\gamma_2/\gamma_1 = 0.2$	0.955	0.568	0.551	0.398	1.000	1.000	0.999	0.986
$\gamma_2/\gamma_1 = 0.5$	0.313	0.100	0.118	0.073	0.988	0.684	0.619	0.439
$\gamma_2/\gamma_1 = 2.0$	0.324	0.100	0.070	0.088	0.988	0.685	0.459	0.443
$\gamma_2/\gamma_1 = 5.0$	0.969	0.649	0.491	0.497	1.000	0.999	0.989	0.992

Table 4.3 **Simulated Power (Nominal Level 0.05)**

Alternative	I-i	I-ii	I-iii	I-iv	II-i	II-ii	II-iii	II-iv
$\gamma_2/\gamma_1 = 0.2$	0.994	0.864	0.839	0.713	1.000	1.000	1.000	0.999
$\gamma_2/\gamma_1 = 0.5$	0.579	0.308	0.321	0.232	0.998	0.874	0.819	0.713
$\gamma_2/\gamma_1 = 2.0$	0.595	0.305	0.227	0.256	0.998	0.879	0.764	0.717
$\gamma_2/\gamma_1 = 5.0$	0.997	0.901	0.799	0.779	1.000	1.000	1.000	0.999

Table 4.4 **Simulated Power (Nominal Level 0.10)**

Alternative	I-i	I-ii	I-iii	I-iv	II-i	II-ii	II-iii	II-iv
$\gamma_2/\gamma_1 = 0.2$	0.999	0.946	0.923	0.846	1.000	1.000	1.000	1.000
$\gamma_2/\gamma_1 = 0.5$	0.702	0.456	0.451	0.364	0.999	0.931	0.887	0.818
$\gamma_2/\gamma_1 = 2.0$	0.719	0.448	0.359	0.382	0.999	0.936	0.864	0.818
$\gamma_2/\gamma_1 = 5.0$	0.999	0.955	0.904	0.879	1.000	1.000	1.000	1.000

apply to this case, because the observations cannot be divided into i.i.d., or even independent, groups.

The true values of parameters are $\mu = \sigma_e^2 = \sigma_u^2 = 1.0$. The value of σ_v^2 varies. First consider the size of the test, so we take $\sigma_v^2 = 1.0$. The simulated sizes corresponding to the nominal levels 0.01, 0.05 and 0.10 are reported in Table 4.1. Next consider the power of the test at the following alternatives: $\sigma_v^2 = 0.2, 0.5, 2, 5$, which correspond to $\gamma_2/\gamma_1 = 0.2, 0.5, 2, 5$, respectively. The simulated powers are reported in Tables 4.2–4.4. All results were based on 10,000 simulation runs.

The numbers seem to follow the same pattern. As the sample size increases, the simulated sizes get closer to the nominal levels, and the simulated powers increase significantly. There does not seem to be difference, in terms of the size, across different distributions. However, the simulated powers appear significantly higher when all the distributions are normal as compared to other cases where the distributions of the random effects are non-normal. Also, the powers are relatively low when the alternatives are close to the null ($\gamma_2/\gamma_1 = 0.5$ or 2.0) but much improved when the alternatives are further away ($\gamma_2/\gamma_1 = 0.2$ and 5.0). Overall, the simulation results are consistent with the theoretical findings of Theorem 4.1.

4.2 Robust versions of classical tests

Robust testing procedures have been studied extensively in the literature. In particular, robust versions of the classical tests, that is, the Wald, score, and likelihood-ratio tests [e.g., Lehmann (1999), §7] have been considered. In the case of i.i.d. observations, see, Foutz and Srivastava (1977), Kent (1982), Hampel *et al.* (1986), Heritier and Ronchetti (1994), among others. In the case of independent but not identically distributed observations, see, for example, Schrader and Hettmansperger (1980), Chen (1985), Silvapulle (1992), and Kim and Cai (1993). In many cases, such as in biomedical research and surveys, the observations are dependent. One important class of models for dependent observations are LMM. Under such a model, the observations are correlated due to the presence of the random effects. Tests with dependent data have been developed in the case of LMM [e.g. Khuri *et al.* (1998)]. Other areas in which dependent data are frequently encountered include time series and stochastic processes, in which tests with dependent data have also been developed [e.g., Basawa and Rao (1980), Dzhaparidze (1986)]. Note that, in these tests with dependent data, it is assumed that the distribution of the data is known up to a set of parameters (therefore the likelihood function is available). In contrast to the independent cases, the literature on robust testing with dependent observations is not extensive. For example, in the case of LMM, Richardson and Welsh (1996) proposed a robust likelihood-ratio test based on restricted maximum likelihood (REML) estimation. A purpose of this section is to provide a unified treatment of these robust versions of classical tests for the case of dependent observations with rigorous proofs. In particular, we apply the results to LMM to obtain robust dispersion tests without the normality assumption.

In many cases, one is interested in testing a hypothesis regarding some parameters associated with the population, but the entire population distribution is unknown given these, and possibly other parameters. In such cases, the likelihood function, based on which the classical tests are defined, is not available, because it relies on full specification of the distribution of the observations given the parameters. However, one may replace the likelihood function by a known objective function, and thus obtain robust versions of the classical tests. In this section, the objective function is assumed to be a function for M-estimation [Huber (1981)].

It should be noted that there has been a quite unified extension of the classical tests in terms of quasi-likelihood method (see Subsection 3.2). The latter is a special case of M-estimation, in which the estimating functions are closer to a (log-)likelihood in the sense that the quasi-score function maintains some properties of the first two moments of a score function. However, it seems that the class of quasi-likelihood functions is not rich enough to include some useful methods. Furthermore, it is not always easy to construct a quasi-likelihood function, especially when the parameter is multivariate, and the observations are dependent [e.g., McCullagh and Nelder (1989), §9.3.2].

The discussions of the current section have the following features: (i) they are for dependent data; (ii) they are based on a general objective function (which is not necessarily a likelihood); and (iii) the hypotheses to be tested are in a general form. The robustness of these tests is in the sense of (ii), that is, the tests require weak distributional assumptions, and therefore are robust to violation of a specific distributional assumption, such as normality. Although many practitioners would apply such tests to dependent data anyway, and assume that the asymptotic results to hold, at least under certain conditions, the exact conditions seem to be unclear for a testing problem that has the characteristics of (i)–(iii).

4.2.1 *Basic idea, assumptions, and examples*

The likelihood-ratio test is known to be applicable to a broad range of testing problems. Let $L(\theta, y)$ be the likelihood function, where y is a vector of observations and θ a vector of parameters. Then, the likelihood-ratio test of the hypothesis

$$H_0: \quad \theta \in \Theta_0 \quad \text{versus} \quad H_1: \quad \theta \notin \Theta_0 , \tag{4.6}$$

where $\Theta_0 \subset \Theta$ and Θ is the parameter space, is based on the likelihood-ratio statistic

$$\frac{\sup_{\theta \in \Theta_0} L(\theta, y)}{\sup_{\theta \in \Theta} L(\theta, y)} = \frac{L(\hat{\theta}_0, y)}{L(\hat{\theta}, y)} , \tag{4.7}$$

where $\hat{\theta}$ is the ML estimator (MLE) of θ, and $\hat{\theta}_0$ the MLE of θ under H_0.

The Wald test, on the other hand, is known to apply to some more restricted testing problems. The original Wald test is for the hypothesis H_0: $\theta = \theta_0$ versus H_1: $\theta \neq \theta_0$, where θ_0 is a known parameter vector. The test statistic can be expressed as $(\hat{\theta} - \theta_0)' I(\theta_0)(\hat{\theta} - \theta_0)$, where $I(\theta)$ is the Fisher information matrix. The method has been extended to test (4.6), provided that H_0 can be expressed as $R(\theta) = 0$, where $R(\cdot)$ is a known (vector valued) function. In such a case, the test statistic has the form

$$R(\hat{\theta})' \left[\left(\left. \frac{\partial R}{\partial \theta} \right|_{\hat{\theta}} \right)' I(\hat{\theta})^{-1} \left(\left. \frac{\partial R}{\partial \theta} \right|_{\hat{\theta}} \right) \right]^{-1} R(\hat{\theta}) \, . \tag{4.8}$$

However, the extension still seems a little restrictive, because it rules out hypotheses that, for example, involve inequalities. Therefore, we consider a Wald test of (4.6) based on the following test statistic

$$(\hat{\theta} - \hat{\theta}_0)' \hat{W}^{-1} (\hat{\theta} - \hat{\theta}_0) \, , \tag{4.9}$$

where $\hat{\theta}$ is an M-estimator of θ, $\hat{\theta}_0$ is an M-estimator of θ under the null hypothesis, and $\hat{W}$ is a "consistent estimator" of the asymptotic covariance matrix of $\hat{\theta} - \hat{\theta}_0$ under the null hypothesis. Such an idea has been used by, e.g., Gourieroux and Monfort (1995) to construct tests that take into account inequality constraints. There is another interesting feature of the proposed Wald test. It is known that, while the likelihood-ratio test involves both $\hat{\theta}$ and $\hat{\theta}_0$, the Wald test based on (4.8) only involves $\hat{\theta}$. However, this is not the case for (4.9).

Finally, The score test is based on the test statistic

$$\left(\left. \frac{\partial l}{\partial \theta} \right|_{\hat{\theta}_0} \right)' I(\hat{\theta}_0)^{-1} \left(\left. \frac{\partial l}{\partial \theta} \right|_{\hat{\theta}_0} \right) , \tag{4.10}$$

where $l = l(\theta, y)$ is the log-likelihood. Note that the score test involves only $\hat{\theta}_0$. Nevertheless, all of the three tests may be thought as involving both $\hat{\theta}$ and $\hat{\theta}_0$, because (4.10) can be written as

$$\left(\left. \frac{\partial l}{\partial \theta} \right|_{\hat{\theta}} - \left. \frac{\partial l}{\partial \theta} \right|_{\hat{\theta}_0} \right)' I(\hat{\theta}_0)^{-1} \left(\left. \frac{\partial l}{\partial \theta} \right|_{\hat{\theta}} - \left. \frac{\partial l}{\partial \theta} \right|_{\hat{\theta}_0} \right) .$$

Note that $(\partial l / \partial \theta)|_{\hat{\theta}} = 0$, under regularity conditions, because $\hat{\theta}$ is the MLE.

We now consider robust versions of Wald, score, and likelihood-ratio tests, which we call W-, S-, and L-tests, respectively. Let $y = (y_k)_{1 \leq k \leq n}$ be a vector of observations. Let θ be a vector of unknown parameters that are associated with the joint distribution of y, but the entire distribution of y may not be known given θ (and possibly other parameters). We are interested in testing the hypothesis:

$$H_0: \quad \theta \in \Theta_0 \tag{4.11}$$

versus H_1: $\theta \notin \Theta_0$, where $\Theta_0 \subset \Theta$, and Θ is the parameter space.

Suppose that there is a new parametrization ϕ such that, under the null hypothesis (4.11), $\theta = \theta(\phi)$ for some ϕ. Here $\theta(\cdot)$ is a map from Φ, the parameter space of ϕ, to Θ. Note that such a reparametrization is almost always possible, but the key is to try to make ϕ unrestricted (unless completely specified, such as in Example 4.3 below). We consider some examples (Exercise 4.2).

Example 4.3. Suppose that, under the null hypothesis, θ is completely specified, i.e.,

$$H_0: \quad \theta = \theta_0 .$$

Then, under H_0, one has $\theta = \phi = \theta_0$.

Example 4.4. Let $\theta = (\theta_1, \dots, \theta_p, \theta_{p+1}, \dots, \theta_q)'$, and suppose that one wishes to test the hypothesis

$$H_0: \quad \theta_j = \theta_{0j} , \quad p+1 \le j \le q ,$$

where θ_{0j}, $p+1 \le j \le q$ are known constants. Then, under the null hypothesis, one has $\theta_j = \phi_j$, $1 \le j \le p$, and $\theta_j = \theta_{0j}$, $p+1 \le j \le q$ for some (unrestricted) $\phi = (\phi_j)_{1 \le j \le p}$.

Example 4.5. Let $y_1, \dots, y_n$ be independent with distribution $N(\mu, \sigma^2)$. Let $\theta = (\mu, \sigma^2)'$. Suppose that one wishes to test the hypothesis H_0: $\sigma^2 = \mu^2$. Then, under the null hypothesis, one has $\theta_1 = \phi$, $\theta_2 = \phi^2$ for some unrestricted ϕ.

Example 4.6. Suppose that the null hypothesis includes inequality constraints:

$$H_0: \quad \theta_j > \theta_{0j} , \quad p_1 + 1 \le j \le p , \quad \text{and} \quad \theta_j = \theta_{0j} , \quad p+1 \le j \le q ,$$

where $p_1 < p < q$. Then, under the null hypothesis, one has $\theta_j = \phi_j$, $1 \le j \le p_1$, $\theta_j = \theta_{0j} + e^{\phi_j}$, $p_1 + 1 \le j \le p$, and $\theta_j = \theta_{0j}$, $p+1 \le j \le q$ for some (unrestricted) $\phi = (\phi_j)_{1 \le j \le p}$.

Let $L(\theta, y)$ be a function of θ and y that takes positive values, and $l(\theta, y) = \log L(\theta, y)$. Also, Let $L_0(\phi, y) = L(\theta(\phi), y)$, and $l_0(\phi, y) = \log L_0(\phi, y)$. Let q and p be the dimensions of θ and ϕ, respectively. Then, we have

$$\frac{\partial l_0}{\partial \phi} = \left(\frac{\partial \theta}{\partial \phi}\right)' \frac{\partial l}{\partial \theta} , \tag{4.12}$$

$$\frac{\partial^2 l_0}{\partial \phi^2} = \left(\frac{\partial \theta}{\partial \phi}\right)' \frac{\partial^2 l}{\partial \theta^2} \left(\frac{\partial \theta}{\partial \phi}\right) + \sum_{i=1}^{q} \frac{\partial l}{\partial \theta_i} \cdot \frac{\partial^2 \theta_i}{\partial \phi^2} , \tag{4.13}$$

where we use the notation

$$\frac{\partial l_0}{\partial \phi} = \left(\frac{\partial l_0}{\partial \phi_j}\right)_{1 \le j \le p}, \frac{\partial l}{\partial \theta} = \left(\frac{\partial l}{\partial \theta_i}\right)_{1 \le i \le q}, \frac{\partial \theta}{\partial \phi} = \left(\frac{\partial \theta_i}{\partial \phi_j}\right)_{1 \le i \le q, 1 \le j \le p};$$

$$\frac{\partial l_0}{\partial \phi^2} = \left(\frac{\partial^2 l_0}{\partial \phi_j \partial \phi_{j'}}\right)_{1 \le j, j' \le p}, \frac{\partial^2 l}{\partial \theta^2} = \left(\frac{\partial^2 l}{\partial \theta_i \partial \theta_{i'}}\right)_{1 \le i, i' \le q},$$

$$\text{and } \frac{\partial^2 \theta_i}{\partial \phi^2} = \left(\frac{\partial^2 \theta_i}{\partial \phi_j \partial \phi_{j'}}\right)_{1 \le j, j' \le p}.$$

Let $\hat{\theta}$ be an estimator of θ, and $\hat{\phi}$ an estimator of ϕ. Note that here we do not require that $\hat{\theta}$ and $\hat{\phi}$ are the (global) maximizers of $l(\theta, y)$ and $l_0(\phi, y)$, respectively. However, we shall require that $\hat{\theta}$ is a solution to $\partial l / \partial \theta = 0$, and $\hat{\phi}$ a solution to $\partial l_0 / \partial \phi = 0$. We use θ_0 to represent the true θ, and ϕ_0 the true ϕ under (4.11).

4.2.2 *The W-, S-, and L-test statistics*

The following quantity is closely related to the W-test:

$$\tilde{\chi}_w^2 = \{\hat{\theta} - \theta(\hat{\phi})\}' G Q_w^- G \{\hat{\theta} - \theta(\hat{\phi})\}, \tag{4.14}$$

where Q_w^- represents the unique Moore-Penrose inverse [e.g., Searle (1971), §1.3] of

$$Q_w = \{A^{-1} - C(C'AC)^{-1}C'\}\Sigma\{A^{-1} - C(C'AC)^{-1}C'\}. \tag{4.15}$$

Note that the expression inside $\{\cdots\}$ in (4.15) can be written in a familiar form: Write $C = A^{-1}D$. Then, we have

$$\begin{aligned} A^{-1} - C(C'AC)^{-1}C &= A^{-1} - A^{-1}D(D'A^{-1}D)^{-1}D'A^{-1} \\ &= K(K'AK)^{-1}K' \equiv P, \end{aligned}$$

where K is any matrix of maximum column rank such that $K'D = 0$ [e.g., Searle *et al.* (1992), pp. 451]. The matrix P is used extensively in linear statistical inference, especially in the context of LMM [e.g., Searle *et al.* (1992), §6]. The definitions of the matrices G, A, C, and Σ will be given in Theorem 4.2 in the sequel, but here we first offer some interpretations: A is the limit of the matrix of second derivatives of l with respect to θ; B is the limit of the matrix of second derivatives of l_0 with respect to ϕ; C is the limit of the matrix of first derivatives of θ with respect to ϕ; and Σ is the asymptotic covariance matrix of $\partial l / \partial \theta$, all after suitable normalizations.

Theorem 4.2 in the sequel shows that, as the symbol indicates, $\tilde{\chi}_w^2$ has an asymptotic χ^2 distribution. Typically, the normalizing matrix G is completely specified. In fact, in many cases G is a diagonal matrix, where the diagonal elements are normalizing constants corresponding to the components of $(\partial l/\partial\theta)|_{\theta_0}$. On the other hand, the matrix Q_w may involve a vector of parameters, say, ϑ. In some cases such as generalized linear models [e.g., McCullagh and Nelder (1989)], ϑ depends only on θ. In other cases, ϑ may involve some additional parameters, which may also need to be estimated. In Section 4.3 we consider a case of estimating the additional parameters in the context of mixed linear models. Let $\hat{Q}_w^-$ be a consistent estimator of Q_w^- in the sense that $\|\hat{Q}_w^- - Q_w^-\| \to 0$ in probability, where for any matrix M, $\|M\| = \{\lambda_{\max}(M'M)\}^{1/2}$ with $\lambda_{\max}$ representing the largest eigenvalue (and similarly $\lambda_{\min}$ the smallest). Note that this also covers the case in which Q_w^- may represent a sequence of matrices (see an Extension of Theorem 4.2). We define the W-test statistic as

$$\hat{\chi}_w^2 = \{\hat{\theta} - \theta(\hat{\phi})\}' G \hat{Q}_w^- G \{\hat{\theta} - \theta(\hat{\phi})\} . \tag{4.16}$$

Similarly, we consider the following:

$$\tilde{\chi}_s^2 = \left\{ \left.\frac{\partial l}{\partial \theta}\right|_{\theta(\hat{\phi})} \right\}' G^{-1} A^{-1/2} Q_s^- A^{-1/2} G^{-1} \left\{ \left.\frac{\partial l}{\partial \theta}\right|_{\theta(\hat{\phi})} \right\} , \tag{4.17}$$

where Q_s^- is the unique Moore-Penrose inverse of

$$Q_s = (I - P) A^{-1/2} \Sigma A^{-1/2} (I - P), \tag{4.18}$$

$P = A^{1/2} C (C'AC)^{-1} C' A^{1/2}$, and the matrices G, A, C are the same as above. Let $\hat{A}$ and $\hat{Q}_s^-$ be consistent estimators of A and Q_s^-, respectively. Note that, quite often, A only depends on θ so that a consistent estimator of A is already available. We define the S-test statistic as

$$\hat{\chi}_s^2 = \left\{ \left.\frac{\partial l}{\partial \theta}\right|_{\theta(\hat{\phi})} \right\}' G^{-1} \hat{A}^{-1/2} \hat{Q}_s^- \hat{A}^{-1/2} G^{-1} \left\{ \left.\frac{\partial l}{\partial \theta}\right|_{\theta(\hat{\phi})} \right\} . \tag{4.19}$$

Finally, the L-ratio for testing the hypothesis (4.11) is defined as

$$R = \frac{L_0(\hat{\phi}, y)}{L(\hat{\theta}, y)} . \tag{4.20}$$

As noted before, here we do not require that $\hat{\theta}$ and $\hat{\phi}$ are, respectively, (global) maximizers of $l(\theta, y)$ and $l_0(\phi, y)$. In particular, if $\hat{\theta}$ and $\hat{\phi}$ are, indeed, the global maximizers, R is equal to $\sup_{\Theta_0} L(\theta, y)/\sup_{\Theta} L(\theta, y)$, which is the original definition of the likelihood-ratio when $L(\theta, y)$ is a likelihood function [see (4.7)]. The L-test statistic is then $-2\log R$.

4.2.3 *Asymptotic theory*

The main results of this subsection are the asymptotic distributions of the W-, S-, and L-test statistics. The following notation will be used: if $a = (a_1, \ldots, a_k)'$ is a vector, then $|a|_v = (|a_1|, \ldots, |a_k|)'$. Here the index v refers to "vector". Also, if $b = (b_1, \ldots, b_k)$, then $a \leq b$ iff $a_i \leq b_i$, $1 \leq i \leq k$; and $a \vee b = (a_1 \vee b_1, \ldots, a_k \vee b_k)$, where $a \vee b = \max(a, b)$ [and similarly $a \wedge b = \min(a, b)$]. For any matrix M, $M > 0$ means that M is positive definite; and $\text{diag}(x_i, 1 \leq i \leq k)$ represents a diagonal matrix with $x_i, 1 \leq i \leq k$ on its diagonal.

Theorem 4.2. Suppose that the following hold:
i) $l(\cdot, y)$ is twice continuously differentiable for fixed y, and $\theta(\cdot)$ is twice continuously differentiable;
ii) with probability $\to 1$, $\hat{\theta}$, $\hat{\phi}$ satisfy $\partial l/\partial\theta = 0$, $\partial l_0/\partial\phi = 0$, respectively;
iii) there are sequences of nonsingular symmetric matrices $\{G\}$ and $\{H\}$ and matrices A, B, C with A, $B > 0$ such that the following $\to 0$ in probability:

$$\sup_{|\theta^{(i)}-\theta_0|_v \leq |\hat{\theta}-\theta_0|_v \vee |\theta(\hat{\phi})-\theta(\phi_0)|_v,\ 1\leq i\leq q} \left\| G^{-1} \left[\left. \frac{\partial^2 l}{\partial\theta_i \partial\theta_j} \right|_{\theta^{(i)}} \right]_{1\leq i,j\leq q} G^{-1} + A \right\|,$$

$$\sup_{|\phi^{(i)}-\phi_0|_v \leq |\hat{\phi}-\phi_0|_v,\ 1\leq i\leq p} \left\| H^{-1} \left[\left. \frac{\partial^2 l_0}{\partial\phi_i \partial\phi_j} \right|_{\phi^{(i)}} \right]_{1\leq i,j\leq p} H^{-1} + B \right\|,$$

$$\sup_{|\phi^{(i)}-\phi_0|_v \leq |\hat{\phi}-\phi_0|_v,\ 1\leq i\leq q} \left\| G \left[\left. \frac{\partial\theta_i}{\partial\phi_j} \right|_{\phi^{(i)}} \right]_{1\leq i\leq q, 1\leq j\leq p} H^{-1} - C \right\|;$$

iv) $D(\partial l/\partial\theta)|_{\theta_0} \to 0$ in probability, where $D = \text{diag}(d_i, 1 \leq i \leq s)$ with $d_i = \|H^{-1}(\partial^2\theta_i/\partial\phi^2)|_{\phi_0} H^{-1}\|$, and

$$G^{-1} \left. \frac{\partial l}{\partial\theta} \right|_{\theta_0} \longrightarrow N(0, \Sigma) \quad \text{in distribution}. \tag{4.21}$$

Then, under the null hypothesis, the asymptotic distribution of $\tilde{\chi}_w^2$ is χ_r^2, where r is the rank of $\Sigma^{1/2}A^{-1/2}(I - P)$ with P given below (4.18). In particular, if Σ is nonsingular, then $r = q - p$.

The theorem may be extended to allow the matrices A, B, etc. to be replaced by sequences of matrices. Such an extension may be useful. For example, suppose G is a diagonal normalizing matrix, then, in many cases, A can be chosen as $-G^{-1}\{\text{E}(\partial^2 l/\partial\theta^2)|_{\theta_0}\}G^{-1}$, but the latter may not have a limit as $n \to \infty$.

Extension of Theorem 4.2. Suppose that, in Theorem 4.2, A, B, C are replaced by sequences of matrices $\{A\}$, $\{B\}$, and $\{C\}$, such that A, B are symmetric satisfying

$$0 < \liminf\{\lambda_{\min}(A) \wedge \lambda_{\min}(B)\} \leq \limsup\{\lambda_{\max}(A) \vee \lambda_{\max}(B)\} < \infty,$$

and $\limsup \|C\| < \infty$. Furthermore, suppose that (4.21) is replaced by

$$\Sigma^{-1/2}G^{-1} \left.\frac{\partial l}{\partial \theta}\right|_{\theta_0} \longrightarrow N(0, I) \quad \text{in distribution}, \tag{4.22}$$

where $\{\Sigma\}$ is a sequence of positive definite matrices such that $0 < \liminf \lambda_{\min}(\Sigma) \leq \limsup \lambda_{\max}(\Sigma) < \infty$, and I is the p-dimensional identity matrix. Then, the asymptotic distribution of $\tilde{\chi}_w^2$ is χ_{q-p}^2.

A proof of Theorem 4.2 is given in Jiang (2012). According to the proof, one has $G[\hat{\theta} - \theta(\hat{\phi})] = O_P(1)$. Thus, we have

$$\hat{\chi}_w^2 = \{\hat{\theta} - \theta(\hat{\phi})\}' G\{Q_w^- + o_P(1)\}G\{\hat{\theta} - \theta(\hat{\phi})\} = \tilde{\chi}_w^2 + o_P(1).$$

Therefore, by Theorem 4.2, we conclude the following.

Corollary 4.1. Under the conditions of Theorem 4.2, the asymptotic distribution of $\hat{\chi}_w^2$ is χ_r^2, where r is the same as in Theorem 4.2. In particular, if Σ is nonsingular, one has $r = q - p$. Under the conditions of Extension of Theorem 4.2, the asymptotic distribution of $\hat{\chi}_w^2$ is χ_{q-p}^2.

Next, we consider asymptotic null distribution of the S-test.

Theorem 4.3. Suppose that the conditions of Theorem 4.2 are satisfied with the following changes: 1) in ii), that $\hat{\theta}$ satisfies $\partial l/\partial \theta = 0$ with probability $\to 1$ is not required; and 2) in iii), the supremum for the first quantity (involving A) is now over $|\theta^{(i)} - \theta_0|_v \leq |\theta(\hat{\phi}) - \theta(\phi_0)|_v$, $1 \leq i \leq q$. Then, under the null hypothesis, the asymptotic distribution of $\tilde{\chi}_s^2$ is χ_r^2, where r is the same as in Theorem 4.2. In particular, if Σ is nonsingular, then $r = q - p$.

In exactly the same way, we have the following results.

Extension of Theorem 4.3. Suppose that, in Theorem 4.3, A, B, and C are replaced by $\{A\}$, $\{B\}$, and $\{C\}$, and (4.21) is replaced by (4.22), where the sequences of matrices $\{A\}$, $\{B\}$, $\{C\}$, and $\{\Sigma\}$ satisfy the conditions of Extension of Theorem 4.2. Then, the asymptotic distribution of $\tilde{\chi}_s^2$ is χ_{q-p}^2.

Corollary 4.2. Under the conditions of Theorem 4.3, the asymptotic distribution of $\hat{\chi}_s^2$ is χ_r^2, where r is the same as in Theorem 4.2. Thus, in particular, if Σ is nonsingular, $r = q - p$. Under the conditions of Extension of Theorem 4.3, the asymptotic distribution of $\hat{\chi}_s^2$ is χ_{q-p}^2.

It is seen that the asymptotic null distributions for W- and S-tests are both χ^2. However, the following theorem states that the asymptotic distribution for L-test is not χ^2 but a "weighted" χ^2 [e.g., Chernoff and Lehmann (1954)]. Let

$$Q_l = [A^{-1} - C(C'AC)^{-1}C']^{1/2}\Sigma[A^{-1} - C(C'AC)^{-1}C']^{1/2} . \quad (4.23)$$

Theorem 4.4. Suppose that the conditions of Theorem 4.2 are satisfied except that the third quantity in iii) (involving C) $\to 0$ in probability is replaced by $G[(\partial\theta/\partial\phi)|_{\phi_0}]H^{-1} \to C$. Then, under the null hypothesis, the asymptotic distribution of $-2\log R$ is the same as $\lambda_1\xi_1^2 + \cdots + \lambda_r\xi_r^2$, where r is the same as in Theorem 4.2; $\lambda_1, \ldots, \lambda_r$ are the positive eigenvalues of Q_l; and $\xi_1, \ldots, \xi_r$ are independent $N(0,1)$ random variables. In particular, if Σ is nonsingular, then $r = q - p$.

It should be pointed out that if $L(\theta, y)$ is, indeed, the likelihood function, in which case L-test is the likelihood-ratio test, the asymptotic null distribution of $-2\log R$ reduces to χ^2 [see Weiss (1975)].

Let $\hat{Q}_l$ be a consistent estimator of Q_l. Then, by Weyl's eigenvalue perturbation theorem [e.g., Bhatia (1997), p. 63], the eigenvalues of $\hat{Q}_l$ are consistent estimators of those of Q_l, and therefore can be used to obtain the asymptotic critical values for the L-test.

In Theorem 4.2, the consistency of $\hat{\theta}$ and $\hat{\phi}$ is not explicitly required. However, condition iii) does implicate some consistent properties of the estimators. In the following, we give sufficient conditions under which $\hat{\theta}$, $\hat{\phi}$ are consistent, and conditions ii), iii) of Theorem 4.2 are satisfied. In the theorem below, the assumption of G and H being diagonal is not essential, but it much simplifies the result.

Theorem 4.5. Suppose that $G = \text{diag}(g_j, 1 \le j \le q)$ and $H = \text{diag}(h_j, 1 \le j \le p)$, where g_j's and h_j's are sequences of positive numbers that $\to \infty$ as sample size increases. Furthermore, suppose that the following conditions are satisfied:

i) $l(\cdot, y)$ is three-times continuously differentiable for fixed y, and $\theta(\cdot)$ is three-times continuously differentiable;

ii) $G^{-1}(\partial l/\partial\theta)_{\theta_0} = O_P(1)$;

iii) there are matrices A, $B > 0$ and C such that

$$G^{-1}\left(\left.\frac{\partial^2 l}{\partial\theta^2}\right|_{\theta_0}\right)G^{-1} + A \longrightarrow 0, \quad H^{-1}\left(\left.\frac{\partial^2 l_0}{\partial\phi^2}\right|_{\phi_0}\right)H^{-1} + B \longrightarrow 0$$

in probability, and $G[(\partial\theta/\partial\phi)|_{\phi_0}]H^{-1} \to C$;

iv) there is $\delta > 0$ such that

$$\frac{1}{g_i g_j} \sup_{|\theta-\theta_0|\le\delta} \left| \frac{\partial^3 l}{\partial\theta_i \partial\theta_j \partial\theta_k} \right|, \quad 1 \le i, j, k \le q\,,$$
$$\frac{1}{h_i h_j} \sup_{|\phi-\phi_0|\le\delta} \left| \frac{\partial^3 l_0}{\partial\phi_i \partial\phi_j \partial\phi_k} \right|, \quad 1 \le i, j, k \le p$$

are bounded in probability, and $g_i h_j^{-1} \sup_{|\phi-\phi_0|\le\delta} |\partial^2\theta_i/\partial\phi_j\partial\phi_k|$, $1 \le i \le q$, $1 \le j, k \le p$ are bounded.
Then, there exist $\hat{\theta}$ and $\hat{\phi}$ such that
1) condition ii) of Theorem 4.2 holds;
2) $G(\hat{\theta} - \theta_0) = O_P(1)$ and $H(\hat{\phi} - \phi_0) = O_P(1)$, therefore, $\hat{\theta}$ and $\hat{\phi}$ are consistent; and
3) condition iii) of Theorem 4.2 holds for the matrices given above.

We illustrate Theorem 4.5 with an example.

Example 4.4 (Continued). For any positive constants g_j, $1 \le j \le q$, let $h_j = g_j$, $1 \le j \le p$. Then,

$$\frac{\partial\theta}{\partial\phi} = \begin{pmatrix} I \\ 0 \end{pmatrix}, \quad \text{hence} \quad G\left(\frac{\partial\theta}{\partial\phi}\right)H^{-1} = \begin{pmatrix} I \\ 0 \end{pmatrix},$$

where I is the $p \times p$ identity matrix. Also, $g_i h_j^{-1} \sup_\phi |\partial^2\theta_i/\partial\phi_j\partial\phi_k| = 0$, $\forall i, j, k$. Furthermore, let $\theta_{(1)} = (\theta_1, \ldots, \theta_p)'$. Then, by (4.13), we have

$$\frac{\partial^2 l_0}{\partial\phi^2} = \left(\frac{\partial\theta}{\partial\phi}\right)' \frac{\partial^2 l}{\partial\theta^2} \left(\frac{\partial\theta}{\partial\phi}\right) = \frac{\partial^2 l}{\partial\theta_{(1)}^2}.$$

Therefore,

$$-G^{-1}\left(\frac{\partial^2 l}{\partial\theta^2}\right)G^{-1} \longrightarrow A = \begin{pmatrix} B & * \\ * & * \end{pmatrix} > 0 \quad \text{in probability}$$

implies that

$$-H^{-1}\left(\frac{\partial^2 l_0}{\partial\phi^2}\right)H^{-1} = -H^{-1}\left(\frac{\partial^2 l}{\partial\theta_{(1)}^2}\right)H^{-1} \longrightarrow B > 0 \quad \text{in probability.}$$

Theorem 4.5 can be easily extended to allow A, B, C to be replaced by sequences of matrices, as in Extensions of Theorem 4.2 and Theorem 4.3. The extension is left as an exercise (Exercise 4.3).

4.3 Robust classical tests for mixed ANOVA model

In this section, we consider the mixed ANOVA model introduced in Section 1.1 [see (1.1)]. However, it is not assumed that the random effects and errors are normally distributed. In fact, the only assumptions regarding the distributions of the random effects and errors are that α_r is an $m_r \times 1$ vector of i.i.d. random variables with mean 0 and variance σ_r^2, $1 \leq r \leq s$; ϵ is an $N \times 1$ vector of i.i.d. random variables with mean 0 and variance τ^2; and $\alpha_1, \dots, \alpha_s, \epsilon$ are independent.

We consider the Hartley-Rao form of variance components, θ (see §3.1.1). Let $\gamma = (\gamma_r)_{1 \leq r \leq s}$, and $\vartheta = (\beta', \theta')' = (\beta', \tau^2, \gamma')'$. Then, without the normality assumption, ϑ is a vector of parameters, which alone may not completely determine the distribution of y. Nevertheless, in many cases, there are still interests in testing the following type of hypotheses:

$$H_0: \quad \theta \in \Theta_0 , \tag{4.24}$$

where $\Theta_0 \subset \Theta = \{\theta : \tau^2 > 0, \gamma_r \geq 0, 1 \leq r \leq s\}$, versus H_1: $\theta \notin \Theta_0$.

When normality is assumed, the use of likelihood-ratio test for potentially complex hypotheses and unbalanced data was first proposed by Hartley and Rao (1967), although a rigorous justification was not given. Welham and Thompson (1997) showed the equivalence of the likelihood ratio, score, and Wald tests under normality. For discussions about other testing procedures in LMM under normality, see Khuri *et al.* (1998). On the other hand, Richardson and Welsh (1996) considered likelihood-ratio test without assuming normality, whose approach is similar to our L-test, but their goal was to select the (fixed) covariates.

Under the normality assumption, the log-likelihood function for estimating ϑ is given by

$$\begin{aligned} l(\vartheta, y) = \text{constant} - \frac{1}{2}\Big\{ & N \log(\tau^2) \\ & + \log(|V|) + \frac{1}{\tau^2}(y - X\beta)'V^{-1}(y - X\beta)\Big\}, \end{aligned} \tag{4.25}$$

where $V = V_\gamma = I + \sum_{r=1}^s \gamma_r V_r$ with I being the N-dimensional identity matrix, $V_r = Z_r Z_r'$, $1 \leq r \leq s$, and $|V|$ the determinant of V. The restricted log-likelihood for estimating θ is given by

$$\begin{aligned} l_R(\theta, y) = \text{constant} - \frac{1}{2}\Big\{ & (N - p)\log(\tau^2) \\ & + \log(|K'VK|) + \frac{y'Py}{\tau^2}\Big\}, \end{aligned} \tag{4.26}$$

where K is any $N\times(N-p)$ matrix such that $\text{rank}(K)=N-p$ and $K'X=0$, and $P=P_\gamma=K(K'VK)^{-1}K'=V^{-1}-V^{-1}X(X'V^{-1}X)^{-1}X'V^{-1}$ (see Section 1.1, but we change the notation A to K to avoid confusion with the A used in the previous section). The restricted log-likelihood is only for estimating the variance components. It is then customary to estimate β by (1.3), where $\hat{V}=V_{\hat{\gamma}}$, and $\hat{\gamma}=(\hat{\gamma}_r)_{1\le r\le s}$ is the REML estimator of γ. Alternatively, one may define the following "restricted log-likelihood":

$$l_R(\vartheta,y)=\text{constant}-\frac{1}{2}\left\{(N-p)\log(\tau^2)\right.$$

$$\left.+\log|K'VK|+\frac{1}{\tau^2}(y-X\beta)'V^{-1}(y-X\beta)\right\}. \quad (4.27)$$

It can be shown (Exercise 4.4) that the maximizer of (4.27) is $\hat{\vartheta}=(\hat{\beta}',\hat{\tau}^2,\hat{\gamma}')'$, where $\hat{\tau}^2$ and $\hat{\gamma}$ are the REML estimators, and $\hat{\beta}$ is given above. The difference is that, unlike $l(\vartheta,y)$, $l_R(\vartheta,y)$ is not a log-likelihood, even if normality holds (Exercise 4.4). Nevertheless, we show that both $l(\vartheta,y)$ and $l_R(\vartheta,y)$ can be used to test (4.24) without the normality assumption.

For LMMs, conditions of Theorems 4.2–4.4 and their extensions correspond to mild restrictions. However, as discussed in Section 3.3, the ACVM of the REML estimator involves higher (i.e., third or fourth) moments of the random effects and errors. More specifically, if the test is only about the fixed effects, no higher moments are involved in Σ; if the test is only about the variance components, the fourth moments, or kurtoses, will appear in Σ; if the test is about both the fixed effects and the variance components, the third moments will be involved, in addition to the kurtoses. The matrices A, B and C in the previous section do not involve the higher moments. To simplify the results, we make the following additional assumption:

$$\text{E}(\epsilon_1^3)=0,\quad\text{and}\quad\text{E}(\alpha_{r1}^3)=0,\quad 1\le r\le s. \quad (4.28)$$

Such conditions hold if, in particular, the distributions of the random effects and errors are symmetric. Another action of simplification is that we shall consider testing a special class of hypotheses (4.24). Let $\vartheta(\cdot)=(\vartheta_j(\cdot))_{1\le j\le q}$, where $q=p+s+1$, be the map such that, under (4.24), $\vartheta=\vartheta(\phi)$ for some $\phi=(\phi_k)_{1\le k\le a}$, where $a\le q$. We assume that there is a subset of indexes $1\le j_1<\cdots<j_a\le q$ such that

$$\begin{cases}\vartheta_{j_k}(\phi)\text{ is a function of }\phi_k\ , & 1\le k\le a\ ,\ \text{ and}\\ \vartheta_j(\phi)\text{ is a constant}\ , & j\in\{1,\dots,q\}\setminus\{j_1,\dots,j_a\}\ .\end{cases} \quad (4.29)$$

Among the examples in Subsection 4.2.1, Examples 4,3, 4,4, and 4.6 satisfy (4.29), while Example 4.5 does not. Also, for notation simplicity, in the rest of this section, ϑ, ϕ, etc. denote the true vectors of parameters. We need the following notation: Write $A_1 = (\mathrm{tr}(V^{-1}V_r)/2\lambda\sqrt{Nm_r})_{1\le r\le s}$, $A_2 = (\mathrm{tr}(V^{-1}V_rV^{-1}V_t)/2\sqrt{m_rm_t})_{1\le r,t\le s}$, and

$$A = \begin{pmatrix} X'V^{-1}X/\tau^2N & 0 & 0 \\ 0 & 1/2\tau^4 & A_1' \\ 0 & A_1 & A_2 \end{pmatrix}. \tag{4.30}$$

Let $b = (I\ \sqrt{\gamma_1}Z_1\ \cdots\ \sqrt{\gamma_s}Z_s)$, $B_0 = b'V^{-1}b$, $B_r = b'V^{-1}V_rV^{-1}b$, $1 \le r \le s$. We define

$$D_{0,rt} = \sum_{l=1}^{N} B_{r,ll}B_{t,ll},$$
$$D_{1,rt} = \sum_{l=N+1}^{N+m_1} B_{r,ll}B_{t,ll}, \dots$$
$$D_{s,rt} = \sum_{l=N+m_1+\cdots+m_{s-1}+1}^{N+m_1+\cdots+m_s} B_{r,ll}B_{t,ll},$$

where $B_{r,kl}$ is the (k,l) element of B_r, $0 \le r \le s$. The kurtoses of the errors and random effects are defined by $\kappa_0 = \{\mathrm{E}(\epsilon_1^4)/\tau^4\} - 3$, and $\kappa_r = \{\mathrm{E}(\alpha_{r1}^4)/\sigma_r^4\} - 3$, $1 \le r \le s$. Let $\Delta_1 = (\Delta_{0r}/\sqrt{Nm_r})_{1\le r\le s}$, $\Delta_2 = (\Delta_{rt}/\sqrt{m_rm_t})_{1\le r,t\le s}$, and

$$\Delta = \begin{pmatrix} 0 & 0 & 0 \\ 0 & \Delta_{00}/N & \Delta_1' \\ 0 & \Delta_1 & \Delta_2 \end{pmatrix}, \tag{4.31}$$

where $\Delta_{rt} = [4\lambda^{1_{(r=0)}+1_{(t=0)}}]^{-1}\sum_{u=0}^{s}\kappa_uD_{u,rt}$, $0 \le r,t \le s$. Let $W = b'V^{-1}X(X'V^{-1}X)^{-1/2}$, and W_l' be the lth row of W, $1 \le l \le N+m$, where $m = m_1 + \cdots + m_s$.

Theorem 4.6. Suppose that
i) $\vartheta(\cdot)$ is three-times continuously differentiable and satisfies (4.29), and $\partial\theta_{j_k}/\partial\phi_k \ne 0$, $1 \le k \le a$;
ii) $\mathrm{E}(\epsilon_1^4) < \infty$, $\mathrm{var}(\epsilon_1^2) > 0$, $\mathrm{E}(\alpha_{r1}^4) < \infty$, $\mathrm{var}(\alpha_{r1}^2) > 0$, $1 \le r \le s$, and (4.28) holds; and
iii) $N \to \infty$, $m_r \to \infty$, $1 \le r \le s$, $0 < \liminf \lambda_{\min}(A) \le \limsup \lambda_{\max}(A) < \infty$, and $\max_{1\le l\le N+m}|W_l| \to 0$;
Then, for $l(\vartheta, y)$ there exist $\hat{\vartheta}$ and $\hat{\phi}$ such that the conditions of Extension of Theorem 4.2 (and hence Extension of Theorem 4.3) are satisfied with $G = \mathrm{diag}(\sqrt{N}, \dots, \sqrt{N}, \sqrt{m_1}, \dots, \sqrt{m_s}) = \mathrm{diag}(g_j, 1 \le j \le q)$, $H = \mathrm{diag}(g_{j_k}, 1 \le k \le a)$, A given by (4.30), $C = \partial\vartheta/\partial\phi$, $B = C'AC$,

and $\Sigma = A + \Delta$, where Δ is given by (4.31). The same conclusion holds for $l_R(\vartheta, y)$ as well.

Note that the jth row of $\partial\vartheta/\partial\phi$ is $\partial\vartheta_j/\partial\phi'$, which, under (4.29), is $(0, \dots, 0)$ if $j \notin \{j_1, \dots, j_a\}$, and is $(0, \dots, 0, \partial\vartheta_{j_k}/\partial\phi_k, 0, \dots, 0)$ (kth component nonzero) if $j = j_k$, $1 \le k \le a$.

Theorem 4.7. Suppose that the conditions of Theorem 4.6 are satisfied except that, in iii), the condition about A is strengthened to that $A \to A_0$, where $A_0 > 0$, and $\Sigma \to \Sigma_0$. Then, the conditions of Theorem 4.4 are satisfied with $A = A_0$, $\Sigma = \Sigma_0$, and everything else given by Theorem 4.6.

It is seen from (4.31) that Δ, and hence Σ, depends on the kurtoses κ_r, $0 \le r \le s$, in addition to the variance components. One already has consistent estimators of the variance components (e.g., REML estimators). As for κ_r, $0 \le r \le s$, they can be estimated by the EMM estimators, introduced in Subsection 3.3.1. However, asymptotic property of the EMM estimator has not yet been studied. Below we given conditions for the consistency of the EMM estimator.

Recall the notation in Subsection 3.3.1. In addition, write $h_r = \|C_r'(y - X\beta)\|_4^4$, $0 \le r \le s$, where C_r is defined below (3.13). The following result shows that, under mild conditions, the EMM estimators are consistent. Note that such a result is not available in Jiang (2003).

Theorem 4.8. Let the C_r's be given below (3.13), and the following hold: i) $\mathrm{E}(\epsilon_1^4) < \infty$, $\mathrm{E}(\alpha_{r1}^4) < \infty$, $1 \le r \le s$; ii) $\hat\beta$, $\hat\theta$ are consistent; iii) $a_{rr}^{-1}\{h_r - \mathrm{E}(h_r)\} = o_{\mathrm{P}}(1)$, $0 \le r \le s$; and iv) the following are bounded: a_{rt}/a_{rr}, $r > t \ge 0$, $a_{rr}^{-1}\sum_{k=1}^{n_r}\left(\sum_{t=0}^{r}|Z_t'c_{rk}|^2\right)^2$, $a_{rr}^{-1}\sum_{k=1}^{n_r}\left(\max_{0\le t\le r}|Z_t'c_{rk}|\right)^w |X'c_{rk}|^{4-w}$, $w = 0, 1, 2, 3$, $0 \le r \le s$. Then, the EMM estimators $\hat\kappa_r$, $0 \le r \le s$ given by (3.14) are consistent.

Proofs of all of the results in this section can be found in Jiang (2012).

4.4 A unified robust goodness-of-fit test

Mixed model diagnostics has been a topic of research and applications of mixed effects models. The topic is of considerable practical interest. For example, mixed effects models have played key roles in small area estimation [SAE; e.g., Rao and Molina (2015)]. It is known that, in case of model misspecification, the traditional empirical best linear unbiased prediction (EBLUP) method may lose efficiency. In fact, in such a case, an alternative method known as observed best prediction (OBP) is likely to be more accurate than the EBLUP. On the other hand, when the underlying model is correctly specified, EBLUP is known to be more efficient than OBP (see the next chapter for details). Therefore, it is important, in practice, to know whether or not the assumed model is appropriate in order to come up with a more efficient SAE strategy.

Another standard assumption in mixed effects models is the normality assumption. For example, in a generalized linear mixed model (GLMM), it is typically assumed that the random effects are normally distributed. If the normality assumption fails, estimators of the fixed effects and variance components derived under the normality assumption may be inconsistent [e.g., Jiang and Nguyen (2009)]. Note that this is very different from the case of a linear mixed model; see Chapter 3.

The literature on mixed model diagnostics is not very extensive. See, for example, Pierce (1982), sec. 2.4.1 of Jiang (2007), Claeskens and Hart (2009). Jiang (2001) proposed a χ^2-type goodness-of-fit test for linear mixed model diagnostics, whose asymptotic null distribution is a weighted χ^2, where the weights are eigenvalues of some nonnegative definite matrix. Claeskens and Hart (2009) proposed an alternative approach to the χ^2 test for checking the normality assumption in LMM. The authors considered a class of distributions that include the normal distribution as a reduced, special case. The test is based on the likelihood-ratio test (LRT) that compares the "estimated distribution" and the null distribution (i.e., normal). A model selection procedure via the information criteria is used to determine the larger class of distributions for the LRT. In particular, the asymptotic null distribution is in the form of the distribution of $\sup_{l \geq 1}\{2Q_l/l(l+3)\}$, where $Q_l = \sum_{q=1}^{l} \chi_{q+1}^2$, and $\chi_2^2, \chi_3^2, \dots$ are independent such that χ_j^2 has a χ^2 distribution with j degrees of freedom, $j \geq 2$.

The χ^2-type tests depend on the choice of cells, based on which the observed and expected cell frequencies are evaluated. As noted by Jiang and Nguyen (2009), performance of the χ^2 test is sensitive to the choice of the cells, and there is no "optimal choice" of such cells known in the literature. On the other hand, the Claeskens-Hart test depends on the choice of the information criterion. As is well known, there are different versions of the information criteria, such as AIC [Akaike (1973)], BIC [Schwarz (1978)], HQ [Hannan and Quinn (1979)]. The difference in the performance of the test by different information criteria is unclear. Furthermore, the weighted-χ^2 asymptotic null distribution of Jiang (2001) depends on eigenvalues of some matrices, whose expressions are complicated, and involve unknown parameters. These parameters need to be estimated in order to obtain the critical values of the tests. Due to such a complication, Jiang (2001) suggests to use a Monte Carlo method to compute the critical value; but, by doing so, the usefulness of the asymptotic result may be undermined. Similarly, the asymptotic distribution of the Claeskens-Hart test is not simple and involves supreme of χ^2 distributions.

It might be argued that, in today's computer era, having a simple asymptotic distribution such as χ^2 is, perhaps, not as important as in the past. However, there are, still, attractive features of the χ^2 limiting distribution that are worth pursuing. First, the χ^2 distribution corresponds to the right standardization–it is the "square" of the multivariate standard normal distribution. In this regard, anything other than χ^2 leaves, at least, some room for improvement. Note that, while there is only one way of a complete standardization, there are many, if not infinitely many, ways of incomplete standardization, so it may not be convincing why one way is chosen over the others. Second, having a computer-determined, non-analytic asymptotic distribution makes it difficult to study properties of the limiting distribution. For example, how does the reduction of complexity of the model under the null hypothesis play a role? It may not be easy to tell if all one gets are a bunch of numbers. A related issue is regarding direction of improvement. This may not be easy to see without a simple analytic expression for the asymptotic distribution.

In what follows, we generalize a method initiated by Fisher (1922) in deriving goodness-of-fit tests (GoFTs) that are guaranteed to have asymptotic χ^2 null distributions. In addition, the proposed test has a robustness feature in that it can test a certain type of model assumption while another aspect of the model may be misspecified. A special case of the method has a connection with the generalized method of moments (GMM), as we shall point out. We then consider application of the test to SAE and present some empirical results, especially in terms of its robustness property. Technical proofs can be found in Jiang and Torabi (2018).

4.4.1 *Tailoring*

In this section, we describe a general approach to obtaining a test statistic that has an asymptotic χ^2 distribution under the null hypothesis. The original idea can be traced back to R. A. Fisher [Fisher (1922)], who used the method to obtain an asymptotic χ^2 distribution for Pearson's χ^2-test, when the so-called minimum chi-square estimator is used. However, Fisher did not put forward the method that he originated under a general framework, as we do here. Suppose that there is a sequence of s-dimensional random vectors, $B(\vartheta)$, which depend on a vector ϑ of unknown parameters such that, when ϑ is the true parameter vector, one has $\mathrm{E}\{B(\vartheta)\} = 0$, $\mathrm{Var}\{B(\vartheta)\} = I_s$, and, as the sample size increases,

$$|B(\vartheta)|^2 \xrightarrow{\mathrm{d}} \chi^2_s, \tag{4.32}$$

where $|\cdot|$ denotes the Euclidean norm. However, because ϑ is unknown, one cannot use (4.32) for GoFT. What is typically done, such as in Pearson's χ^2-test, is to replace ϑ by an estimator, $\hat{\vartheta}$. Question is: what is $\hat{\vartheta}$? The ideal scenario would be that, after replacing ϑ by $\hat{\vartheta}$ in (4.32), one has a reduction of degrees of freedom (d.f.), which leads to

$$|B(\hat{\vartheta})|^2 \xrightarrow{\mathrm{d}} \chi^2_\nu, \tag{4.33}$$

where $\nu = s - r > 0$ and $r = \dim(\vartheta)$. This is the famous "subtract one degree of freedom for each parameter estimated" rule taught in many elementary statistics books [e.g., Rice (1995), p. 242]. However, as is well known [e.g., Moore (1978)], depending on what $\hat{\vartheta}$ is used, (4.33) may or may not hold, regardless of what d.f. is actually involved. In fact, the only method that is known to achieve (4.33) without restriction on the distribution of the data is Fisher's minimum χ^2 method. In a way, the method allows one to "cut-down" the d.f. of (1) by r, and thus convert an asymptotic χ^2_s to an asymptotic χ^2_ν. For such a reason, we have coined the method, under the more general setting below, *tailoring*. We develop the method with a heuristic derivation, with the rigorous justification given in Jiang and Torabi (2018).

The "right" estimator of ϑ for tailoring is supposed to be the solution to an estimating equation of the following form:

$$C(\vartheta) \equiv A(\vartheta)B(\vartheta) = 0, \tag{4.34}$$

where $A(\vartheta)$ is an $r \times s$ non-random matrix that plays the role of tailoring the s-dimensional vector, $B(\vartheta)$, to the r-dimensional vector, $C(\vartheta)$. The specification of A will become clear at the end of the derivation. Throughout the derivation, ϑ denotes the true parameter vector. For notation simplicity, we use A for $A(\vartheta)$, $\hat{A}$ for $A(\hat{\vartheta})$, etc. Under regularity conditions, one has the following expansions, which can be derived from the Taylor series expansion and large-sample theory [e.g., Jiang (2010)]:

$$\hat{\vartheta} - \vartheta \approx -\left\{\mathrm{E}_\vartheta\left(\frac{\partial C}{\partial \vartheta'}\right)\right\}^{-1} C, \tag{4.35}$$

$$\hat{B} \approx B - \mathrm{E}_\vartheta\left(\frac{\partial B}{\partial \vartheta'}\right)\left\{\mathrm{E}_\vartheta\left(\frac{\partial C}{\partial \vartheta'}\right)\right\}^{-1} C. \tag{4.36}$$

Because $\mathrm{E}_\vartheta\{B(\vartheta)\} = 0$ [see above (4.32)], one has

$$\mathrm{E}_\vartheta\left(\frac{\partial C}{\partial \vartheta'}\right) = A\mathrm{E}_\vartheta\left(\frac{\partial B}{\partial \vartheta'}\right). \tag{4.37}$$

Combining (4.36), (4.37), we get

$$\hat{B} \approx \{I_s - U(AU)^{-1}A\}B, \tag{4.38}$$

where $U = \mathrm{E}_\vartheta(\partial B/\partial \vartheta')$. We assume that A is chosen such that

$$U(AU)^{-1}A \text{ is symmetric.} \tag{4.39}$$

Then, it is easy to verify that $I_s - U(AU)^{-1}A$ is symmetric and idempotent. If we further assume that the following limit exists:

$$I_s - U(AU)^{-1}A \longrightarrow P, \tag{4.40}$$

then P is also symmetric and idempotent. Thus, assuming that $B \stackrel{\mathrm{d}}{\to} N(0, I_s)$, which is typically the argument leading to (4.32), one has, by (4.38), $\hat{B} \stackrel{\mathrm{d}}{\to} N(0, P)$, hence [e.g., Searle (1971), p. 58] $|\hat{B}|^2 \stackrel{\mathrm{d}}{\to} \chi^2_\nu$, where $\nu = \mathrm{tr}(P) = s - r$. This is exactly (4.33).

It remains to answer one last question: Is there such a non-random matrix $A = A(\vartheta)$ that satisfies (4.39) and (4.40)? We show that, not only the answer is yes, there is an optimal one. Let $A = N^{-1}U'W$, where W is a symmetric, non-random matrix to be determined, and N is a normalizing constant that depends on the sample size. By (4.35) and the fact that $\mathrm{Var}_\vartheta(B) = I_s$ [see above (4.32)], we have

$$\mathrm{var}_\vartheta(\hat{\vartheta}) \approx (U'WU)^{-1}U'W^2U(U'WU)^{-1} \geq (U'U)^{-1}, \tag{4.41}$$

by, for example, Lemma 5.1 of Jiang (2010). The equality on the right side of (4.41) holds when $W = I_s$, giving the optimal A:

$$A = A(\vartheta) = \frac{U'}{N} = \frac{1}{N}\mathrm{E}_\vartheta\left(\frac{\partial B'}{\partial \vartheta}\right). \tag{4.42}$$

The A given by (4.42) clearly satisfy (4.39) [which is equal to $U(U'U)^{-1}U'$]. It will be seen in the next section that, with $N = m$, (4.40) is expected to be satisfied. It should be noted that the solution to (4.34), $\hat{\vartheta}$, does not depend on the choice of N.

A basic assumption for the tailoring method to work is that $\mathrm{E}\{B(\vartheta)\} = 0$ when ϑ is the true parameter vector. However, from the proof of the result [see Jiang and Torabi (2018)] it is seen that the condition "ϑ is the true parameter vector" is not critical. For example, in case there is a model misspecification, a "true parameter vector" may not exist. Nevertheless, what is important is that there is some parameter vector, ϑ, which is not necessarily the true parameter vector, such that the equation

$$A(\vartheta)\mathrm{E}\{B(\vartheta)\} = 0 \tag{4.43}$$

holds. This equation holds, of course, when ϑ is the true parameter vector, but it can also hold when the true parameter vector does not exist, such as under model misspecification. In fact, in the latter case, one may define the "true parameter vector" as the unique ϑ, assumed exist, that satisfies (4.43). Note that the number of equations in (4.43) is the same as the dimension of ϑ; thus, one expect that a solution exists and is unique, under some regularity conditions. To see that (4.43) is the key, note that under (4.42), (4.34) is equivalent to $A(\vartheta)[B(\vartheta) - \mathrm{E}\{B(\vartheta)\}] = 0$, where the expectation is with respect to the true underlying distribution. It follows that one can replace $B(\vartheta)$ by $B(\vartheta) - \mathrm{E}\{B(\vartheta)\}$, which has mean zero, and all of the arguments in the proof go through. This property has given tailoring some unexpected robustness feature, that is, it can work correctly inspite of some model misspecification. We illustrate this point more specifically in the next section.

It should be noted that the special case of (4.42) is closely related to the specification test (ST) in GMM [e.g., Newey (1985)], where the estimator $\hat{\vartheta}$ is defined as the minimizer of the left side of (4.32). This is the same as Fisher's minimum chi-square estimator. Thus, the ST may be viewed as an extension of Fisher's idea, too. Although the minimizer of $|B(\vartheta)|^2$ and the solution to (4.34), with A given by (4.42), are not the same, they are asymptotically equivalent under some regularity conditions. Nevertheless, tailoring is more general than ST in that A does not have to be given by (4.42); it is not exactly the same as ST even if (4.42) holds. We also prefer the name *tailoring* over ST as it is more intuitive.

4.4.2 *Application to SAE*

A standard assumption in SAE models, including the Fay-Herriot model [Fay and Herriot (1979)], the nested-error regression (NER) model [Battese *et al.* (1988)], and the mixed logistic model [Jiang and Lahiri (2001)], is that the random effects are normally distributed. This assumption has had substantial impact on many aspects of the inference. For example, estimation of the mean squared prediction error (MSPE) is an important issue in SAE [e.g., Rao and Molina (2015)]. The well-known Prasad-Rao method (Prasad and Rao 1990) depends on the normality assumption and may not be accurate if the assumption fails. Also, prediction intervals obtained via parametric bootstrap methods [e.g., Chatterjee *et al.* (2008)] depends heavily on the normality assumption. Although there are strategies that are less dependent on the normality assumption, those strategies are often less efficient than the normality-based method when the normality assumption holds, even approximately. Thus, it is important to check the validity of the normality assumption so that an appropriate method can be used for the inference.

In this section, we develop two goodness-of-fit tests that are applicable to SAE using the tailoring method. Although the focus is testing for the normality assumption of the random effects, the method can be easily extended to testing other aspects of the assumed model. Also, we shall focus on the Fay-Herriot model; extension of the method to other types of SAE models, such as the NER model, is fairly straightforward. The first test is based on the existing ML method; while the second test is developed based on a new inference method, which may be of interest on its own.

The Fay-Herriot model may be expressed as that (i) $(y_i, \theta_i), i = 1, \dots, m$ are independent; (ii) $y_i|\theta_i \sim N(\theta_i, D_i)$; and (iii) $\theta_i \sim N(x_i'\beta, A)$. Here, y_i is the direct estimate from the ith area, θ_i is the small area mean, x_i is a vector of observed covariates, β is a vector of unknown parameters, A is an unknown variance, and D_i is a sampling variance that is assumed known. The normality assumption has to do with (iii). The reason that this is not an issue with (ii) is because, in practice, y_i is typically a sample summary such as a sample mean or proportion; as a result, the normality assumption in (ii) often holds approximately due to the central limit theorem (CLT). However, there is no obvious reason to believe that the CLT should hold for (iii). Thus, we consider a broader class of distributions, namely, the skewed normal distribution [SN; e.g., Azzalini and Capitanio (2014)], which includes the normal distribution as a special case. Under the SN distribution, (iii) is replaced by (iii) $\theta_i \sim \mathrm{SN}(x_i'\beta, A, \alpha)$, which denotes the SN distribution with mean $x_i'\beta$, variance A, and skewness parameter α (see below). Noting that $\alpha = 0$ leads to the normal distribution. We denote the model parameters as $\psi = (\beta', A, \alpha)'$.

Suppose that, under the null hypothesis, there is a reduction in the dimension of the parameter vector such that $\gamma = \gamma_0$ under the null hypothesis, where γ is a subvector of ψ and γ_0 is known. Let ϑ denote the vector of parameters in ψ other than γ. In this section, notation such as E_ϑ, etc. will be understood as expectation, etc. under the null hypothesis.

1. Maximum likelihood. Under the Fay-Herriot model, one can show that $y_i \sim SN(x_i'\beta, A + D_i, \alpha_1)$ where $\alpha_1 = \alpha\sqrt{A/(A + D_i + \alpha^2 D_i)}$. We can then write the pdf of y_i as:

$$f(y_i, \psi) = \frac{2}{\sqrt{A + D_i}}\phi\Big(\frac{y_i - x_i'\beta}{\sqrt{A + D_i}}\Big)\Phi\Big(\alpha_1 \frac{y_i - x_i'\beta}{\sqrt{A + D_i}}\Big), \quad i = 1, ..., m,$$

where $\psi = (\beta', A, \alpha)'$, $\phi(.)$ and $\Phi(.)$ are the pdf and cdf of $N(0,1)$, respectively. The likelihood function is $L = \prod_{i=1}^m f(y_i, \psi)$. Under this setting, testing normality is equivalent to testing H_0: $\alpha = 0$. Let $z_i(y_i, \vartheta) = \{(\partial/\partial\psi)\log f(y_i, \psi)\}|_{\alpha=0}$, where $\vartheta = (\beta', A)'$,

$$\begin{aligned}
\frac{\partial \log f(y_i, \psi)}{\partial \beta}\bigg|_{\alpha=0} &= \frac{x_i(y_i - x_i'\beta)}{A + D_i}, \\
\frac{\partial \log f(y_i, \psi)}{\partial A}\bigg|_{\alpha=0} &= \frac{1}{2}\left\{\left(\frac{y_i - x_i'\beta}{A + D_i}\right)^2 - \frac{1}{A + D_i}\right\}, \\
\frac{\partial \log f(y_i, \psi)}{\partial \alpha}\bigg|_{\alpha=0} &= \sqrt{\frac{2\ A}{\pi}}\left(\frac{y_i - x_i'\beta}{A + D_i}\right).
\end{aligned}$$

If the model is correctly specified, let ϑ denote the true ϑ. Then, under the null hypothesis, $\sum_{i=1}^m z_i(y_i, \vartheta)$ is a sum of independent random vectors with mean zero.

If, somehow, the model is misspecified in its mean function that there is no true β, hence no true ϑ, we define the "true ϑ" as the unique solution to (4.43). Then, in the proof of the asymptotic null distribution [see Jiang and Torabi (2018)], one can replace $\sum_{i=1}^m z_i(y_i, \vartheta)$ by $\sum_{i=1}^m [z_i(y_i, \vartheta) - \mathrm{E}\{z_i(y_i, \vartheta)\}]$, which is a sum of independent random vectors with mean zero, and all of the arguments go through.

Furthermore, it can be shown that

$$V_z(\vartheta) = \mathrm{Var}_\vartheta\left\{\sum_{i=1}^m z_i(y_i, \vartheta)\right\} = \sum_{i=1}^m \mathrm{Var}_\vartheta\{z_i(y_i, \vartheta)\},$$

and this is true regardless of the definition of the true ϑ, where

$$\begin{aligned}
&\mathrm{Var}_\vartheta\{z_i(y_i, \vartheta)\} \\
&= \begin{bmatrix} x_i x_i'/(A + D_i) & 0_p & x_i\sqrt{2A}/\sqrt{\pi}(A + D_i) \\ 0_p' & 1/2(A + D_i)^2 & 0 \\ x_i'\sqrt{2A}/\sqrt{\pi}(A + D_i) & 0 & 2A/\pi(A + D_i) \end{bmatrix}
\end{aligned}$$

(Exercise 4.6).

Therefore, if we let $B(\vartheta) = V_z^{-1/2}(\vartheta)\sum_{i=1}^m z_i(y_i, \vartheta)$, we have $B(\vartheta) \xrightarrow{\mathrm{d}} N(0, I_s)$, where $s = \dim(\psi)$, it follows that (4.32) holds. Because $r =$

$\dim(\vartheta) < s$, the tailoring method applies to yield (4.33) with $\nu = s - r = 1$. In particular, we have $A(\vartheta) = m^{-1}\{\sum_{i=1}^{m} \mathrm{E}_\vartheta(\partial z_i'/\partial\vartheta)\}V_z^{-1/2}(\vartheta)$, where z_i is define above and

$$\mathrm{E}_\vartheta\left(\frac{\partial z_i'}{\partial\vartheta}\right) = -\begin{bmatrix} x_i x_i'/(A+D_i) & 0_p & x_i\sqrt{2A}/\sqrt{\pi}(A+D_i) \\ 0_p' & 1/2(A+D_i)^2 & 0 \end{bmatrix}.$$

This gives $A(\vartheta)$ for solving the tailoring equation (4.34) (Exercise 4.6).

2. Maple. The likelihood-ratio test (LRT) is often used in the context of goodness-of-fit. However, because, in SAE, the primary interest is prediction of mixed effects [e.g., Jiang (2007), Rao and Molina (2015)], it is reasonable to develop something that is closely related to the special interest of SAE. To motivate something that is in the same spirit of LRT, but takes into account the SAE interest, let us consider the problem from a "Bayesian" perspective.

Let θ be a vector of unobserved quantities that one wishes to predict, and y a vector of observations. The likelihood function may be viewed, using a Bayesian term, as a marginal likelihood with the distribution of θ, $f(\theta)$, treated as *a prior*, that is,

$$f(y) = \int f(y|\theta)f(\theta)d\theta. \tag{4.44}$$

The likelihood is used for estimation of fixed parameters, which are associated with either $f(\theta)$ or $f(y|\theta)$ or both. To come up with a predictive version of the likelihood, we may simply replace the prior in (4.44) by its "posterior", that is, the conditional pdf of θ given y, $f(\theta|y)$. With this replacement, we obtain

$$f(y|y) \equiv \int f(y|\theta)f(\theta|y)d\theta. \tag{4.45}$$

We call (4.45) the *predictive likelihood*, or PL. The reason is that, if parameter estimation is of primary interest, one uses the prior, $f(\theta)$, to obtain the (marginal) likelihood (4.44). Now, because we replace $f(\theta)$ by $f(\theta|y)$, which is the main outcome for the prediction of θ, and then go through the same operation, the output (4.45) should be called a predictive likelihood. It should be noted that the predictive likelihood is not necessarily a likelihood, as it does not always possess some of the well-known properties of the likelihood. However, we can, at least, adjust the score equation of the PL to make it unbiased.

Let ψ be the vector of parameters involved in either $f(y|\theta)$ or $f(\theta)$. Accordingly, write $f(y|y) = f(y|y,\psi)$. The adjusted PL score is given by

$$s_{\mathrm{a}}(\psi) = \frac{\partial}{\partial\psi}\log f(y|y,\psi) - \mathrm{E}_\psi\left\{\frac{\partial}{\partial\psi}\log f(y|y,\psi)\right\}. \tag{4.46}$$

We call the estimator of ψ obtained by solving the adjusted PL score equation, $s_{\mathrm{a}}(\psi) = 0$, or, equivalently, the following equation:

$$\frac{\partial}{\partial \psi} \log f(y|y, \psi) = \mathrm{E}_{\psi} \left\{ \frac{\partial}{\partial \psi} \log f(y|y, \psi) \right\} \tag{4.47}$$

maximum adjusted PL estimator, or Maple, in view of its analogy to MLE.

Under the Fay-Herriot model, it is easy to show that $f(\theta|y) = \prod_{i=1}^{m} f(\theta_i|y_i)$. Thus, we have

$$\begin{aligned} f(y|y) &= \int \prod_{i=1}^{m} f(y_i|\theta_i) f(\theta_i|y_i) d\theta \\ &= \prod_{i=1}^{m} \int f(y_i|\theta_i) f(\theta_i|y_i) d\theta_i \\ &= \prod_{i=1}^{m} f(y_i|y_i), \end{aligned} \tag{4.48}$$

where $f(y_i|y_i) = \int f(y_i|\theta_i) f(\theta_i|y_i) d\theta_i$. Under the Fay-Herriot model, we have $y_i|\theta_i \sim N(\theta_i, D_i)$ and $\theta_i \sim \mathrm{SN}(x_i'\beta, A, \alpha)$. It is then shown in Jiang and Torabi (2018) that (Exercise 4.7)

$$\begin{aligned} & f(y_i|y_i) \\ &= \frac{1}{\sqrt{D_i(1+B_i)}} \phi \left[\frac{y_i - x_i'\beta}{\sqrt{D_i(1+B_i)/(1-B_i)}} \right] \frac{\Phi[\alpha_{2i}(y_i - x_i'\beta)]}{\Phi[\alpha_{3i}(y_i - x_i'\beta)]}, \end{aligned} \tag{4.49}$$

where $\alpha_{si} = (4-s)\sqrt{A}\alpha/\sqrt{\{(4-s)A + D_i\}\{(4-s)A + (1+\alpha^2)D_i\}}$, $s = 2, 3$. Note that, when $\alpha = 0$, (4.49) reduces to that under normality. Also note that $f(y_i|y_i) \neq f(y_i)$.

By (4.48), the PL can be expressed as $\prod_{i=1}^{m} f(y_i|y_i, \psi)$. To test $\mathrm{H}_0 : \alpha = 0$, let

$$b_i(y_i, \vartheta) = \{(\partial/\partial\psi) \log f(y_i|y_i, \psi)\}|_{\alpha=0} - E[\{(\partial/\partial\psi) \log f(y_i|y_i, \psi)\}|_{\alpha=0}].$$

One can derive the adjusted PL equation, (4.47), as follows:

$$\begin{aligned} & \left.\frac{\partial \log f(y_i|y_i, \psi)}{\partial \beta}\right|_{\alpha=0} - \mathrm{E}\left\{ \left.\frac{\partial \log f(y_i|y_i, \psi)}{\partial \beta}\right|_{\alpha=0} \right\} = a_i(A) x_i (y_i - x_i'\beta), \\ & \left.\frac{\partial \log f(y_i|y_i, \psi)}{\partial A}\right|_{\alpha=0} - \mathrm{E}\left\{ \left.\frac{\partial \log f(y_i|y_i, \psi)}{\partial A}\right|_{\alpha=0} \right\} \\ & = b_i(A)(y_i - x_i'\beta)^2 - c_i(A), \\ & \left.\frac{\partial \log f(y_i|y_i, \psi)}{\partial \alpha}\right|_{\alpha=0} - \mathrm{E}\left\{ \left.\frac{\partial \log f(y_i|y_i, \psi)}{\partial \alpha}\right|_{\alpha=0} \right\} = d_i(A)(y_i - x_i'\beta), \end{aligned}$$

where $a_i(A) = (1-B_i)^2/D_i(1+B_i)$, $b_i(A) = (1-B_i)^3(3+B_i)/2D_i^2(1+B_i)^2$, $c_i(A) = (1-B_i)^2(3+B_i)/2D_i(1+B_i)^2$, $d_i(A) = \sqrt{2A/\pi}(1-B_i)^2/D_i(1+B_i)$.

Let ϑ denote the true ϑ. If the model is correctly specified under the null hypothesis, then, under the null hypothesis, $\sum_{i=1}^m b_i(y_i, \vartheta)$ is a sum of independent random vectors with mean zero. On the other hand, if there is some misspecification in the mean function that the true β, hence the true ϑ, does not exist (under the null hypothesis), we again define the "true ϑ" as the unique solution to (4.43). Then, similar to the ML case, all of the arguments in the proof go through by replacing $\sum_{i=1}^m b_i(y_i, \vartheta)$ with $\sum_{i=1}^m [b_i(y_i, \vartheta) - \mathrm{E}\{b_i(y_i, \vartheta)\}]$. Furthermore, we have $V_b(\vartheta) = \mathrm{Var}_\vartheta\{\sum_{i=1}^m b_i(y_i, \vartheta)\} = \sum_{i=1}^m \mathrm{Var}_\vartheta\{b_i(y_i, \vartheta)\}$, where

$$\mathrm{Var}_\vartheta\{b_i(y_i, \vartheta)\} = \begin{bmatrix} g_i(A)x_i x_i' & 0_p & g_i(A)(x_i\sqrt{2A/\pi}) \\ 0_p' & h_i(A) & 0 \\ g_i(A)(x_i'\sqrt{2A/\pi}) & 0 & g_i(A)(2A/\pi) \end{bmatrix}$$

with $g_i(A) = (1-B_i)^3/D_i(1+B_i)^2$ and $h_i(A) = (1-B_i)^4(3+B_i)^2/2D_i^2(1+B_i)^4$. Thus, if we let $B(\vartheta) = V_b^{-1/2}(\vartheta)\sum_{i=1}^m b_i(y_i, \vartheta)$, we have $B(\vartheta) \xrightarrow{\mathrm{d}} N(0, I_s)$, where $s = \dim(\psi)$. It follows that (4.32) holds. Because $r = \dim(\vartheta) < s$, the tailoring method applies to yield (4.33) with $\nu = s - r = 1$. In particular, we have $A(\vartheta) = m^{-1}\{\sum_{i=1}^m \mathrm{E}_\vartheta(\partial b_i'/\partial\vartheta)\}V_b^{-1/2}(\vartheta)$, where b_i is define above and (Exercise 4.7)

$$\mathrm{E}_\vartheta\left(\frac{\partial b_i'}{\partial\vartheta}\right) = -\begin{bmatrix} a_i(A)x_i x_i' & 0_p & d_i(A)x_i \\ 0_p' & b_i(A) & 0 \end{bmatrix}.$$

4.4.3 *Empirical results*

A simulation study was carried out to evaluate performance of the tailoring methods based on ML and Maple, described in the previous subsection, and compare them with existing methods. Specifically, we compare the tailoring methods with those of Pierce (1982), Jiang (2001), and Claeskens and Hart (2009). For Pierce (1982), the test statistic under the Fay-Herriot model for $\mathrm{H}_0 : \alpha = 0$ is given by $\hat{F} \equiv m\hat{T}_m^2/V$, where

$$\hat{T}_m = \frac{1}{m}\sum_{i=1}^m \frac{\sqrt{D_i}(y_i - x_i'\hat{\beta})}{\hat{A} + D_i},$$

$$V = \frac{1}{m}\sum_{i=1}^m \frac{D_i}{D_i + A} - mP\{\mathrm{var}(\hat{\psi} - \psi)\}P'$$

with $P = \lim \mathrm{E}(\partial T_m / \partial \psi)$. The asymptotic null distribution of the test statistic is χ_1^2. In the current case, it can be shown that

$$P = -\frac{1}{m}\begin{bmatrix} \sum_{i=1}^m \sqrt{D_i} x_i' / (D_i + A) \\ 0 \end{bmatrix}.$$

In the case of Jiang (2001), one has the test statistic

$$\hat{\chi}_{\mathrm{J}}^2 = \frac{1}{m}\sum_{k=1}^K \{N_k - p_k(\hat{\psi})\}^2,$$

where $N_k = \sum_{i=1}^m 1_{(y_i \in C_k)} = \#\{1 \leq i \leq m : y_i \in C_k\}$, and $p_k(\psi) = \sum_{i=1}^m P_\psi(y_i \in C_k) = \sum_{i=1}^m p_{ik}(\psi)$. More specifically, the cells, C_k, $1 \leq k \leq K$ are defined as follows: $C_1 = (-\infty, c_1], C_k = (c_{k-1}, c_k], 2 \leq k \leq K-1$, and $C_K = (c_{K-1}, \infty)$. Regarding the choice of K and c_k's, by Jiang (2001), we may choose $K = \max(p+2, [m^{1/5}])$, where p is the dimension of β. Once K is chosen, the c_k's are chosen so that there are equal number of y_i's within each $C_k, 1 \leq k \leq K$. It then follows that $N_k = m/K, 1 \leq k \leq K$. Finally, the $p_{ik}(\psi)$ have the following expressions:

$$\begin{aligned} p_{i1}(\psi) &= \Phi\left(\frac{c_1 - x_i'\beta}{\sqrt{A + D_i}}\right), \\ p_{ik}(\psi) &= \Phi\left(\frac{c_k - x_i'\beta}{\sqrt{A + D_i}}\right) - \Phi\left(\frac{c_{k-1} - x_i'\beta}{\sqrt{A + D_i}}\right), \quad 2 \leq k \leq K-1, \\ p_{iK}(\psi) &= 1 - \Phi\left(\frac{c_{K-1} - x_i'\beta}{\sqrt{A + D_i}}\right), \end{aligned}$$

where $\Phi(\cdot)$ is the cdf of $N(0,1)$. We then use a Monte Carlo method (e.g., bootstrapping) to compute the critical values, as suggested by Jiang (2001).

In the case of Claeskens and Hart (2009), one uses the test statistic

$$\hat{\chi}_{\mathrm{CH}}^2 = \max_{1 \leq l \leq M} \frac{2\{\log L_l - \log L_{M=0}\}}{l(l+3)/2},$$

where M is the order of polynomial which plays the role of a smoothing parameter. The test is based on the LRT which compares the estimated distribution ($M > 0$) and the null distribution ($M = 0$; i.e., normal). Similar to Jiang (2001), one needs to use replications from the test statistic above to approximate the critical values. Here we consider $M = 1$.

To evaluate and compare performance the aforementioned methods, let $\hat{B}_{\mathrm{ML}}^2$, $\hat{B}_{\mathrm{PL}}^2$, $\hat{F}$, $\hat{\chi}_{\mathrm{J}}^2$, and $\hat{\chi}_{\mathrm{CH}}^2$ represent the test statistics for the tailoring tests with ML, Maple, and the tests of Pierce, Jiang, and Claeskens-Hart, respectively [for notation simplicity we write $|B(\hat{\vartheta})|^2$ as $\hat{B}^2$]. We consider the following Fay-Herriot model:

$$y_i = \beta_1 x_i + v_i + e_i, \quad 1 \le i \le n,$$
$$y_i = \beta_2 x_i + v_i + e_i, \quad n+1 \le i \le m,$$

where $m = 2n$, $D_i = D_{i1}$ for $1 \le i \le n$ and $D_i = D_{i2}$ for $n+1 \le i \le m$. We choose $A = 10$. Recall that $v_i \sim \mathrm{SN}(0, A, \alpha)$ and $e_i \sim N(0, D_i)$. The D_{i1} are generated form the uniform distribution between 3.5 and 4.5. There are two scenarios for D_{i2}: one generated from (i) Uniform$(3.5, 4.5)$ and the other from (ii) Uniform$(0.5, 1.5)$. The x_i's are generated from the uniform distribution between 0 and 1. The x_i's and D_i's are fixed throughout the simulation study.

The actual model that we are fitting is the Fay-Herriot model with $\beta_1 = \beta_2 \equiv \beta$, that is,

$$y_i = \beta x_i + v_i + e_i, \quad 1 \le i \le m$$

with the same assumptions for v_i and e_i. Two scenarios are considered. In the first scenario, we let $\beta_1 = \beta_2 = 1$; in the second scenario, we let $\beta_1 = 1$ and $\beta_2 = 3$. Thus, under the first scenario, the mean function is correctly specified with a true $\beta = 1$; under the second scenario, the mean function is misspecified, therefore, there is no true β. However, it is easy to see that, even under the second scenario, there is a unique $\vartheta = (\beta, A)'$ that satisfies (4.43). Therefore, according to the theory developed in the previous sections, the tailoring methods apply under both scenarios.

To test $\mathrm{H}_0 : \alpha = 0$, we consider four different sample sizes $m = 50, 100, 200,$ and 500. We carry out $R = 5{,}000$ simulation runs to calculate $\hat{B}^2_{\mathrm{ML}}, \hat{B}^2_{\mathrm{PL}}, \hat{F}, \hat{\chi}^2_{\mathrm{J}}$, and $\hat{\chi}^2_{\mathrm{CH}}$ [for notation simplicity we write $|B(\hat{\vartheta})|^2$ as $\hat{B}^2$]. Note that for the tailoring methods, A is estimated by solving (4.34), while for $\hat{F}$ and $\hat{\chi}^2_{\mathrm{J}}$ we use the Prasad-Rao method [Prasad and Rao (1990)] to estimate the parameters, and for $\hat{\chi}^2_{\mathrm{CH}}$ we use ML for parameter estimation. Also, we use 1000 replications to obtain the critical values (in each simulation run) for $\hat{\chi}^2_{\mathrm{J}}$, and 100 replications for $\hat{\chi}^2_{\mathrm{CH}}$.

To compare the powers of $\hat{B}^2_{\mathrm{ML}}, \hat{B}^2_{\mathrm{PL}}, \hat{F}, \hat{\chi}^2_{\mathrm{J}}$, and $\hat{\chi}^2_{\mathrm{CH}}$, we repeat similar steps as above but now generate data in each simulation run r under the alternative model: $y_i^{(r)} = \beta_1 x_i + v_i^{(r)} + e_i^{(r)}, (1 \le i \le n)$; and $y_i^{(r)} = \beta_2 x_i + v_i^{(r)} + e_i^{(r)}, (n+1 \le i \le m)$ for $r = 1, \dots, R$, where $v_i^{(r)} \sim \mathrm{SN}(0, A, \alpha = 0.5)$ and $e_i^{(r)}$ are simulated the same way as under H_0.

The simulated size and power are reported in Table 4.5. It seems that with the increasing sample size (m), we get the correct size and high power

Table 4.5 **Size (Power) under Different Sample Sizes (m) and Scenarios**

D_{i2}	β_2	m	$\hat{B}^2_{\text{ML}}$	$\hat{B}^2_{\text{PL}}$	$\hat{F}$	$\hat{\chi}^2_{\text{J}}$	$\hat{\chi}^2_{\text{CH}}$
(i)	1	50	0.048 (0.853)	0.048 (0.861)	0.046 (0.898)	0.049 (0.023)	0.054 (0.389)
		100	0.053 (0.966)	0.052 (0.966)	0.052 (0.981)	0.051 (0.017)	0.044 (0.398)
		200	0.051 (0.999)	0.051 (0.999)	0.049 (0.999)	0.045 (0.004)	0.007 (0.108)
		500	0.055 (1.000)	0.055 (1.000)	0.056 (1.000)	0.048 (0.0002)	0.007 (0.096)
	3	50	0.044 (0.840)	0.044 (0.851)	0.040 (0.894)	0.053 (0.021)	0.044 (0.377)
		100	0.050 (0.962)	0.050 (0.961)	0.049 (0.980)	0.060 (0.013)	0.046 (0.332)
		200	0.046 (0.999)	0.046 (0.999)	0.046 (0.999)	0.053 (0.003)	0.003 (0.149)
		500	0.053 (1.000)	0.052 (1.000)	0.052 (1.000)	0.049 (0.000)	0.000 (0.155)
(ii)	1	50	0.046 (0.848)	0.045 (0.810)	0.049 (0.855)	0.051 (0.025)	0.060 (0.461)
		100	0.053 (0.964)	0.048 (0.934)	0.048 (0.970)	0.052 (0.017)	0.034 (0.372)
		200	0.050 (0.999)	0.047 (0.994)	0.050 (0.998)	0.048 (0.006)	0.008 (0.112)
		500	0.054 (1.000)	0.054 (1.000)	0.054 (1.000)	0.048 (0.0002)	0.004 (0.159)
	3	50	0.042 (0.838)	0.043 (0.799)	0.092 (0.765)	0.061 (0.021)	0.062 (0.467)
		100	0.050 (0.958)	0.048 (0.922)	0.129 (0.932)	0.074 (0.011)	0.047 (0.329)
		200	0.046 (0.999)	0.044 (0.994)	0.212 (0.991)	0.070 (0.003)	0.009 (0.114)
		500	0.052 (1.000)	0.053 (1.000)	0.406 (1.000)	0.081 (0.000)	0.004 (0.370)

for $\hat{B}^2_{\text{ML}}$ and $\hat{B}^2_{\text{PL}}$ under different scenarios. However, in the cases of $\hat{F}$ and $\hat{\chi}^2_{\text{CH}}$ we do not get the right size, if we have both misspecification in the mean and different sampling variances for the small areas; we also do not have the right size for $\hat{\chi}^2_{\text{CH}}$ with the increasing sample size. Furthermore, we observe poor performance in terms of power for $\hat{\chi}^2_{\text{J}}$ and $\hat{\chi}^2_{\text{CH}}$ compared to $\hat{B}^2_{\text{ML}}$ and $\hat{B}^2_{\text{PL}}$. Overall, the simulation results demonstrate a robustness feature of tailoring as suggested by the theoretical result that the other methods do not possess.

4.5 Real-data examples

4.5.1 *TVSFP data*

We use data from the Television School and Family Smoking Prevention and Cessation Project (TVSFP) to illustrate the robust likelihood-ratio test discussed in the previous sections. For a complete description of the TVSFP study, see Hedeker *et al.* (1994). The original study was designed to test independent and combined effects of a school-based social-resistance curriculum and a television-based program in terms of tobacco use prevention and cessation. The subjects were seventh-grade students from Los Angeles (LA) and San Diego in the State of California in the United States. The students were pretested in January 1986 in an initial study. The same students completed an immediate postintervention questionnaire in April 1986, a one-year follow-up questionnaire (in April 1987), and a two-year follow-up (in April 1988). In this analysis, we consider a subset of the TVSFP data involving students from 28 LA schools, where the schools were randomized to one of four study conditions: (a) a social-resistance classroom curriculum (CC); (b) a media (television) intervention (TV); (c) a combination of CC and TV conditions; and (d) a no-treatment control. A tobacco and health knowledge scale (THKS) score was one of the primary study outcome variables, and the one used for this analysis. The THKS consisted of seven questionnaire items used to assess student tobacco and health knowledge. A student's THKS score was defined as the sum of the items that the student answered correctly. Only data from the pretest and immediate postintervention are available for the current analysis. More specifically, the data only involved subjects who had completed the THKS at both of these time points. In all, there were 1,600 students from the 28 schools, with the number of students from each school ranging from 18 to 137.

Hedeker *et al.* (1994) carried out a mixed-model analysis based on a number of NER models to illustrate maximum likelihood estimation for the analysis of clustered data. Jiang *et al.* (2015b) considered the same data small area estimation [e.g., Rao and Molina (2015)]. The following NER model [Battese *et al.* (1988)] was proposed, which is a special case of linear mixed model:

$$y_{ij} = \beta_0 + \beta_1 x_{i1} + \beta_2 x_{i2} + \beta_3 x_{i1} x_{i2} + v_i + e_{ij}, \tag{4.50}$$

$i = 1, \dots, 28, j = 1, \dots, n_i$, where y_{ij} is the difference between the immediate postintervention and pretest THKS scores; x_{i1}, x_{i2} are indicators (1 or 0) of CC and TV programs, respectively; $\beta_j, j = 0, 1, 2, 3$ are unknown fixed effects, with β_3 corresponding to the interaction between CC and TV; v_i is a school-specific random effect, and e_{ij} is an additional error. It is assumed that the random effects and errors are independent, with $v_i \sim N(0, \sigma_v^2)$ and $e_{ij} \sim N(0, \sigma_e^2)$, where σ_v^2, σ_e^2 are unknown variances (Exercise 4.5).

The analysis of Jiang *et al.* (2015b) suggested that the CC program is more effective than the TV program. Also, the sign of the estimated β_3 was negative, indicating that the interaction may have negative effect. These findings have led to consideration of a testing problem, with the following null hypothesis:

$$\mathrm{H}_0 : \ \beta_2 = 0, \ \beta_1 > 0, \ \beta_3 < 0. \tag{4.51}$$

In a way, the null hypothesis is complex, especially because of the inequality constraints. Therefore, it is natural to consider the likelihood-ratio test. However, the data is clearly not normal in this case. Note that the possible values of y_{ij} are integers between -7 and 7. Therefore, the Gaussian likelihood function is not a likelihood function. However, we can still use it for the L-test, discussed in the previous sections.

The maximizer of the objective function, which is the negative Gaussian log-likelihood, is given by $\hat{\beta}_0 = 0.211$, $\hat{\beta}_1 = 0.707$, $\hat{\beta}_2 = 0.240$, $\hat{\beta}_3 = -0.319$,

$\hat{\sigma}_v = 0.069$, and $\hat{\sigma}_e = 1.550$. The maximizer under the null hypothesis is $\hat{\beta}_{00} = 0.222$, $\hat{\beta}_{01} = 0.584$, $\hat{\beta}_{03} = -0.082$, $\hat{\sigma}_{0v} = 0.120$, and $\hat{\sigma}_{0e} = 1.550$. The L-test statistic is $-2\log R = 3.923$. According to Theorem 4.4, the asymptotic null distribution of the L-test is χ_1^2, which, for the level of significance 0.05, has a critical value of 3.841. Thus, the null hypothesis is rejected. The conclusion is that, in spite of being less effective than the CC program, the TV program still has a positive impact on improving the schools' THKS scores (whether the improved THKS score means improved tobacco use prevention and cessation is a different matter though).

4.5.2 *Median income data*

It is known that the income data are typically not normal. In this application, our goal is to check the normality assumption for median incomes of four-person families at the state level in the USA Ghosh *et al.* (1996). The data has been analyzed by several researchers using different set-ups. In this analysis, the response variable y_i is sample survey of four-person family median income at state i in year 1989, and x_i is the census four-person family median income at state i in year 1979 ($i = 1, ..., 51$). An inspection of the scatter plot (see Fig. 4.1) suggests that a quadratic model would fit the data. Thus, the following model is considered:

$$y_i = \beta_1 x_i + \beta_2 x_i^2 + v_i + e_i, i = 1, ..., m = 51, \tag{4.52}$$

where the v_i's are state-specific random effects and e_i's are sampling errors. It is assumed that v_i and e_i are independent with $v_i \sim \mathrm{SN}(0, A, \alpha)$ and $e_i \sim N(0, D_i)$ with known D_i.

We considering testing $\mathrm{H}_0 : \alpha = 0$ vs $\mathrm{H}_1 : \alpha \neq 0$. First, we apply the Maple and ML approaches in conjunction with tailoring. In the case of tailoring/Maple, the parameter estimates are $\hat{\beta}_1 = 2.07, \hat{\beta}_2 = -1.2e - 5, \hat{A} = 1.9 \times 10^7$, which result in rejecting the H_0 at 10% significance level $[\hat{B}^2_{\mathrm{PL}} = 4.67 > 2.70 = \chi_1^2(0.90)]$. In the case of tailoring/ML, the parameter estimates are $\hat{\beta}_1 = 2.07, \hat{\beta}_2 = -1.3e - 5, \hat{A} = 1.9 \times 10^7$, which leads to the same conclusion ($\hat{B}^2_{\mathrm{ML}} = 3.68 > 2.70$). Thus, the tailoring methods suggest that the normality assumption is not valid.

We also applied the methods of Pierce (1982), Jiang (2001), and Claeskens and Hart (2009) to test the hypothesis. None of these tests were able to reject the normality assumption. This seems to be consistent with the pattern observed in our simulation study in Section 4.4.3 that these tests appear to be less powerful than our tests.

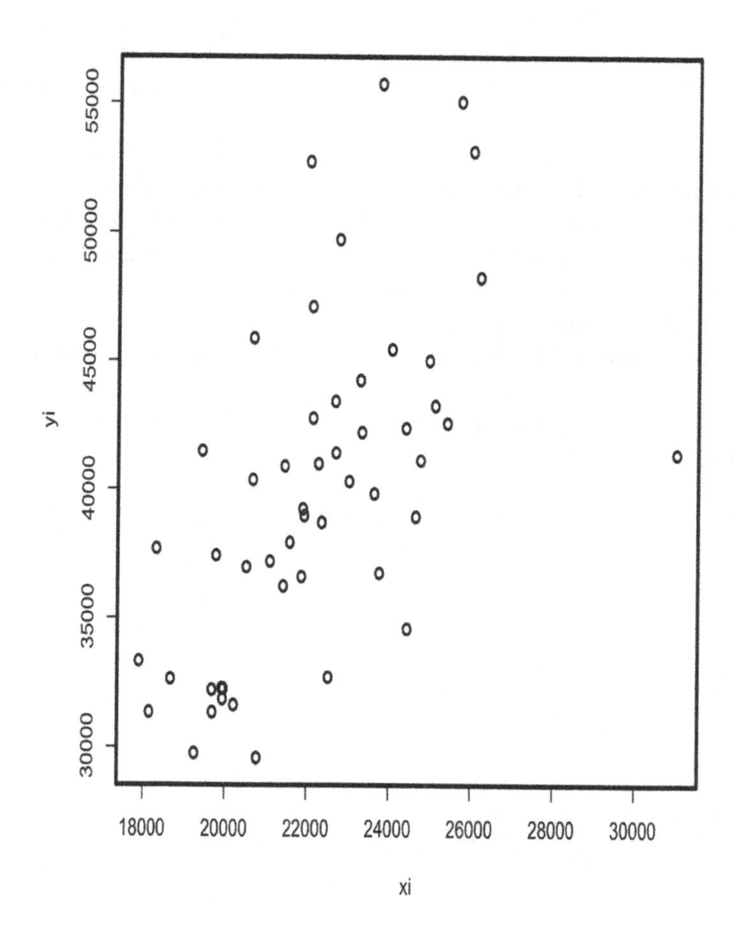

Fig. 4.1 *Plot of Census Four-person Family Median Income in 1979 (x) vs Sample Survey of Four-person Family Median Income in 1989 (y) for Different States.*

4.6 Exercises

4.1. Show that, for the test considered in Subsection 3.3.3, the test statistic (4.4) reduces to $\hat{\chi}^2 = (\hat{\gamma}_1 - 1)^2/\hat{\Sigma}_{R,11}$, where $\hat{\Sigma}_{R,11}$ is given by (3.26), and the asymptotic null distribution is χ_1^2.

4.2. Write out specifically the parameter space Θ_0 in the null hypothesis (4.11) for the cases of Examples 4.3–4.6.

4.3. Extend Theorem 4.5 in the same was as the Extensions of Theorem 4.2 and Theorem 4.3 to allow A, B, C to be replaced by sequences of matrices.

4.4. Show that the maximizer of (4.27) is $\hat{\vartheta} = (\hat{\beta}', \hat{\tau}^2, \hat{\gamma}')'$, where $\hat{\tau}^2$ and $\hat{\gamma}$ are the REML estimators, and $\hat{\beta}$ is given above (4.27). However, $l_R(\vartheta, y)$ is not a log-likelihood function, even if normality holds. Please explain why.

4.5. Derive an expression of the Gaussian log-likelihood under the nested-error regression model of Section 4.4. The expression should not involve an matrix products.

4.6. Verify the expressions for $\text{Var}_\vartheta\{z_i(y_i, \vartheta)\}$ and $\text{E}_\vartheta(\partial z_i'/\partial\vartheta)$ given in §4.4.2.1.

4.7. Verify expression (4.49). Also verify the expressions for $\text{Var}_\vartheta\{b_i(y_i, \vartheta)\}$ and $\text{E}_\vartheta(\partial b_i'/\partial\vartheta)$ given in §4.4.2.2.

Chapter 5

Observed Best Prediction

So far, we have been considering the types of inference such as estimation and testing. There is another type of statistical inference that has not been touched: Prediction. As noted in Jiang (2007, sec. 2.3), there are two kinds of prediction problems associated with the mixed effects models. The first kind, which is encountered more often in practice, is prediction of mixed effects; the second kind is prediction of future observation. Here in this chapter, we shall focus on prediction of the mixed effects. In this regard, the best known method is best linear unbiased prediction, or BLUP. So, before we present a new method, let us first review basic elements of BLUP.

5.1 Best linear unbiased prediction

A (linear) mixed effect is a linear combination of fixed and random effects, expressed as $\eta = b'\beta + a'\alpha$, where b, a are known vectors, and α and β are the vectors of fixed and random effects. Prediction of a mixed effect has a long history, dating back to C. R. Henderson in his early work in the field of animal breeding [e.g., Henderson (1948). Robinson (1991) gives a wide-ranging account of prediction of mixed effects using BLUP with examples and applications. Jiang and Lahiri (2006) offers another review of prediction of mixed effects, in which the authors used the term *mixed model prediction* for such kind of prediction problems, with focus on small area estimation. When the fixed effects and variance components are known, the best predictor (BP) of η, in the sense of minimum mean squared prediction error (MSPE), is its conditional expectation given the data:

$$\tilde{\eta} = \mathrm{E}(\xi|y) = b'\beta + a'\mathrm{E}(\alpha|y). \tag{5.1}$$

Now assume that y satisfies the following LMM:

$$y = X\beta + Z\alpha + \epsilon, \tag{5.2}$$

where X, Z are known matrices; β is a vector of fixed effects; α, ϵ are vectors of random effects and errors, respectively, such that $\alpha \sim N(0, G)$, $\epsilon \sim N(0, R)$, and α, ϵ are independent. Under such LMM assumptions, one has the well-known expression for the conditional expectation in (5.1):

$$\mathrm{E}(\alpha|y) = GZ'V^{-1}(y - X\beta), \tag{5.3}$$

where $V = \mathrm{Var}(y) = ZGZ' + R$. Thus, if β, G, R are known, the BP of η is given by (5.1) and (5.3).

In practice, the β, G, R involved in the BP are usually unknown. Before we handle the most practical situation, let us first make a small step, by assuming that β is unknown but G, R are known. In such a case, it is customary to estimate β by its MLE, which is given by (1.3) with $\hat{V}$ replaced by the V given below (5.3). This leads to the BLUP. In other words, the BLUP of η is $\tilde{\eta} = b'\tilde{\beta} + a'\tilde{\alpha}$, where

$$\tilde{\beta} = (X'V^{-1}X)^{-1}X'V^{-1}y, \tag{5.4}$$

$$\tilde{\alpha} = GZ'V^{-1}(y - X\tilde{\beta}). \tag{5.5}$$

Note that (5.4) is the same as the BLUE introduced earlier [see (2.30)] while (5.5) is the BLUP of α.

As for the most general situation, where G, R are also unknown, one needs to replace G, R in (5.4), (5.5) by their estimators. For example, it is often assumed that G, R are known functions of some vector of variance components, θ, that is, $G = G(\theta)$ and $R = R(\theta)$. Then, if $\hat{\theta}$ is a consistent estimator of θ, G, R are estimated by $\hat{G} = G(\hat{\theta}), \hat{R} = R(\hat{\theta})$, respectively. For example, $\hat{\theta}$ may be the REML estimator, discussed in Section 1.1 (also see Section 3.2). Once the G, R in BLUP are replaced by their estimators, the result is called empirical BLUP, or EBLUP. In other words, the EBLUP of η is $\hat{\eta} = b'\hat{\beta} + a'\hat{\alpha}$, $\hat{\beta}$ is given by (1.3),

$$\hat{\alpha} = \hat{G}Z'\hat{V}^{-1}(y - X\hat{\beta}), \tag{5.6}$$

and $\hat{V} = Z\hat{G}Z' + \hat{R}$.

5.2 Observed best prediction: Another look at the BLUP

If one thinks more carefully, the BLUP may be viewed as a hybrid of optimal prediction, that is, BP and optimal estimation, that is, ML. More specifically, up to (5.3) one has the BP, but after that there is, perhaps unnotably, a shift of focus to best estimation, that is, β by its MLE, assuming that G, R are known. However, if prediction of the mixed effect is of main interest, it may be wondered why one has to hybridize.

To make it simpler, let us stay, for now, with the assumption G, R are known. Can we estimate β in a way that is also to the best interest of prediction (of the mixed effect)? Note that the MLE is known to be (asymptotically) optimal in terms of estimation, but not necessarily in terms of prediction (in fact, it is not, as we shall show). This idea leads to a new approach for mixed model prediction, proposed by Jiang *et al.* (2011a). The approach has a surprising bonus, as it turns out, that is, it leads to a predictor that is more robust to model misspecification than the BLUP, or EBLUP. A key to the new approach is to entertain two models. One is called assumed model, which is (5.2); the other is called broader model, or true model. For simplicity, let us consider, for now, the case that the potential model misspecification is only in terms of the mean function; in other words, the true model is

$$y = \mu + Z\alpha + \epsilon, \tag{5.7}$$

where $\mu = \mathrm{E}(y)$. Here, E denotes expectation under the true distribution of y, which may be unknown, but is not model-dependent. Our interest is prediction of a vector of mixed effects, expressed as

$$\zeta = F'\mu + C'\alpha, \tag{5.8}$$

where F, C are known matrices. We illustrate with an example.

Example 5.1 (Fay–Herriot model). Fay and Herriot [Fay and Herriot (1979)] proposed the following model to estimate per-capita income of small places with population size less than 1,000: $y_i = x_i'\beta + v_i + e_i$, $i = 1, \dots, m$, where y_i is a direct estimate (sample mean) for the ith area, x_i is a vector of known covariates, β is a vector of unknown regression coefficients, v_i's are area-specific random effects and e_i's are sampling errors. It is assumed that v_i's, e_i's are independent with $v_i \sim N(0, A)$ and $e_i \sim N(0, D_i)$. The variance A is unknown, but the sampling variances D_i's are assumed known (in practice, the D_i's can be estimated with accuracy, hence can be assumed known, at least approximately). Here, the interest is to estimate the small area mean $\zeta = (\zeta_i)_{1 \le i \le m} = \mu + v$, where $\mu = (\mu_i)_{1 \le i \le m}$ with $\mu_i = \mathrm{E}(y_i)$, and $v = (v_i)_{1 \le i \le m}$. In other words, $\zeta = \mathrm{E}(y|v)$, which can be expressed as (5.8) with $F = C = I_m$.

Similar to (5.1), (5.3), the BP of ζ is given by

$$\begin{aligned} \mathrm{E}_M(\zeta|y) &= F'\mu + C'\mathrm{E}_M(\alpha|y) \\ &= F'X\beta + C'GZ'V^{-1}(y - X\beta) \\ &= F'y - \Gamma(y - X\beta), \end{aligned} \tag{5.9}$$

where $\Gamma = F' - B$ with $B = C'GZ'V^{-1}$. Here, unlike E, E_M denotes conditional expectation under the assumed model, (5.2), denoted by M. Thus, the BP is model-based in the sense that its expression depends on the assumed LMM, (5.2), including the normality assumption.

Let $\check{\zeta}$ denote the right side of (5.9) with a fixed, arbitrary β. Then, by (5.7), (5.8), we have $\check{\zeta} - \zeta = H'\alpha + F'\epsilon - \Gamma(y - X\beta)$, where $H = Z'F - C$. Thus, we have

$$\begin{aligned} \mathrm{MSPE}(\check{\zeta}) &= \mathrm{E}(|\check{\zeta} - \zeta|^2) \\ &= \mathrm{E}(|H'\alpha + F'\epsilon|^2) - 2\mathrm{E}\{(\alpha'H + \epsilon'F)\Gamma(y - X\beta)\} \\ &\quad + \mathrm{E}\{(y - X\beta)'\Gamma'\Gamma(y - X\beta)\} \\ &= I_1 - 2I_2 + I_3. \end{aligned}$$

It is easy to see that I_1 does not depend on β. In fact, I_2 does not depend on β either, because $I_2 = \mathrm{E}\{(\alpha'H + \epsilon'F)\Gamma(y - \mu)\} + \{\mathrm{E}(\alpha'H + \epsilon'F)\}\Gamma(\mu - X\beta) = \mathrm{E}\{(\alpha'H + \epsilon'F)\Gamma(Z\alpha + \epsilon)\}$, by (5.7). Thus, we can express the MSPE as

$$\mathrm{MSPE}(\check{\zeta}) = \mathrm{E}\{(y - X\beta)'\Gamma'\Gamma(y - X\beta) + \cdots\}, \tag{5.10}$$

where $\cdots$ does not depend on β. The idea is to estimate β by minimizing the observed MSPE, which is the expression inside the expectation in (5.10). Then, because $\cdots$ does not depend on β, this is equivalent to minimizing $(y - X\beta)'\Gamma'\Gamma(y - X\beta)$. The solution is what we call best predictive estimator, or BPE, given by

$$\hat{\beta} = (X'\Gamma'\Gamma X)^{-1}X'\Gamma'\Gamma y, \tag{5.11}$$

assuming nonsingularity of $\Gamma'\Gamma$ and that X is full rank. The resulting predictor of ζ is called *observed best predictor*, or OBP, given by the right side of (5.9) with β replaced by $\hat{\beta}$. The term OBP is due to the fact that the BPE is obtained by minimizing the observed MSPE (if the observed MSPE were the true MSPE, the same procedure would lead to the BP). We consider an example.

Example 5.1 (continued). It is easy to verify (Exercise 5.1) that the BPE of β is given by

$$\hat{\beta} = \left\{\sum_{i=1}^m \left(\frac{D_i}{A + D_i}\right)^2 x_i x_i'\right\}^{-1} \sum_{i=1}^m \left(\frac{D_i}{A + D_i}\right)^2 x_i y_i. \tag{5.12}$$

Also, the BP of ζ_i, the ith component of ζ, is given by

$$\tilde{\zeta}_i = \frac{A}{A + D_i} y_i + \frac{D_i}{A + D_i} x_i' \beta, \tag{5.13}$$

where β, A are the true parameters. Thus, assuming that A is known, the OBP of ζ_i is given by (5.13) with β replaced by $\hat{\beta}$ of (5.12).

On the other hand, it can be shown that the MLE of β, assuming that A is known, is

$$\tilde{\beta} = \left(\sum_{i=1}^{m} \frac{x_i x_i'}{A + D_i} \right)^{-1} \sum_{i=1}^{m} \frac{x_i y_i}{A + D_i}. \tag{5.14}$$

The BLUP of ζ_i is given by (5.13) with β replaced by $\tilde{\beta}$ of (5.14).

It is interesting to note that both the MLE, (5.14), and BPE, (5.12), are in the forms of weighted averages, the only difference being the weights. In particular, the BPE gives more weights to areas with larger sampling variances, D_i, while the MLE does just the opposite–assigning more weights to areas with smaller sampling variances. A question is: Who is right?

To answer this question, first note that, from the expression of the BP, (5.13), it is evident that the assumed model is involved only through $x_i'\beta$, whose corresponding weight, $D_i/(A + D_i)$, is increasing with D_i. In other words, the model-based BP is more relevant to those areas with larger D_i. Imagine that there is a meeting of representatives from the different small areas to discuss what estimate of β is to be used in the BP. The areas with larger D_i think that their "voice" should be heard more (i.e., they should receive more weights), because the BP is more relevant to their "business". Their request is reasonable (although, politically, this may not work out within a democratic voting system).

Not only the OBP has a intuitive explanation, it also has an attractive property that it is more robust to model misspecification than BLUP, in terms of MSPE. This is not surprising, given the way that OBP is derived, but the following theorem, proved in Jiang *et al.* (2011a), makes this story precise. Consider an empirical best predictor (EBP), $\check{\zeta}$, which is the BP [i.e., right side of (5.9)] with β replaced by a weighted least squares (WLS) estimator (recall that G, R are assumed known),

$$\check{\beta} = (X'WX)^{-1} X'Wy, \tag{5.15}$$

where W is a positive definite weighting matrix. The BPE and MLE are special cases of the WLS estimator, with $W = \Gamma'\Gamma$ for the former and $W = V^{-1}$ for the latter. Thus, the OBP and BLUP are special cases of the EBP. Also recall that $\mu = \mathrm{E}(y)$.

Theorem 5.1. The MSPE of EBP can be expressed as

$$\text{MSPE}(\check{\zeta}) = a_0 + \mu' A_1(W)\mu + \text{tr}\{A_2(W)\}, \tag{5.16}$$

where a_0 is a nonnegative constant that does not depend on W; $A_j(W), j = 1, 2$ are nonnegative definite matrices that depend on W. Furthermore, $\mu' A_1(W)\mu$ is minimized by the OBP, with $W = \Gamma'\Gamma$, and $\text{tr}\{A_2(W)\}$ is minimized by the BLUP, with $W = V^{-1}$.

The expressions of $a_0, A_j(W), j = 1, 2$ are given below. Define

$$L = F' - \Gamma\{I - X(X'WX)^{-1}X'W\},$$

where I is the identity matrix, and recall $B = C'GZ'V^{-1}$. Then, we have

$$\begin{aligned} a_0 &= \text{tr}\{C'(G - GZ'V^{-1}ZG)C\}, \\ A_1(W) &= (F' - L)'(F' - L), \\ A_2(W) &= (L - B)V(L - B)'. \end{aligned}$$

Here are some of the important implications of Theorem 5.1. First note that the second terms disappear when the mean of y is correctly specified, that is, $\mu = X\beta$ for some β. So, when the underlying model is correct, BLUP usually wins the battle of MSPE, because it minimizes the only (remaining) term that depends on W. On the other hand, when the mean is misspecified, the second term is usually of higher order than the third term. Thus, in this situation, OBP is likely to win the battle, because it minimizes a higher order term. We use an example to illustrate.

5.3 Example

We use a very simple example to show that the potential gain of OBP over BLUP can be substantial, if the underlying model is misspecified. Consider a special case of Example 5.1, in which $x_i'\beta = \beta$, an unknown mean. To make it even simpler, suppose that A is known, so that one can actually compute the BLUP. Furthermore, suppose that $m = 2n$, $D_i = a, 1 \le i \le n$, and $D_i = b, n + 1 \le i \le m$, where a, b are positive known constants. Now suppose that, actually, the underlying model is

$$y_i = \begin{cases} c + v_i + e_i, \ \le i \le n, \\ d + v_i + e_i, \ n + 1 \le i \le m, \end{cases}$$

where $c \neq d$; in other words, we have a model misspecification by assuming $c = d$. Consider $\zeta = (\zeta_i)_{1 \le i \le m}$, where

$$\zeta_i = \begin{cases} c + v_i, \ 1 \le i \le n, \\ d + v_i, \ n + 1 \le i \le m. \end{cases}$$

For this special case, we can actually derive the exact expressions of the MSPEs. It can be shown (Exercise 5.2) that

$$\begin{aligned}
\text{MSPE}(\hat{\zeta}_{\text{OBP}}) &= \left\{\left(\frac{a}{A+a}+\frac{b}{A+b}\right)A+\frac{a^2b^2(c-d)^2}{a^2(A+b)^2+b^2(A+a)^2}\right\}n \\
&\quad +\frac{a^4(A+b)^3+b^4(A+a)^3}{(A+a)(A+b)\{a^2(A+b)^2+b^2(A+a)^2\}} \\
&= g_1 n + h_1, \qquad (5.17) \\
\text{MSPE}(\hat{\zeta}_{\text{BLUP}}) &= \left\{\left(\frac{a}{A+a}+\frac{b}{A+b}\right)A+\frac{(a^2+b^2)(c-d)^2}{(2A+a+b)^2}\right\}n \\
&\quad +\frac{a^2(A+b)^2+b^2(A+a)^2}{(A+a)(A+b)(2A+a+b)} \\
&= g_2 n + h_2. \qquad (5.18)
\end{aligned}$$

If $c \neq d$, then $g_1 \leq g_2$ and $h_1 \geq h_2$ with equality holding in both cases if and only if the following identity holds:

$$a^2(A+b) = b^2(A+a). \qquad (5.19)$$

Now suppose that (5.19) does not hold, then we have $g_1 < g_2$, and $h_1 > h_2$, but the latter is not important when n is large. In fact, we have

$$\lim_{n\to\infty}\frac{\text{MSPE}(\hat{\zeta}_{\text{OBP}})}{\text{MSPE}(\hat{\zeta}_{\text{BLUP}})} = \frac{g_1}{g_2} < 1.$$

For example, suppose that $A/(c-d)^2 \approx 0$, $A/b \approx 0$ and $b/a \approx 0$. Then it is easy to show that $g_1/g_2 \approx 0.5$ (Exercise 5.2). Thus, in this case, the MSPE of the OBP is asymptotically about half of that of the BLUP. On the other hand, if $c = d$, that is, if the underlying model is correctly specified, then, we have $g_1 = g_2$ while, still, $h_1 \geq h_2$. Therefore, in this case, $\text{MSPE}(\hat{\zeta}_{\text{OBP}}) \geq \text{MSPE}(\hat{\zeta}_{\text{BLUP}})$; however, we have

$$\lim_{n\to\infty}\frac{\text{MSPE}(\hat{\zeta}_{\text{OBP}})}{\text{MSPE}(\hat{\zeta}_{\text{BLUP}})} = 1;$$

hence, the MSPEs of the OBP and BLUP are asymptotically the same.

5.4 OBP for two classes of SAE models

In this section, we derive the OBP for two important classes of LMMs that are frequently used in small area estimation (SAE).

5.4.1 *Fay-Herriot model*

Let us now refer back to the Fay-Herriot model, introduced in Example 5.1, but with A unknown. Again, we begin with the left side of (5.10), and note that the expectations involved are with respect to the true underlying distribution that is unknown, but not model-dependent. By (5.13), the BP of ζ can be expressed, in matrix form, as $\tilde{\zeta} = y - \Gamma(y - X\beta)$, where Γ is defined below (5.9). Thus, by (5.7) and (5.8) (specialized to this particular case), it can be shown (Exercise 5.3) that

$$\text{MSPE}(\tilde{\zeta}) = \text{E}\{(y - X\beta)'\Gamma^2(y - X\beta) + 2A\text{tr}(\Gamma) - \text{tr}(D)\}. \quad (5.20)$$

The BPE of $\psi = (\beta', A)'$ is obtained by minimizing the expression inside the expectation on the right side of (5.20), which is equivalent to minimizing

$$Q(\psi) = (y - X\beta)'\Gamma^2(y - X\beta) + 2A\text{tr}(\Gamma). \quad (5.21)$$

Note that, in (5.20), ψ is treated as an unknown parameter vector, rather than the true parameter vector. Let $\tilde{Q}(A)$ be $Q(\psi)$ with $\beta = \tilde{\beta}$ given by the right side of (5.12), considered as a function of A. It can be shown (Exercise 5.3) that

$$\tilde{Q}(A) = y'\Gamma P_{(\Gamma X)^{\perp}}\Gamma y + 2A\text{tr}(\Gamma), \quad (5.22)$$

where for any matrix M, $P_{M^{\perp}} = I - P_M$ with $P_M = M(M'M)^{-1}M'$ (assuming nonsingularity of $M'M$), hence, we have

$$P_{(\Gamma X)^{\perp}} = I_m - \Gamma X(X'\Gamma^2 X)^{-1}X'\Gamma$$

and I_m is the m-dimensional identity matrix. The BPE of A is the minimizer of $\tilde{Q}(A)$ with respect to $A \geq 0$, denoted by $\hat{A}$. Once $\hat{A}$ is obtained, the BPE of β, $\hat{\beta}$, is given by (5.12) with A replaced by $\hat{A}$. Given the BPE of ψ, $\hat{\psi} = (\hat{\beta}', \hat{A})'$, the OBP of ζ is given by the BP (5.13) with $\psi = \hat{\psi}$.

5.4.2 *Nested-error regression model*

Consider sampling from a finite population. To link the population to a nested-error regression (NER) model, consider a super-population NER model. Suppose that the subpopulations of responses $\{Y_{ik}, k = 1, \dots, N_i\}$ and auxiliary data $\{X_{ikl}, k = 1, \dots, N_i\}, l = 1, \dots, p$ are realizations from corresponding super-populations that are assumed to satisfy

$$Y_{ik} = X'_{ik}\beta + v_i + e_{ik}, \; i = 1, \dots, m, k = 1, \dots, N_i, \quad (5.23)$$

where β is a vector of unknown regression coefficients, v_i is a subpopulation-specific random effect, and e_{ik} is an error. It is assumed that the v_i's and e_{ik}'s are independent with $v_i \sim N(0, \sigma_v^2)$ and $e_{ik} \sim N(0, \sigma_e^2)$. Under the finite-population setting, the true small area mean is $\theta_i = \bar{Y}_i = N_i^{-1} \sum_{k=1}^{N_i} Y_{ik}$ (as opposed to $\theta_i = \bar{X}_i'\beta + v_i$ under the infinite-population setting) for $1 \leq i \leq m$. Furthermore, write $r_i = n_i/N_i$. Now suppose that simple random samples [e.g., Fuller (2009)] $(y_{ij}, x_{ij}), 1 \leq j \leq n_i$ are drawn from the finite population, $\{(Y_{ik}, X_{ik}), 1 \leq k \leq N_i\}$, $1 \leq i \leq m$. Then, the finite-population version of the BP has the expression (Exercise 5.4)

$$\tilde{\theta}_i = \mathrm{E}_M(\theta_i|y_i) = \bar{X}_i'\beta + \left\{ r_i + (1 - r_i)\frac{n_i\gamma}{1 + n_i\gamma} \right\} (\bar{y}_{i\cdot} - \bar{x}_{i\cdot}'\beta), \quad (5.24)$$

where $y_i = (y_{ij})_{1 \leq j \leq n_i}$, $\bar{y}_{i\cdot} = n_i^{-1} \sum_{j=1}^{n_i}$, $\bar{x}_{i\cdot} = n_i^{-1} \sum_{j=1}^{n_i} x_{ij}$, E_M denotes (conditional) expectation under the assumed super-population NER model, and β and $\gamma = \sigma_v^2/\sigma_e^2$ are the true parameters. Again, note that the BP is model-dependent.

In practice, an assumed model is subject to misspecification. Jiang *et al.* (2011a) considers misspecification of the mean function, while assuming that the variance-covariance structure of the data is correctly specified. However, the latter, too, may be misspecified in practical. In this subsection, we extend the potential model misspecification to both the mean function and the variance-covariance structure. One possible misspecification of the variance-covariance structure is heteroscedasticity, defined in terms of $\mathrm{var}(e_{ij}) = \sigma_i^2$ for area i, $1 \leq i \leq m$, where the σ_i^2's are unknown and possibly different. For example, Jiang and Nguyen (2012) used the well-known Iowa crops data [Battese *et al.* (1988)] to show that the homoscedastic variance assumption of the NER model may not hold in practice.

On the other hand, in spite of the potential model misspecification, there are reasons that one cannot "abandon" the assumed model, and the model-based BP. First, the assumed model and BP are relatively simple to use, and therefore, attractive to practitioners. In particular, they explore simple (linear) relationship between the response and auxiliary variables. For example, in contrast to (5.23), one may assume $Y_{ik} = \mu_{ik} + v_i + e_{ik}$, where the μ_{ik} are completely unspecified, unknown constants. The latter model is almost always correct, but is nevertheless useless, because it does not utilize any relationship between Y and X at all. In fact, in practice, if auxiliary data are available, it is often "politically incorrect" not to use them. Secondly, even though there is a concern about the model misspecification, it often lacks (statistical) evidence on why something else is more reasonable, or whether a complication is necessary. For example, sometimes there is a concern about the normality assumption, as discussed in Section 4.4.2, but no indication on why an alternative distribution, say, t_5, is more reasonable. As another example, suppose that one fits a quadratic model and finds that the coefficient of the quadratic term is insignificant. Then, one is not sure whether the complication of quadratic modeling is necessary as opposed to linear modeling. Thus, as far as this section is concerned, we are not attempting to change the assumed model, or the BP based on the assumed model. In particular, we assume a single parameter, γ, for the ratio σ_v^2/σ_e^2, rather than considering a heteroscedastic NER model such as in Jiang and Nguyen (2012).

Our goal is to find a better way to estimate the parameters, $\psi = (\beta', \gamma)'$, under the assumed model, so that the resulting BP, (5.24), is more robust against model misspecifications. We do so by considering an objective MSPE that is not model-dependent, defined as follows. Let $\theta = (\theta_i)_{1\leq i\leq m}$ denote the vector of small area means, and $\tilde{\theta} = [\tilde{\theta}_i]_{1\leq i\leq m}$ the vector of BPs. Note that $\tilde{\theta}_i$ depends on ψ, that is, $\tilde{\theta}_i = \tilde{\theta}_i(\psi)$. The design-based [e.g., Fuller (2009)] MSPE is

$$\text{MSPE}(\tilde{\theta}) = \text{E}(|\tilde{\theta} - \theta|^2) = \sum_{i=1}^{m} \text{E}\{\tilde{\theta}_i(\psi) - \theta_i\}^2. \tag{5.25}$$

Note that the E in (5.25) is different from the E_M in (5.24) in that E is completely model-free; namely, the expectation in (5.25) is with respect to the simple random sampling from the finite populations, which has nothing to do with the assumed model. It can be shown (Exercise 5.5) that the MSPE in (5.25) has an alternative expression. Namely, we have $\text{MSPE}(\tilde{\theta}) = \text{E}\{Q(\psi) + \cdots\}$, where $\cdots$ does not depend on ψ, and

$$Q(\psi) = \sum_{i=1}^{m} \left\{\tilde{\theta}_i^2(\psi) - 2\frac{1 - r_i}{1 + n_i\gamma}\bar{y}_{i\cdot}\bar{X}_i'\beta + b_i(\gamma)\hat{\mu}_i^2\right\} = \sum_{i=1}^{m} Q_i. \tag{5.26}$$

In (5.26), ψ is considered as a parameter vector, rather than the true parameter vector, $b_i(\gamma) = 1 - 2a_i(\gamma)$ with $a_i(\gamma) = r_i + (1 - r_i)n_i\gamma(1 + n_i\gamma)^{-1}$. Furthermore, $\hat{\mu}_i^2$ is a design-unbiased estimator of $\bar{Y}_i^2$ that has the following expression (Exercise 5.6):

$$\hat{\mu}_i^2 = \frac{1}{n_i}\sum_{j=1}^{n_i} y_{ij}^2 - \frac{N_i - 1}{N_i(n_i - 1)}\sum_{j=1}^{n_i}(y_{ij} - \bar{y}_{i\cdot})^2, \tag{5.27}$$

assuming that $n_i > 1$. The BPE of ψ, $\hat{\psi}$, is the minimizer of $Q(\psi)$ with respect to ψ. Also note that the BP is based on the (model-based) area-specific MSPE (so it is optimal for every small area, if the assumed model is correct), while the BPE is based on the (design-based) overall MSPE. This is because, here, we do not want the estimator of ψ to be area-dependent. One reason is that area-dependent estimators are often unstable due to the small sample size from the area, while an estimator obtained by utilizing all of the areas, such as the BPE defined above, tends to be much more stable. However, area-dependent estimator could be explored, if one can solve the stability problem.

Once the BPE is obtained, the OBP of θ_i is given by (5.24) with ψ replaced by its BPE.

5.5 OBP for small area counts

So far OBP has been applied to linear models, which may be appropriate when the response variable is continuous. However, in many SAE problems, the data for the response variables are counts. See, for example, Münnich *et al.* (2009), Ferrante and Trivisano (2010). Such cases are often treated under the framework of GLMM (see, e.g., Section 2.1). Such models have been studied in the context of SAE. See, for example, Malec *et al.* (1997), Ghosh *et al.* (1998), Jiang *et al.* (2001). In particular, Poisson mixed models, a special class of GLMM that is considered in this paper, have appeared in recent literature in SAE. For example, Münnich *et al.* (2009) used both the binomial and Poisson mixed models for estimating population counts in the 2011 German census; Ferrante and Trivisano (2010) proposed a Poisson log-normal model, which is similar to the Poisson mixed model except formulated as a hierarchical model, for estimation of the number of recruits by firms; in a similar development, Hajarisman (2013) proposed a two-level hierarchical Poisson model for estimation of infant mortality rates using a Bayesian approach. In this section, we extend the OBP method to SAE with count data.

5.5.1 *Best prediction under a two-stage model*

For simplicity, suppose that responses are counts at the area-level, denoted by y_i, and that, in addition, a vector of covariates, x_i, is also available at the area-level. The model of interest has two stages. In the first stage, it assumes that, given the area-specific means, $\mu_i, 1 \leq i \leq m$, where m is the number of small areas, $y_i, 1 \leq i \leq m$ are independent such that $y_i \sim$ Poisson(μ_i). In the second stage, a model is assumed for the distribution of the μ_i's. In this section, we mainly focus on two of these models. The first is a mixed-effect log-normal (LN) model, expressed as

$$\log(\mu_i) = x_i'\beta + v_i, \tag{5.28}$$

where β is a vector of fixed effects, and v_i is an area-specific random effect that is assumed to be independent, for different i's, and distributed as $N(0, \sigma^2)$ for some variance σ^2. The second is a Gamma (GM) model, which assumes that the μ_i's are independent with

$$\mu_i \sim \text{Gamma}(x_i'\beta/\phi, \phi), \tag{5.29}$$

where Gamma(α, ϕ) is the Gamma distribution with shape parameter α, scale parameter ϕ, and probability density function (pdf) given by

$$f(u|\alpha, \phi) = \frac{1}{\Gamma(\alpha)\phi^\alpha} u^{\alpha-1} e^{-u/\phi}, \quad u > 0.$$

Under either model, the BP of μ_i can be expressed as

$$\text{E}_{M,\psi}(\mu_i|y) = g_i(\psi, y_i), \tag{5.30}$$

where $\text{E}_{M,\psi}$ denotes conditional expectation under the assumed model, M, and parameter vector, ψ, under M, and $g_i(\cdot, \cdot)$ is a function that depends on the assumed model. Note that, from a Bayesian point of view, the BP is simply the posterior mean of μ_i under the model $y_i|\mu_i \overset{\text{ind}}{\sim} \text{Poisson}(\mu_i)$ and treating the distribution of μ_i as a prior. However, the latter is potentially misspecified which is a main concern here. Throughout this section, we assume that the conditional model of y_i given μ_i is correctly specified.

The LN model, also known as the log-linear model, is, perhaps, the most popular in practice. It is a special case of GLMM [e.g., Breslow and Clayton (1993), Jiang (2007)]. In particular, the random effect v_i often has an interpretation, which is attractive to practitioners. Of course, it may also be expressed without using the random effects, in a way similar to the GM model, that is

$$\mu_i \sim \text{LN}(x_i'\beta, \sigma^2), \tag{5.31}$$

where LN stands for the log-normal distribution [$\xi \sim \text{LN}(\mu, \sigma^2)$ iff $\log(\xi) \sim N(\mu, \sigma^2)$]. On the other hand, the BP under the LN model does not have an analytic expression. In fact, if we use the expression $v_i = \sigma\xi_i$, where $\xi_i \sim N(0, 1)$, and let $\psi = (\beta', \sigma)'$, then it is easy to derive that, in this case,

$$g_i(\psi, y_i) = e^{x_i'\beta} \frac{\int \exp\{(y_i + 1)\sigma u - \mu_i(u) - u^2/2\} du}{\int \exp\{y_i \sigma u - \mu_i(u) - u^2/2\} du}, \tag{5.32}$$

where $\mu_i(u) = \exp(x_i'\beta + \sigma u)$. The one-dimensional integrals in (5.32) can be approximated by numerical integration. Alternatively, if y_i is large, the integrals in (5.32) can be approximated by Laplace approximation [e.g., Jiang (2007), sec. 3.5.1], leading to the following approximation: $g_i(\psi, y_i) = a_i$ if $\sigma = 0$ (this result is exact rather than approximate);

$$\begin{aligned} & g_i(\psi, y_i) \\ & \approx a_i \sqrt{\frac{\sigma^2 y_i + 1}{\sigma^2(y_i+1)+1}} \exp\left[(y_i+1)\log\left(\frac{y_i+1}{a_i}\right)\right. \\ & -y_i \log\left(\frac{y_i}{a_i}\right) - 1 + \frac{y_i}{2(\sigma^2 y_i + 1)}\left\{\log\left(\frac{y_i}{a_i}\right)\right\}^2 \\ & \left. -\frac{y_i+1}{2\{\sigma^2(y_i+1)+1\}}\left\{\log\left(\frac{y_i+1}{a_i}\right)\right\}^2\right] \end{aligned} \tag{5.33}$$

if $\sigma > 0$, where $a_i = e^{x_i'\beta}$ (Exercise 5.7). The Laplace approximation is computationally much more efficient.

As for the GM model, it has a major advantage in that, in this case, the BP has a simple analytic expression as a result of Gamma being the conjugate prior. In fact, if we let $\psi = (\beta', \phi)'$, then we have

$$g_i(\psi, y_i) = \frac{\phi y_i + x_i'\beta}{\phi + 1}. \tag{5.34}$$

The derivation of (5.34) is left as an exercise (Exercise 5.8). The conjugate Gamma prior also has its attractiveness from a Bayesian perspective.

5.5.2 *Derivation of OBP*

As noted in the previous sections, the difference between OBP and BLUP or EBLUP is not the BP, but how the parameters involved in the BP are estimated. The same is true here. Following the earlier approach, we evaluate the performance of the BP under a broader model, which is simply the first-stage model, but the second-stage model about the distribution of the μ_i is not assumed. In other words, under the broader model, the μ_i's are completely unspecified. Write $\mu = (\mu_i)_{1\le i\le m}$, and $\tilde{\mu} = (\tilde{\mu}_i)_{1\le i\le m}$, where $\tilde{\mu}_i$ is the right side of (5.30), where ψ is considered as an unknown parameter vector. Consider

$$
\begin{aligned}
\text{MSPE} &= \mathrm{E}(|\tilde{\mu} - \mu|^2) \\
&= \sum_{i=1}^{m} \mathrm{E}\{g_i(\psi, y_i) - \mu_i\}^2 \\
&= \mathrm{E}\left\{\sum_{i=1}^{m} g_i^2(\psi, y_i)\right\} - 2\sum_{i=1}^{m} \mathrm{E}\{g_i(\psi, y_i)\mu_i\} + \sum_{i=1}^{m} \mathrm{E}(\mu_i^2) \\
&= I_1 - 2I_2 + I_3, \qquad (5.35)
\end{aligned}
$$

where E denotes expectation under the broader model. Note that I_3 does not involve ψ, even though it may be completely unknown. Also note that

$$
\begin{aligned}
\mathrm{E}\{g_i(\psi, y_i)\mu_i\} &= \mathrm{E}[\mu_i \mathrm{E}\{g_i(\psi, y_i)|\mu\}] \\
&= \sum_{k=0}^{\infty} g_i(\psi, k) \mathrm{E}\left(\frac{e^{-\mu_i}\mu_i^{k+1}}{k!}\right) \\
&= \sum_{k=0}^{\infty} g_i(\psi, k)(k+1) \mathrm{E}\left\{\frac{e^{-\mu_i}\mu_i^{k+1}}{(k+1)!}\right\} \\
&= \sum_{k=0}^{\infty} g_i(\psi, k)(k+1) \mathrm{E}\{1_{(y_i=k+1)}\},
\end{aligned}
$$

where 1_A is the indicator of event A (= 1 is A occurs, and 0 otherwise). Thus, if we define $g_i(\psi, -1) = 0$, we have

$$
\begin{aligned}
\mathrm{E}\{g_i(\psi, y_i)\mu_i\} &= \mathrm{E}\left\{\sum_{k=0}^{\infty} g_i(\psi, k)(k+1) 1_{(y_i=k+1)}\right\} \\
&= \mathrm{E}\{g_i(\psi, y_i - 1)y_i\}.
\end{aligned}
$$

Therefore, combined with (5.35), we have

$$
\text{MSPE} = \mathrm{E}\left\{\sum_{i=1}^{m} g_i^2(\psi, y_i) - 2\sum_{i=1}^{m} g_i(\psi, y_i - 1)y_i + \cdots\right\}, \qquad (5.36)
$$

where $\cdots$ does not depend on ψ, which leads to the BPE of ψ, $\hat{\psi}$, by minimizing the expression inside the expectation on the right side of (5.36) without $\cdots$, that is,

$$
Q(\psi, y) = \sum_{i=1}^{m} g_i^2(\psi, y_i) - 2\sum_{i=1}^{m} g_i(\psi, y_i - 1)y_i. \qquad (5.37)
$$

Once the BPE of ψ is obtained, the OBP of the small area mean count, μ_i, is obtained by (5.30) with ψ replaced by $\hat{\psi}$, that is,

$$
\hat{\mu}_i = g_i(\hat{\psi}, y_i), \quad 1 \le i \le m. \qquad (5.38)
$$

It is easy to show (Exercise 5.9) that, under the GM model, the BPE has a closed-form expression:

$$\hat{\beta} = \left(\sum_{i=1}^{m} x_i x_i'\right)^{-1} \sum_{i=1}^{m} x_i y_i = (X'X)^{-1}X'y, \tag{5.39}$$

$$\hat{\phi} = \frac{\sum_{i=1}^{m}(y_i - x_i'\hat{\beta})^2}{\sum_{i=1}^{m} y_i} - 1 = \frac{\text{RSS}}{y_\cdot} - 1. \tag{5.40}$$

(5.39) is the well-known least squares (LS) estimator, where $X = (x_i')_{1\le i\le m}$ and $y = (y_i)_{1\le i\le m}$, while RSS is the residual sum of squares, and $y_\cdot = \sum_{i=1}^{m} y_i$. Under the LN model, the BPE does not have an analytic expression, so a numerical procedure is needed to compute the BPE.

It is fairly straightforward to extend the above derivation to more complicated area-level models [see Chen (2012)]. For example, in Subsection 5.7.3, we consider a case where interest is demographic groups within the areas, but the random effect is at the area-level. Some further extensions are considered in the next section.

5.5.3 *Extensions*

A further extension would be to consider the one-parameter exponential family [e.g., McCullagh and Nelder (1989), p. 28] with the pdf

$$\mu_i \sim \exp\left\{\frac{u\theta_i - b(\theta_i)}{a_i(\phi)} + h_i(u,\phi)\right\}, \tag{5.41}$$

where u is the variable for the pdf, θ_i is the natural parameter for the exponential family, ϕ is an additional dispersion parameter, and $b(\cdot)$, $a_i(\cdot)$, and $h_i(\cdot,\cdot)$ are known functions. Furthermore, the natural parameter, θ_i, is associated with the linear predictor, $\eta_i = x_i'\beta$, through a link function. It can be shown that, under (5.41), the model-based conditional expectation, (5.30), is given by

$$\mathrm{E}_{M,\psi}(\mu_i|y) = a_i(\phi)\frac{\partial}{\partial\theta_i}\log d_i(\theta_i, y_i, \phi), \tag{5.42}$$

where $\psi = (\beta', \phi)'$, and

$$d_i(\theta_i, y_i, \phi) = \int \exp\left\{\frac{\mu_i\theta_i}{a_i(\phi)} + c_i(\mu_i, y_i, \phi)\right\} d\mu_i \tag{5.43}$$

with $c_i(\mu_i, y_i, \phi) = y_i \log\mu_i - \mu_i + h_i(\mu_i, \phi)$. However, the right side of (5.42) does not necessarily have an analytic expression; it reduces to (5.34), of course, in the Gamma case.

One may also consider adding a (sampling) weight in the first-stage model by assuming $y_i|\mu_i \sim \text{Poisson}(w_i\mu_i)$, where w_i is a known weight, while maintaining the second-stage model for μ_i. Again, denote the model-based BP of μ_i by (5.30). A very similar derivation to that in Subsection 5.5.2 leads to

$$\mathrm{E}\{g_i(\psi, y_i)\mu_i\} = \mathrm{E}\left\{\frac{g_i(\psi, y_i - 1)}{w_i} y_i\right\}.$$

Thus, similar to (5.36), we have

$$\text{MSPE} = \mathrm{E}\left\{\sum_{i=1}^m g_i^2(\psi, y_i) - 2\sum_{i=1}^m \frac{g_i(\psi, y_i - 1)}{w_i} y_i + \cdots\right\}, \quad (5.44)$$

where $\cdots$ does not depend on ψ. The BPE of ψ is then obtained by minimizing the expression inside the expectation on the right side of (5.44) without $\cdots$, that is,

$$Q(\psi, y, w) = \sum_{i=1}^m g_i^2(\psi, y_i) - 2\sum_{i=1}^m \frac{g_i(\psi, y_i - 1)}{w_i} y_i, \quad (5.45)$$

where $w = (w_i)_{1\leq i\leq m}$. It is clear that, when $w_i = 1, 1 \leq i \leq m$, $Q(\psi, y, w)$ reduces to $Q(\psi, y)$ of (5.37), leading to the BPE without the weights.

It is easy to show that, under the GM model, the BP has a closed-form expression:

$$g_i(\psi, y_i) = \frac{\phi y_i + x_i'\beta}{1 + \phi w_i}. \quad (5.46)$$

The BPE of ψ does not have a closed-form expression, in general. However, it can be computed via the following simple numerical algorithm: First solve

$$\sum_{i=1}^m \frac{y_i/w_i}{(1 + \phi w_i)^2} - \sum_{i=1}^m w_i \frac{(y_i/w_i - x_i'\tilde{\beta})^2}{(1 + \phi w_i)^3} = 0, \quad (5.47)$$

where

$$\tilde{\beta} = \left\{\sum_{i=1}^m \frac{x_i x_i'}{(1 + \phi w_i)^2}\right\}^{-1} \sum_{i=1}^m \frac{x_i(y_i/w_i)}{(1 + \phi w_i)^2}. \quad (5.48)$$

Let the solution to (5.47) be $\hat{\phi}$, which is the BPE of ϕ. The BPE of β is (5.48) with ϕ replaced by $\hat{\phi}$. It is easy to verify that, when $w_i = 1, 1 \leq i \leq m$, this leads to (5.39) and (5.40).

On the other hand, under the LN model, neither the BP nor the BPE have closed-form expressions, and one does not have a simple algorithm like (5.47) and (5.48) for the computation.

5.6 Asymptotic property of BPE

There have been studies examining the asymptotic behaviors of estimators under model misspecifications. For example, in the context of maximum likelihood estimation with i.i.d. observations, White (1982) showed that, when the underlying model is misspecified (therefore the true parameter vector does not exist), the MLE under the assumed model still converges almost surely to "something", and the "something" is the (unique) maximizer of the expected "log-likelihood" (considered as a function of the parameters), under some regularity conditions. Here "log-likelihood" refers to the logarithm of the misspecified likelihood function, which White called quasi-likelihood function. A similar approach can be used to study the asymptotic behavior of the BPE under a misspecified model.

We consider a general setting which includes the cases considered in the previous sections as special cases. Suppose that the BP of θ, a vector of mixed effects, under the assumed model, $\theta_{\rm BP}$, depends on ψ, a vector of parameters that may include β as well as some variance components. Also, suppose that MSPE($\theta_{\rm BP}$) can be expressed as $\mathrm{E}\{Q(y,\psi)\} \equiv M(\psi)$, where E denotes expectation under the true distribution of y, either with respect to a true model (e.g., the case of Fay-Herriot model), or with respect to the sampling design (e.g., the case of nested-error regression model). Throughout this section, E, and later P, are understood in the same sense. We call $M(\psi)$ the *MSPE function*. The BPE of ψ, denoted by $\tilde{\psi}$, is the minimizer of $Q(y,\psi)$ over $\psi \in \Psi$, the parameter space for ψ. The OBP of θ, $\tilde{\theta}$, is $\theta_{\rm BP}$ with ψ replaced by $\tilde{\psi}$. Furthermore, we assume

$$Q(y,\psi) = \sum_{i=1}^{m} Q_i(y_i,\psi), \tag{5.49}$$

where m is the number of small areas; y_i is the subvector of y corresponding to data collected from the ith small area such that $y_1,\dots,y_m$ are independent; and $Q_i(y_i,\psi)$ is three-times continuously differentiable with respect to ψ. In addition, the following regularity conditions are assumed, where $\lambda_{\min}$ denotes the smallest eigenvalue.

A1. There exists a unique $\psi_* \in \Psi^o$, the interior of Ψ, such that $M(\psi_*) = \inf_{\psi\in\Psi} M(\psi)$. Note that ψ_* may depend on the (joint) distribution of $y_1,\dots,y_m$ as well as other quantities that may be involved in the definition of $M(\psi)$ (such as the x_i's in the Fay-Herriot model).

A2. One can differentiate $\mathrm{E}\{Q(y,\psi)\}$ with respect to ψ under the expectation, that is, $(\partial/\partial\psi)\mathrm{E}\{Q(y,\psi)\} = \mathrm{E}\{(\partial/\partial\psi)Q(y,\psi)\}$.

A3. As $m \to \infty$, we have
(i) $\liminf \lambda_{\min}[m^{-1}\sum_{i=1}^m (\partial^2/\partial\psi\partial\psi')\mathrm{E}\{Q_i(y_i, \psi_*)\}] > 0$;
(ii) $\limsup m^{-1}\sum_{i=1}^m \mathrm{E}\{(\partial/\partial\psi_r)Q_i(y_i, \psi_*)\}^2 < \infty$;
(iii) $m^{-2}\sum_{i=1}^m \mathrm{E}\{(\partial^2/\partial\psi_r\partial\psi_s)Q_i(y_i, \psi_*)\}^2 \to 0$; and
(iv) $m^{-3/2}\sum_{i=1}^m \mathrm{E}\{\sup_{\psi\in\bar{S}_\rho(\psi_*)} |\partial^3 Q_i(y_i, \psi)/\partial\psi_r\partial\psi_s\partial\psi_t|\} \to 0$ for some $\rho > 0$ and any $1 \le r, s, t \le q = \dim(\psi)$, where $\bar{S}_\rho(\psi_*) = \{\psi \in \Psi : |\psi - \psi_*| \le \rho\}$.

Condition *A1* requires identifiability of the parameter vector, ψ_*, which assumes the role of the "true parameter vector". Condition *A2* is a regularity condition that is often required in asymptotic theory [e.g., Lehmann and Casella (1998), pp. 441]. Condition (i) of *A3* corresponds to an "information assumption", which is an extension of that (in the i.i.d. case) the sample size goes to infinity. The rest of the conditions in *A3* are essentially moment conditions. These conditions are fairly mild in that they are satisfied in typical situations. We use an example to illustrate how to verify these conditions.

Example 5.1 (continued). Consider the Fay-Herriot model with A known, for simplicity. In this case, the minimizer of $M(\psi)$ has an explicit expression (Exercise 5.10),

$$\psi_* = (X'\Gamma^2 X)^{-1}X'\Gamma^2\mathrm{E}(y). \tag{5.50}$$

Clearly, ψ_* lies in the interior of the parameter space, R^p, where p is the dimension of β. Thus, condition *A1* is satisfied. Note that, in this case, $Q(y, \psi)$ is given by the expression inside the expectation in (5.10). Thus, it is easy to verify that condition *A2* is satisfied. Similarly,

$$Q_i(y_i, \psi) = \mathrm{E}(B_i y_i - \theta_i)^2 + (1 - B_i)^2(x_i'\beta)^2 - 2(1 - B_i)^2 x_i'\beta y_i,$$

where $B_i = A/(A + D_i)$. It follows that

$$\frac{1}{m}\sum_{i=1}^m \left(\frac{\partial^2}{\partial\psi\partial\psi'}\right)\mathrm{E}\{Q_i(y_i, \psi_*)\} = \frac{2}{m}\sum_{i=1}^m (1 - B_i)^2 X_i X_i'.$$

Suppose that the D_i's are bounded away from zero. Then, it is easy to show that condition (i) of *A3* is satisfied if $\liminf \lambda_{\min}(m^{-1}X'X) > 0$. The latter is a regularity condition often required in studying large sample properties of the least squares estimator [e.g., sec. 6.7 of Jiang (2010) and the references therein]. For example, if $X_i = (1, x_i)'$, where x_i is a scalar, then the latter condition is equivalent to

$$\liminf \frac{1}{m}\sum_{i=1}^m (x_i - \bar{x})^2 > 0,$$

where $\bar{x} = m^{-1}\sum_{i=1}^{m} x_i$. The rest of the conditions in *A3* are easy to verify; in particular, we have $\partial^3 Q_i/\partial\beta_j\partial\beta_k\partial\beta_l = 0$ for all j, k, l.

The following theorem, proved in Jiang *et al.* (2011a), states that, under a possibly misspecified model, the BPE is "$\sqrt{m}$-consistent" with respect to ψ_*, the minimizer of the MSPE function.

Theorem 5.2. Under the assumptions *A1–A3*, there exists with probability tending to one, as $m \to \infty$, a local minimizer, $\tilde{\psi}$, of $Q(y, \psi)$ in a neighborhood of ψ_*, such that $\sqrt{m}(\tilde{\psi} - \psi_*) = O_{\mathrm{P}}(1)$. Thus, in particular, we have $\tilde{\psi} - \psi_* \xrightarrow{\mathrm{P}} 0$ as $m \to \infty$.

Note that we do not write $\tilde{\psi} \xrightarrow{\mathrm{P}} \psi_*$, because ψ_*, although is nonrandom, may depend on m (see assumption *A1*).

5.7 Examples with simulations

In this section, we present three examples, each supported by a simulation study, to demonstrate the properties of OBP and its comparison to the standard methods in SAE. The first is a simple example used to demonstrate the theoretical properties of OBP established in Sections 5.2 and 5.3. The second is a more practical example used to illustrate the design-based predictive performance of OBP. The last example is used to illustrate the performance of OBP for estimating small area counts.

5.7.1 *A simple example*

The example is the same as the simple Fay-Herriot model considered in Section 5.3 except that A is unknown. Note that the assumed model can be written as $y_i = \beta + v_i + e_i, i = 1, \dots, m$, where the v_i's and e_i's are independent such that $v_i \sim N(0, A)$ with A unknown, and $e_i \sim N(0, D_i)$ with D_i given in Section 5.3 and further specified below.

We consider four different estimators of A that have been traditionally used. These are the ML, REML, Fay-Herriot [F-H; Fay and Herriot (1979); also see Datta *et al.* (2005)] and Prasad-Rao [P-R; Prasad and Rao (1990)] estimators. Given one of the estimators of A, denoted by $\hat{A}$, the EBLUP is obtained by the following steps: (i) computing $\hat{\beta}$ of (5.14) with $x_i = 1$ and A replaced by $\hat{A}$; and (ii) computing the EBLUP $\hat{\theta} = (\hat{\theta}_i)_{1 \le i \le m}$ of

Table 5.1 **Empirical MSPE (% increase over OBP)**

m	d	EBLUP-1	EBLUP-2	EBLUP-3	EBLUP-4	OBP
50	1	28.76 (28%)	27.94 (25%)	25.00 (11%)	25.87 (15%)	22.43
100	1	51.05 (26%)	50.20 (24%)	47.74 (18%)	49.02 (21%)	40.42
200	1	94.22 (26%)	93.95 (26%)	92.52 (24%)	93.87 (25%)	74.86
50	5	95.83 (42%)	95.05 (41%)	93.55 (39%)	93.12 (38%)	67.41
100	5	189.93 (44%)	189.22 (43%)	186.51 (41%)	185.61 (41%)	132.01
200	5	372.59 (44%)	371.92 (44%)	366.96 (42%)	365.21 (41%)	258.60

$\theta = (\theta_i)_{1\le i\le m}$, where

$$\begin{aligned}\hat{\theta}_i &= \frac{\hat{A}}{\hat{A}+a}y_i + \frac{a}{\hat{A}+a}\hat{\beta}, \quad 1 \le i \le n;\\ \hat{\theta}_i &= \frac{\hat{A}}{\hat{A}+b}y_i + \frac{b}{\hat{A}+b}\hat{\beta}, \quad n+1 \le i \le m \end{aligned} \tag{5.51}$$

[note that the $\hat{A}$ in (5.51) is the same $\hat{A}$ used to compute $\hat{\beta}$]. Depending on whether $\hat{A}$ is the ML, REML, F-H, or P-R estimator, the corresponding EBLUPs are denoted by EBLUP-1, EBLUP-2, EBLUP-3, and EBLUP-4, respectively. The OBP of θ, $\tilde{\theta} = (\tilde{\theta}_i)_{1\le i\le m}$, is given by the right sides of (5.51) with $\hat{A}$ and $\hat{\beta}$ replaced by $\tilde{A}$ and $\tilde{\beta}$, respectively, where $\tilde{\psi} = (\tilde{\beta}', \tilde{A})'$ is the BPE of $\psi = (\beta', A)'$.

The empirical MSPEs are reported in Table 5.1, taken from Jiang *et al.* (2011a). The % in the parentheses is the percentage increase in MSPE by the corresponding EBLUP over the OBP. It is seen that all of the EBLUPs perform very similarly, while the OBP is a distance away from the EBLUPs. The MSPEs of the EBLUPs are somewhere between 11% to 44% higher than that of the OBP. The percentage increase in MSPE by the EBLUPs is more substantial for $d = 5$ than for $d = 1$. This makes sense because $d = 5$ features a more serious model misspecification than $d = 1$ does, and the OBP shines when the underlying model is misspecified. A more explicit explanation of this pattern may be seen from the comparison of the formulae for the exact MSPEs, that is, (5.17) and (5.18) (with $c = 0$), although the latter are derived for the case that A is known. The formulae show that the difference in MSPE between the OBP and any of the EBLUPs gets larger, even proportionally, as d increases.

We next compare the OBP and EBLUPs in terms of area-specific MSPEs. Although the OBP is defined by minimizing the overall (observed) MSPE, there is no guarantee that its area-specific MSPEs are minimal. On the other hand, area-specific MSPEs are often of main interest in SAE. Therefore, a comparison of the area-specific MSPEs of the OBP with those of the EBLUP is important, especially from a practical point of view. Such a comparison may also provide further details for the overall MSPEs reported in Table 5.1. The empirical area-specific MSPE is evaluated by

$$\text{MSPE}_i^* = \frac{1}{K}\sum_{k=1}^{K}\{\breve{\theta}_i^{(k)} - \theta_i^{(k)}\}^2$$

with $K = 500$, where $\breve{\theta}_i^{(k)}$ and $\theta_i^{(k)}$ are the predictor and true small area mean, respectively, for the ith small area in the kth simulation run, $1 \le i \le m, 1 \le k \le K$. Due to the fairly large number of small areas involved ($m =$ 50, 100 or 200 in our simulations), we summarize the results using boxplots and histograms, as shown in Figures 5.1 and 5.2. The figures reveal some untold stories by the overall MSPEs. First, the boxplots show a significant difference in the distributions of the empirical MSPEs between the OBP and EBLUPs. Not only does the OBP have smaller median empirical MSPE in each case, what is more apparent is the range, or variation, of the empirical MSPEs overwhelmingly in favor of the OBP. Second, the histograms exhibit quite different shapes between the OBP and EBLUPs. A closer look at the numbers shows that the empirical MSPEs of the EBLUPs are somewhere between slightly to moderately smaller than those of the OBP for half of the small areas; but for the other half, the empirical MSPEs of the EBLUPs are much larger than those of the OBP. The pattern can also be seen from the histograms. Recall the assumed model has a common mean for all of the small areas, while the true model has one mean for half of the areas and another mean for the other half. Apparently, what the EBLUP does is to "side with" one mean while "abandoning" the other. The OBP, on the other hand, uses a rather different strategy, by "staying in the middle" or "balancing" between the two means. This explains the bimodal histograms for the EBLUPs, compared to the fairly normal-look-like histograms for the OBP (with much narrower spreads). Overall, the simulation results show a much more robust performance of the OBP in terms of the area-specific MSPE as compared to the EBLUPs.

5.7.2 *Cases when the assumed model is correct, or partially correct*

The situation considered in Subsection 5.7.1 might be a little extreme. In practice, the assumed model may not be completely wrong, or may be close to be correct. In this subsection we first consider a case where the assumed model is "partially correct". It is a case of finite-population setting

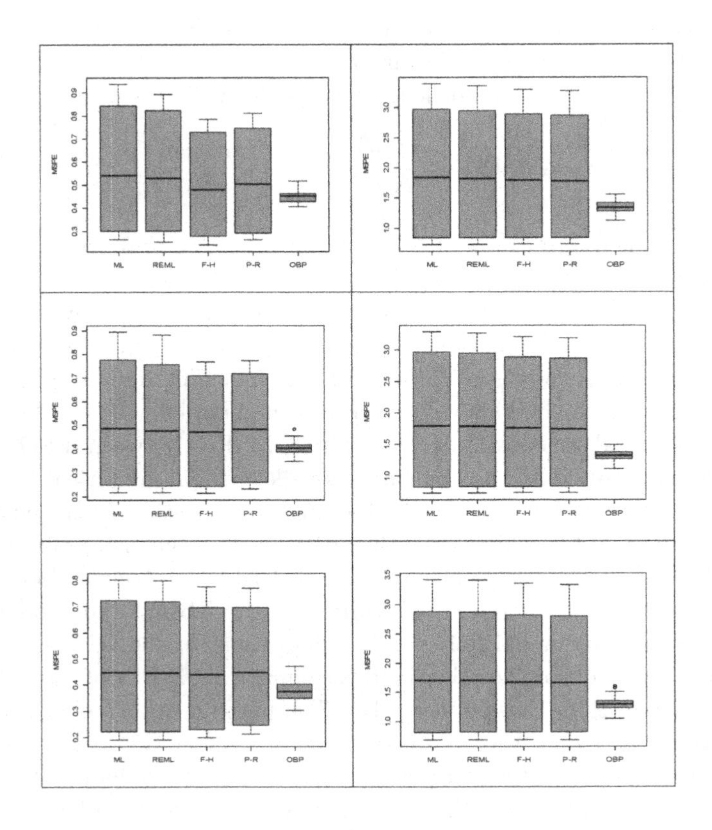

Fig. 5.1 *Boxplots of* $\mathrm{MSPE}_i^*, i = 1, \ldots, m$. *Upper Left:* $m = 50, d = 1$; *Upper Right:* $m = 50, d = 5$; *Middle Left:* $m = 100, d = 1$; *Middle Right:* $m = 100, d = 5$; *Lower Left:* $m = 200, d = 1$; *Lower Right:* $m = 200, d = 5$. *Within each plot from left to right: EBLUP-1, EBLUP-2, EBLUP-3, EBLUP-4, OBP.*

considered in Subsection 5.4.2. A single covariate, x_{ij}, is thought to be linearly associated with the response, y_{ij}, through the NER model

$$y_{ij} = \beta x_{ij} + v_i + e_{ij}, \quad i = 1, \ldots, m, \; j = 1, \ldots, 5 \tag{5.52}$$

(so we have $n_i = 5, 1 \leq i \leq m$ in this case), where β is an unknown coefficient, and v_i, e_{ij} are the same as in (5.23). Thus, in particular, there is a belief that the mean response should be zero when the value of the

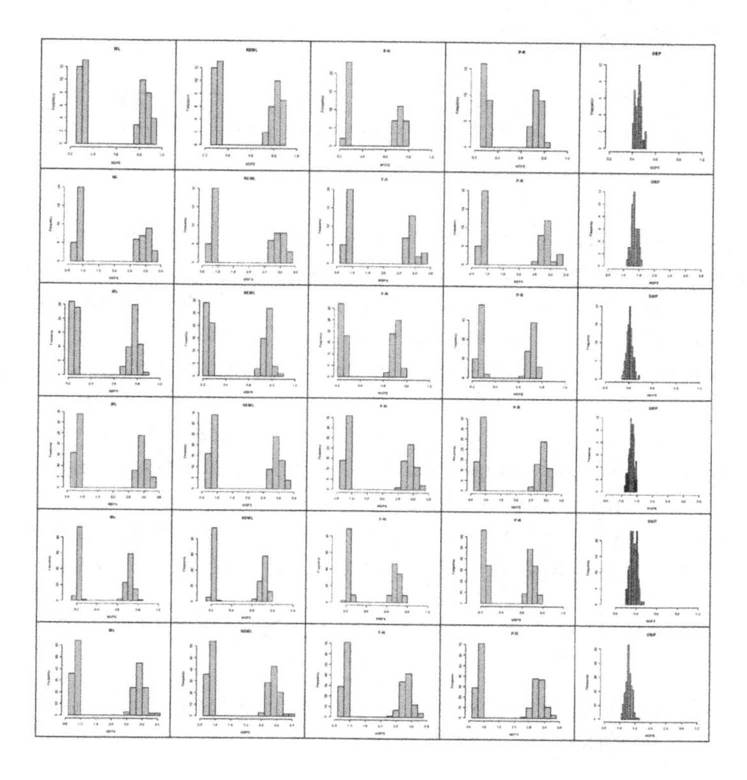

Fig. 5.2 *Histograms of* MSPE_i^*, $i = 1, \dots, m$. *The rows, from top to bottom, correspond to* $m = 50, d = 1$; $m = 50, d = 5$; $m = 100, d = 1$; $m = 100, d = 5$; $m = 200, d = 1$; *and* $m = 200, d = 5$, *respectively. Within each row from left to right: EBLUP-1, EBLUP-2, EBLUP-3, EBLUP-4, OBP. Within each row the* x*-axes have the same range.*

covariate is zero. The truth is that the slope in (5.52) is nonzero (so the assumed model is correct in this regard); but, there is a nonzero intercept, although its value is much smaller compared to, say, those considered in Subsection 5.7.1 (so the assumed model is wrong, but not "terribly wrong"). More specifically, the true underlying model is

$$Y_{ij} = b_0 + b_1 X_{ik} + v_i + e_{ik}, \quad i = 1, \dots, m, \; k = 1, \dots, 1000, \quad (5.53)$$

Table 5.2 **Overall Empirical MSPE (bias, variance contribution):** Assumed model is partially correct; % Increase is MSPE of EBLUP over MSPE of OBP (negative number means decrease).

m	OBP	EBLUP	% Increase
50	0.421 (0.224, 0.197)	0.405 (0.238, 0.167)	-4.0
100	0.733 (0.448, 0.285)	0.748 (0.457, 0.291)	2.1
400	2.745 (1.847, 0.899)	2.848 (1.878, 0.971)	3.8

as opposed to (5.52), where the X subpopulation is generated from the normal distribution with mean equal to 1 and standard deviation equal to $\sqrt{0.1} \approx 0.32$; $b_0 = 0.2$, $b_1 = 0.1$; the v_i are generated independently from the normal distribution with mean 0 and standard deviation 0.1; the e_{ik} are generated from the heteroscedastic normal distribution so that $e_{ik} \sim N(0, \sigma_i^2)$, where σ_i^2 are generated independently from the Uniform$[0.05, 0.15]$ distribution (so that range for σ_i is approximately from 0.22 to 0.39); and the v_i's and e_{ik}'s are generated independently. In addition to the overall MSPE, we also report contribution to the MSPE due to "bias" and "variance". Let $d_i = \hat{\theta}_i - \theta_i$, and $d_i^{(k)}$ be d_i based on the kth simulated data set, $1 \leq k \leq K$. We define the empirical bias and variance for the ith small area as

$$\bar{d}_i = \frac{1}{K}\sum_{k=1}^{K} d_i^{(k)} \quad \text{and} \quad v_i^2 = \frac{1}{K-1}\sum_{k=1}^{K}\{d_i^{(k)} - \bar{d}_i\}^2,$$

respectively. Let MSPE_i denote the empirical MSPE for the ith small area. It is easy to show that the overall empirical MSPE is

$$\sum_{i=1}^{m} \text{MSPE}_i = \frac{K-1}{K}\sum_{i=1}^{m} v_i^2 + \sum_{i=1}^{m} (\bar{d}_i)^2.$$

Thus, the bias and variance contribution to the overall MSPE are defined as $\sum_{i=1}^{m}(\bar{d}_i)^2$ and $\sum_{i=1}^{m} v_i^2$, respectively. Results based on $K = 1000$ simulation runs are presented in Table 5.2. As we can see, for the smaller m, $m = 50$, OBP performs (slightly) worse than the EBLUP, but for the larger m, $m = 100$ and $m = 400$, OBP performs (slightly) better, and its advantage increases with m. As for the bias, variance contribution, OBP seems to have smaller bias, and smaller variance for larger m ($m = 100, 400$).

Next, we consider a case where the assumed model is actually correct. Namely, the true underlying model is (5.53) with $b_0 = 0$; the errors e_{ik} are homoscedastic with variance equal to 0.1, and everything else is the same as the case considered above. Results based on $K = 1000$ simulation runs are presented in Table 5.3. This time, we see that the EBLUP performs slightly better than OBP under different m, but the difference is diminishing as the

Table 5.3 **Overall Empirical MSPE (bias, variance contribution):** Assumed model is correct; % Increase is MSPE of EBLUP over MSPE of OBP (negative number indicates decrease).

m	OBP	EBLUP	% Increase
50	0.335 (0.204, 0.131)	0.330 (0.205, 0.125)	-1.4
100	0.749 (0.457, 0.292)	0.746 (0.456, 0.290)	-0.4
400	2.796 (1.800, 0.997)	2.794 (1.799, 0.996)	-0.1

sample size increases. As for the bias, variance contribution, EBLUP seems to have smaller variance, and smaller bias for larger m ($m = 100, 400$), but its advantages in both bias and variance shrink as m increases.

The results reported in Tables 5.2 and 5.3 are copied from Jiang *et al.* (2015b).

In summary, the simulation results suggest that, when the assumed model is slightly misspecified, OBP may not outperform EBLUP when m, the number of small areas, is relatively small; however, OBP is expected to outperform EBLUP when m is relatively large, and the advantage of OBP over EBLUP increases with m (recall the definition of the overall MSPE). On the other hand, when the assumed model is correct, EBLUP is expected to perform better than OBP, although the difference may be ignorable; and the advantage of EBLUP over OBP is disappearing as m increases. These findings, along with those in Subsection 5.7.2, are very much in line with those of Jiang *et al.* (2011a) for the Fay-Herriot model.

5.7.3 *Comparison of OBP and EBP under an LN model*

Finally, we consider a simulated example regarding OBP under an LN model for count data, discussed in Section 5.5. As noted (see the end of Subsection 5.5.2), the OBP method can be extended to more complicated area-level models. Here we consider one of such cases, with $y_{ij}|\mu_{ij} \sim \text{Poisson}(\mu_{ij})$ and $\mu_{ij} = \exp(\eta_{ij} + v_i)$, $i = 1, \dots, m, j = 1, \dots, s$, where the y_{ij}'s are conditionally independent, and the v_i's are independent $N(0, 1)$. Here the index i corresponds to the area (e.g., county), and the index j to a classification of demographic groups (e.g., age groups) within the area. The random effect, v_i, is at the area-level, but the interest is the area/demographic mean counts, μ_{ij}. The assumed model has $\eta_{ij} = x_{ij}\beta$, where the x_{ij}'s are generated independently from the Uniform$(0, 2)$ distribution, and then fixed throughout the simulation study.

We consider three true underlying distributions: (i) $\eta_{ij} = x_{ij} + 1$, (ii) $\eta_{ij} = 2x_{ij}^2/3 + 1$; and (iii) $\eta_{ij} = x_{ij} + z_{ij}$, where the z_{ij}'s are generated

Table 5.4 **OBP vs EBP under an LN model.** Reported are simulated MSPEs (based on $K = 500$ simulations) under different true models and sample sizes.

	Assumed Model		Model (i)		Model (ii)		Model (iii)	
m	OBP	EBP	OBP	EBP	OBP	EBP	OBP	EBP
8	64	56	241	258	527	649	5500	5739
32	218	204	811	841	2481	3277	35640	36445
128	856	839	3421	4090	11049	14979	178534	183780

Table 5.5 **Ratio of Empirical MSPE (OBP over EBP) from Table 5.4**

m	Assumed Model	Model (i)	Model (ii)	Model (iii)
8	1.14	0.93	0.81	0.96
32	1.07	0.96	0.76	0.98
128	1.02	0.84	0.74	0.97

independently from the Uniform$(0, 2)$ distribution, and fixed. It follows that, under each of the three cases, the assumed model is misspecified. Three different samples sizes are considered: $m = 8$, $m = 32$, and $m = 128$, with $s = 6$ in all of the cases. A similar measure of (overall) simulated MSPE, is considered:

$$\text{MSPE} = \frac{1}{K}\sum_{k=1}^{K}\sum_{i=1}^{m}\sum_{j=1}^{s}\{\hat{\mu}_{ij}^{(k)} - \mu_{ij}^{(k)}\}^2,$$

where $K = 500$, $\mu_{ij}^{(k)}$ and $\hat{\mu}_{ij}^{(k)}$ are the true small area mean count and the corresponding OBP, or EBP, for the kth simulation run. We compare the OBP with the EBP under the cases where the assumed model holds, and where the true underlying model is (i), (ii), or (iii). The results, rounded to the nearest integers and copied from Chen *et al.* (2015), are presented in Table 5.4.

It is seen that, when the underlying model is correctly specified, the EBP performs better than the OBP, but the (proportional) difference is diminishing as m increases. On the other hand, when the underlying model is misspecified, the OBP performs better than the EBP, and the (proportional) difference is not going away as m increases. These patterns, as illustrated by Table 5.5, are consistent with those found in Jiang *et al.* (2011a) and Jiang *et al.* (2015b) (see the previous subsection).

5.8 Estimation of area-specific MSPE

An important and challenging problem in SAE is to obtain measures of uncertainty for small-area estimators, or predictors. This is typically done, under the frequentist framework, by estimating the area-specific MSPE for the small-area predictors. A desirable property for the MSPE estimator is that the estimator is nearly unbiased or, more precisely, second-order unbiased. This means that the bias of the MSPE estimator is of the order $o(m^{-1})$, where m is the number of small areas.

The Prasad-Rao method [Prasad and Rao (1990)] is well-known in deriving second-order unbiased MSPE estimator for the EBLUP. It is known that the naive estimator of the MSPE (of the EBLUP), which simply replaces the unknown variance components in the (analytic) expression of the MSPE of BLUP by their estimators, underestimates the MSPE of the EBLUP, and is only first-order unbiased, that is, the bias is of the order $O(m^{-1})$, if the parameter estimators are consistent. Prasad and Rao (1990) used Taylor expansions to obtain a second-order approximation to the MSPE, and then bias-corrected the plug-in estimator based on the approximation, again to the second order, to obtain an estimator of the MSPE whose bias is $o(m^{-1})$. The method has since been used extensively in SAE and several extensions have been given [e.g., Lahiri and Rao (1995), Datta and Lahiri (2000), Jiang and Lahiri (2001), and Datta *et al.* (2005)].

The Prasad-Rao method is based on the assumption that the underlying model is correct, hence the existence of the true parameters. In this section, however, such an assumption is not made, which makes estimation of the area-specific MSPE much more difficult. Consider, for example, the Fay-Herriot model. It is known (see Section 5.6) that $\tilde{\psi}$, the BPE, is $\sqrt{m}$-consistent to ψ_* that is defined in Section 5.6, but ψ_* is not necessarily the true parameter vector. Even in the much simpler case in which ψ_* is known, hence one can replace $\tilde{\psi}$ by ψ_* in the OBP, it is easy to see that the MSPE of $\tilde{\theta}_i$ is not a known function of ψ_* but depends on the unknown μ_i and $\mathrm{E}(y_i^2)$. It follows that, under the weak model assumption, the MSPE cannot be consistently estimated.

On the other hand, it is still possible to obtain a nearly unbiased estimator of the MSPE of $\tilde{\theta}_i$. Jiang *et al.* (2011a) derived a second-order unbiased MSPE estimator for the case of Fay-Herriot model. First note that, instead of making the Taylor expansion at the point of the true parameter vector, as in the Prasad-Rao method, the expansion can be made at ψ_*. However, the current weak assumption about the mean functions makes it difficult to carry out the standard arguments of Prasad and Rao (1990), which make use of such assumptions as $\mathrm{E}(y_i) = x_i'\beta$ and $\mathrm{E}\{(y_i - x_i'\beta)^2\} = A + D_i$, $1 \le i \le m$, where $\psi = (\beta', A)'$ is the true parameter vector. Nevertheless, with the asymptotic property of the BPE (Theorem 5.2), and using a different technique, the authors obtained a second-order unbiased estimator of the area-specific MSPE. First introduce some notation. Let $\tilde{B}_i$ be B_i (defined in Section 5.6) with A replaced by $\tilde{A}$. Define $\tilde{M}_i = \mathrm{diag}\{x_i, (A+D_i)^{-1}\}$, $\tilde{U}_i = (u_i, u_{m+i})'$, where $u_i = y_i - x_i'\tilde{\beta}$ and $u_{m+i} = (y_i - x_i'\tilde{\beta})^2 - (\tilde{A} + D_i)$. Let $\tilde{W}_i = (\tilde{W}_{i,ab})_{a,b=1,2}$ with $\tilde{W}_{i,11} = x_i x_i'$,

$$\tilde{W}_{i,12} = 2(\tilde{A} + D_i)^{-1}(y_i - x_i'\tilde{\beta})x_i,$$

$\tilde{W}_{i,21} = \tilde{W}_{i,12}'$ and

$$\tilde{W}_{i,22} = (\tilde{A} + D_i)^{-2}\{3(y_i - x_i'\tilde{\beta})^2 - (\tilde{A} + D_i)\}.$$

Let $\tilde{f}_i = -2(1 - \tilde{B}_i)^2 \tilde{M}_i \tilde{U}_i$ and

$$\tilde{G}_2 = 2\sum_{j=1}^{m}(1 - \tilde{B}_j)^2(\tilde{W}_j - \tilde{\Delta}_j),$$

where $\tilde{\Delta}_j = \mathrm{diag}\{0, \dots, 0, (\tilde{A} + D_j)^{-1}\}$ (p zeros; p is the dimension of β), and $\tilde{h}_2'$ be the last row of $\tilde{G}_2^{-1}$. Jiang *et al.* (2011a) showed that the following is a second-order unbiased estimator of the MSPE of $\tilde{\theta}_i$, the OBP of θ_i,

$$\begin{aligned}\widetilde{\mathrm{MSPE}}(\tilde{\theta}_i) &= (\tilde{\theta}_i - y_i)^2 + D_i(2\tilde{B}_i - 1) \\ &\quad + 2(1 - \tilde{B}_i)^2 \tilde{h}_2' \tilde{f}_i + 4D_i(1 - \tilde{B}_i)^3 \mathrm{tr}(\tilde{G}_2^{-1}\tilde{W}_i). \quad (5.54)\end{aligned}$$

In a simulation study carried out in Jiang *et al.* (2011a), the authors showed that the MSPE estimator (5.54) performed better in reducing the bias, compared to a bootstrap MSPE estimator (see below for more detail). However, the estimator (5.54) is not guaranteed nonnegative. Note that it does not make sense to use a negative number as an estimate for a nonnegative quantity, such as the MSPE. Furthermore, it is seen that the leading term on the right side of (5.54), $(\tilde{\theta}_i - y_i)^2 + D_i(2\tilde{B}_i - 1)$, is $O_{\mathrm{P}}(1)$, and it depends on the area-specific data, y_i. [The remaining terms are of the order $O(m^{-1})$]. Because y_i is an observation from a single small area, it has a relatively large variance. On the other hand, the term $D_i(2\tilde{B}_i - 1)$ can be negative. Thus, as a result of the high variation of $(\tilde{\theta}_i - y_i)^2$, there is a non-vanishing probability (as m increases) that the leading term, hence the estimated MSPE, is negative.

Jiang *et al.* (2011a) also proposed an alternative, bootstrap approach to the MSPE estimation. The proposed bootstrap MSPE estimator is guaranteed nonnegative, but its bias appears to be larger than the estimator (5.54). Note that, typically, a bootstrap MSPE estimator without bias correction is only first order unbiased [e.g., Hall and Maiti (2006)]. Also, the justification of the bootstrap approach is questionable given the potential model misspecification.

A different, nonparametric bootstrap approach was proposed by Jiang *et al.* (2015b). Unlike Jiang *et al.* (2011a), there is a better justification of the nonparametric bootstrap approach in terms of Efron's original idea [Efron (1979)]. Consider the finite-population situation of Subsection 5.4.2. Suppose that the small area subpopulations, or the N_i's, are large enough, so that the sampling from the subpopulations can be treated approximately as with replacement. Let $z_{ij} = (x_{ij}', y_{ij})', j = 1, \dots, n_i$ denote the (original) samples from the ith small area, $1 \leq i \leq m$. We then draw samples, $z_{ij}^{(a)} = [\{x_{ij}^{(a)}\}', y_{ij}^{(a)}]', j = 1, \dots, n_i$, with replacement, from $\{z_{ij}, j = 1, \dots, n_i\}$, independently for $1 \leq i \leq m$. Suppose that B bootstrap samples are drawn, yielding samples $z^{(a)} = \{z_{ij}^{(a)}, 1 \leq j \leq n_i, 1 \leq i \leq m\}$, $1 \leq a \leq B$. The bootstrapped version of the BP (5.24) is

$$\tilde{\theta}_i^{(a)} = \bar{X}_{i\cdot}'\beta + \left\{r_i + (1 - r_i)\frac{n_i\gamma}{1 + n_i\gamma}\right\}[\bar{y}_{i\cdot}^{(a)} - \{\bar{x}_{i\cdot}^{(a)}\}'\beta], \tag{5.55}$$

where β and γ are the same population parameters for β and γ, respectively, as for the original population. Note that the original samples of z_{ij} are assumed to satisfy the same NER model, (5.23), with X_{ik} (Y_{ik}) replaced by x_{ij} (y_{ij}). Because the original samples are treated as the bootstrap population, following Efron's original idea, the population parameters, β, γ, for the bootstrap samples are the same as those for the original samples. Nevertheless, as mentioned, the proposed bootstrap procedure is nonparametric in the sense that the assumed model, (5.23), plays no role in drawing the bootstrap samples. In particular, the BPE of β and γ, based on the original samples, are not used anywhere in the bootstrapping; and the population quantities of interest are $\bar{Y}_i$, $1 \leq i \leq m$, whose bootstrap analogies are $\bar{y}_{i\cdot}, 1 \leq i \leq m$. This is different from the parametric bootstrap of Jiang *et al.* (2011a), where the BPE of the model parameters, based on the original samples, are used to draw bootstrap samples under the assumed model. Also note that, because the $\bar{X}_i$ are known, they are treated as known constants, and therefore do not change during the bootstrapping (it does not make sense to "estimate" something that one already knows). Other than those, the procedure follows closely the standard procedure [e.g., Efron and Tibshirani (1993)]. The bootstrap estimator of $\text{MSPE}(\hat{\theta}_i) = \text{E}(\hat{\theta}_i - \bar{Y}_i)^2$ is

$$\widehat{\text{MSPE}}(\hat{\theta}_i) = \frac{1}{B}\sum_{a=1}^{B}\{\hat{\theta}_i^{(a)} - \bar{y}_{i\cdot}\}^2, \tag{5.56}$$

where $\hat{\theta}_i^{(a)}$ is (5.55) with β, γ replaced by their BPE based on the bootstrapped samples.

One might be concerned that, because the n_i's may be small in typical SAE problems, there may not be many distinct bootstrap samples for each small area. However, the data consist of not just one, but many small areas. When all of the small areas are combined, there are, still, a lot of distinct bootstrap samples, even if the n_i's are small. As an example, below we report the results of a simulation study by Jiang *et al.* (2015b).

Example 5.2. Consider the NER model (5.52). The setting is the same as in Subsection 5.7.2 except that the true underlying model is

$$Y_{ik} = b + v_i + e_{ik}, \quad i = 1, \dots, m, \; k = 1, \dots, 1000 \tag{5.57}$$

(so the subpopulation size is $N_i = 1000$, $1 \le i \le m$), and the sample sizes are smaller. More specifically, we have $b = 0.5$; $m = 10$ or 20, and $n_i = 5$ or 10. We first consider the design-based bias of $\widehat{\text{MSPE}}(\hat{\theta}_i)$. Two finite populations are generated, and then fixed, so that the finite population for $m = 10$ is a subpopulation of the finite population for $m = 20$. Table 5.6 reports, for the first 10 small areas (these are all the small areas that are common under different values of m), the simulated true MSPE (MSPE), obtained the same way as in Section 5.7, the simulated mean of $\widehat{\text{MSPE}}(\hat{\theta}_i)$ ($\widehat{\text{MSPE}}$), and the percentage relative bias (%RB) defined as

$$100 \times \left\{\frac{\text{E}(\widehat{\text{MSPE}}) - \text{True MSPE}}{\text{True MSPE}}\right\},$$

where the expectation is based on the simulations. Another measure of performance is the square root of the mean squared error (RMSE) over the simulations, defined as

$$\sqrt{\frac{1}{K}\sum_{k=1}^{K}(\widehat{\text{MSPE}}_{i,k} - \text{MSPE}_i)^2}$$

for the ith small area, where MSPE_i is the true MSPE for the ith small area (which does not depend on k), evaluated over the simulations, and $\widehat{\text{MSPE}}_{i,k}$ is the MSPE estimate based on the kth simulated data set. We

Table 5.6 **Empirical Performance of $\widehat{\text{MSPE}}$**

m	n_i	i	MSPE	$\widehat{\text{MSPE}}$	%RB	RMSE	i	MSPE	$\widehat{\text{MSPE}}$	%RB	RMSE
10	5	1	.041	.042	4.5	.103	6	.034	.043	26.3	.070
10	10	1	.036	.036	-0.4	.068	6	.034	.036	6.4	.070
20	5	1	.031	.032	4.1	.051	6	.028	.031	12.5	.046
10	5	2	.046	.038	-16.1	.078	7	.032	.040	25.4	.078
10	10	2	.035	.033	-4.1	.078	7	.033	.034	2.7	.068
20	5	2	.031	.029	-7.2	.050	7	.030	.031	3.6	.055
10	5	3	.038	.042	10.2	.121	8	.042	.042	-0.4	.150
10	10	3	.037	.036	-1.7	.091	8	.033	.035	7.5	.067
20	5	3	.031	.032	4.4	.052	8	.030	.031	4.1	.058
10	5	4	.056	.052	-7.6	.121	9	.050	.042	-15.0	.074
10	10	4	.037	.040	6.3	.072	9	.034	.034	-1.0	.063
20	5	4	.040	.035	-11.3	.068	9	.034	.030	-11.1	.049
10	5	5	.033	.037	11.8	.066	10	.041	.043	3.1	.082
10	10	5	.032	.033	2.5	.066	10	.034	.033	-2.9	.073
20	5	5	.024	.025	2.9	.052	10	.035	.033	-7.9	.062

consider $B = 100$ as the number of bootstrap samples used to evaluate the MSPE estimator, (5.56). All results are based on 1,000 simulation runs.

It is seen that, overall, the results improve when either n_i or m increase, but, in terms of %RB, the improvement is more universal, or effective, when n_i increases. This is mainly due to the fact that, as n_i increases, the sample provides a better approximation to the population; hence, the bootstrap distribution better approximates to the population distribution. Also note that, depending on the area, the sign of the RB can be either positive or negative. This is mainly due to the area-to-area difference (recall that the populations are fixed) as well as the bootstrap errors. To obtain some overall measures, we report the mean and standard deviation (s.d.) of the %RBs over the 10 small areas as follows: $m = 10, n_i = 5$: mean $= 4.2\%$, s.d. $= 14.8\%$; $m = 10, n_i = 10$: mean $= 1.5\%$, s.d. $= 4.2\%$; $m = 20, n_i = 5$: mean $= -0.6\%$, s.d. $= 8.1\%$. The boxplots of the %RBs are presented in Figure 5.3. The plots further illustrate the pattern of improvement. On the other hand, in terms of RMSE, the improvement is much more significant when m increases than when n_i increases. This is because having a larger m reduces the MSPEs, in general; hence, naturally, the corresponding MSPE estimates also drop. In other words, both the estimator and the parameter (the MSPE) decrease, which typically results in a reduction in RMSE.

Finally, we consider measure of uncertainty in the case of count data, discussed in Section 5.5. Chen *et al.* (2015) proposed a bootstrap approach for estimating the conditional MSPE of the OBP given the small area

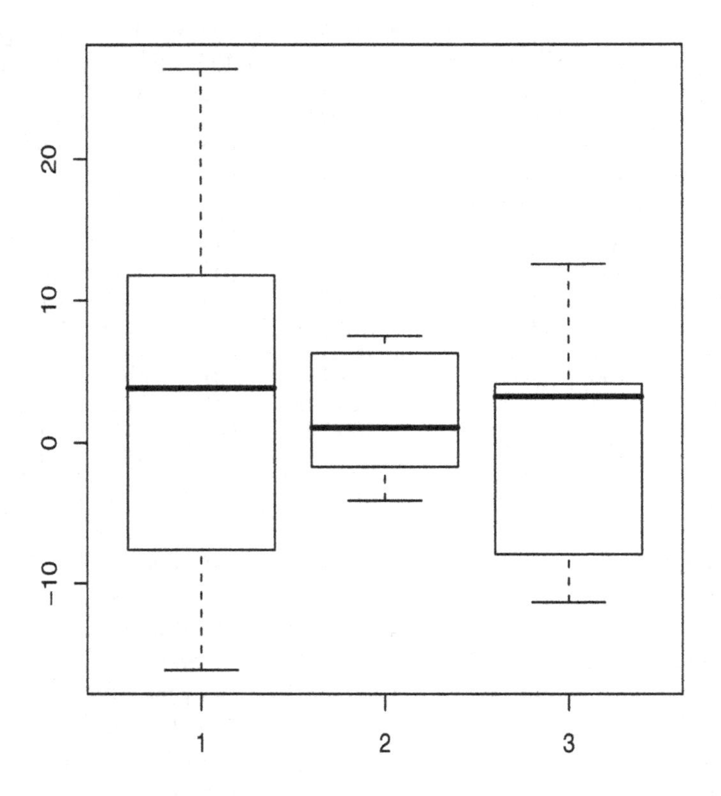

Fig. 5.3 *Boxplots of %RB. 1: $m = 10, n_i = 5$; 2: $m = 10, n_i = 10$; 3: $m = 20, n_i = 5$.*

means, $\mu_i, 1 \leq i \leq m$ [e.g., Datta *et al.* (2011)]. The method is surprisingly simple. For each $1 \leq i \leq m$, draw a bootstrap sample, y_i^*, from the Poisson distribution with mean y_i. Let $y_{i,b}^*, 1 \leq i \leq m$ be the bth bootstrap sample, $1 \leq b \leq B$, and $\hat{\mu}_{i,b}^*$ be the OBP computed based on the bth bootstrap sample for area i. Then, for $1 \leq i \leq m$, the conditional MSPE of the OBP for area i, $\text{MSPE}(\hat{\mu}_i|\mu_i) = \text{E}\{(\hat{\mu}_i - \mu_i)^2|\mu_i\}$, is estimated by

$$\widehat{\text{MSPE}}_i = \frac{1}{B}\sum_{b=1}^{B}(\hat{\mu}_{i,b}^* - y_i)^2. \tag{5.58}$$

Although the proposal might sound simple, it actually carries an idea on how to use the original non-parametric bootstrap idea of Efron (1979) to justify something that might sound like parametric bootstrap with one observation. A key idea is called Poisson approximation to binomial, detailed below, from which, one is able to create an imaginative "box", and draw samples from the box as in Efron's (1979) bootstrap. After the bootstrap samples are drawn, one "throws out" the box and go back to the original Poisson observation (this is why the box is called imaginative).

Consider the case that the (conditional) Poisson distribution can be viewed as an approximation to a binomial distribution. The latter is, of course, well known in probability theory. Part of the theory states that, if $Y \sim \text{Binomial}(n, p)$, where n is large and p is small such that $np \approx \lambda$, the distribution of Y is approximately Poisson(λ). See, for example, Jiang (2010, sec. 10.3), for a discussion as well as some (historical) examples. Suppose that the observation, Y, is a Poisson count such that, approximately, $Y = \sum_{i=1}^{n} Y_i$, where the Y_i's are i.i.d. Bernoulli. If the Y_i's were observed, it would be straightforward to apply Efron's (1979) bootstrap by drawing samples, with replacement, from $\{Y_1, \dots, Y_n\}$, say, $Y_1^*, \dots, Y_n^*$, and then compute $Y^* = \sum_{i=1}^{n} Y_i^*$ as the bootstrapped Y. The point is that, actually, the Y_i's need not to be observed, because Y^* is the same as a random sample from the Binomial($n, \bar{Y}$) distribution, where $\bar{Y} = Y/n$. Here, we assume that the sample size, n, is known. Thus, once again, using the Poisson approximation to binomial, Y^* is approximately a random sample from the Poisson(Y) distribution. It turns out that, after all, one does not need to really do anything with the "box", nor does one need to know the sample size, n, for the Poisson approximation to Binomial except that n is large.

Note that, under the broader model, and conditioning on μ_i, y_i is a Poisson count. It follows that the bootstrapped y_i, y_i^*, is approximately a random sample from the Poisson(y_i) distribution, conditioning on μ_i.

Alternatively, y_i^* may also be viewed as a parametric bootstrap sample in that, conditioning on μ_i, the distribution of y_i is Poisson(μ_i). The parameter μ_i is then estimated by y_i, resulting the bootstrap distribution, Poisson(y_i).

The performance of the latest bootstrap method was evaluated empirically by Chen *et al.* (2015), who showed that the method performs reasonably well, especially when the small area means, μ_i, are relatively large.

Table 5.7 **The Hospital Data, EBLUP-4, OBP with Measure of Uncertainty**

Area	y_i	x_i	$\sqrt{D_i}$	OBP $\left(\widehat{\text{RMSPE}}\right)$	EBLUP-4 $\left(\widehat{\text{RMSPE}}\right)$
1	.302	.112	.055	.239 (.060)	.226 (.026)
2	.140	.206	.053	.181 (.019)	.181 (.027)
3	.203	.104	.052	.220 (.017)	.209 (.026)
4	.333	.168	.052	.249 (.085)	.238 (.026)
5	.347	.337	.047	.347 (.047)	.347 (.049)
6	.216	.169	.046	.234 (.016)	.224 (.026)
7	.156	.211	.046	.172 (.020)	.175 (.028)
8	.143	.195	.046	.197 (.045)	.193 (.026)
9	.220	.221	.044	.162 (.058)	.170 (.031)
10	.205	.077	.044	.180 (.017)	.176 (.028)
11	.209	.195	.042	.206 (.015)	.203 (.026)
12	.266	.185	.041	.228 (.031)	.222 (.026)
13	.240	.202	.041	.201 (.031)	.200 (.026)
14	.262	.108	.036	.234 (.024)	.224 (.027)
15	.144	.204	.036	.180 (.032)	.179 (.027)
16	.116	.072	.035	.154 (.040)	.152 (.029)
17	.201	.142	.033	.236 (.032)	.224 (.027)
18	.212	.136	.032	.238 (.020)	.226 (.027)
19	.189	.172	.031	.223 (.031)	.214 (.026)
20	.212	.202	.029	.199 (.014)	.198 (.026)
21	.166	.087	.029	.187 (.017)	.181 (.026)
22	.173	.177	.027	.212 (.039)	.204 (.025)
23	.165	.072	.025	.165 (.017)	.163 (.027)

5.9 Real-data examples

We illustrate the OBP method using three real-data examples. The first example is regarding a case of Fay-Herriot model; the second is regarding the TVSFP data, discussed in Section 4.4, but here used to illustrate the OBP under an NER model; the last email is about a case of count data.

5.9.1 *Hospital data*

Morris and Christiansen (1995) presented a data set involving 23 hospitals (out of a total of 219 hospitals) that had at least 50 kidney transplants during a 27 month period. See Table 5.7. The y_i's are graft failure rates for kidney transplant operations, that is, y_i = number of graft failures/n_i, where n_i is the number of kidney transplants at hospital i during the period of interest. The variance for the graft failure rate, D_i, is approximated by $(0.2)(0.8)/n_i$, where 0.2 is the observed failure rate for all hospitals. Thus,

D_i is assumed known. In addition, a severity index x_i is available for each hospital, which is the average fraction of females, blacks, children and extremely ill kidney recipients at hospital i. Ganesh (2009) proposed a Fay-Herriot model for the graft failure rates, which is Example 5.1 with $x_i'\beta = \beta_0 + \beta_1 x_i$. Note that the graft failure rates are binomial proportion of fairly large denominators (at least 50). Thus, a normal distribution for the y_i's is not unreasonable, at least from an approximation point of view, by the central limit theorem. However, inspections of the raw data suggest some nonlinear trends in the mean function (see Figure 5.4). Jiang *et al.* (2010) proposed a cubic model for the same data. On the other hand, there has been a concern that the point at the upper right corner might be an "outlier", in some way, so a cubic model to accommodate a single point might overfit. Due to such a concern, Jiang *et al.* (2011a) proposed a quadratic-outlying (Q-O) model. Suppose that there is an abrupt "jump" in the mean response when x_i is greater than 0.3; otherwise, the mean response is a quadratic function of the covariate. This can be expressed as

$$y_i = \beta_0 + \beta_1 x_i + \beta_2 x_i^2 + d1_{(x_i > 0.3)} + v_i + e_i, \tag{5.59}$$

$i = 1, \dots, m$, where the x_i's are the severity index given in Table 5.7 and so are the D_i's.

Jiang *et al.* (2011a) considered OBP and EBLUP-4 (see Subsection 5.7.1) under the Q-O model. The latter appears to perform the best among the EBLUPs in this case based on the simulation results of Jiang *et al.* (2011a). Furthermore, the MSPE estimates (5.54) for the OBP and the Prasad-Rao MSPE estimates for EBLUP-4 [Prasad and Rao (1990)] were obtained. For 6 out of the 23 areas (area #3, 6, 7, 11, 20, and 23) the estimates (5.54) are negative. Following the recommendation of Jiang *et al.* (2011a), for those areas the bootstrap MSPE estimates are used, with the bootstrap sample size equal to 100. The BPE for the parameters are, approximately, $\tilde{\beta}_0 = -0.084$, $\tilde{\beta}_1 = 4.614$, $\tilde{\beta}_2 = -16.045$, $\tilde{d} = 0.698$ and $\tilde{A} = 3.4 \times 10^{-4}$; the corresponding estimates for EBLUP-4 are, approximately, $\hat{\beta}_0 = -0.040$, $\hat{\beta}_1 = 3.723$, $\hat{\beta}_2 = -12.745$, $\hat{d} = 0.580$ and $\hat{A} = 3.6 \times 10^{-4}$. The OBP and EBLUP-4 are reported in Table 5.7 with the square roots of the corresponding MSPE estimates (RMSPE; in the parentheses) used as measures of uncertainty.

If the assumed Q-O model is correct, then according to the properties of OBP, EBLUP-4 may work better in this case; on the other hand, if the Q-O model is misspecified, the OBP may have an edge. While we may never know for sure whether the Q-O model is correct, it is very possible, as is

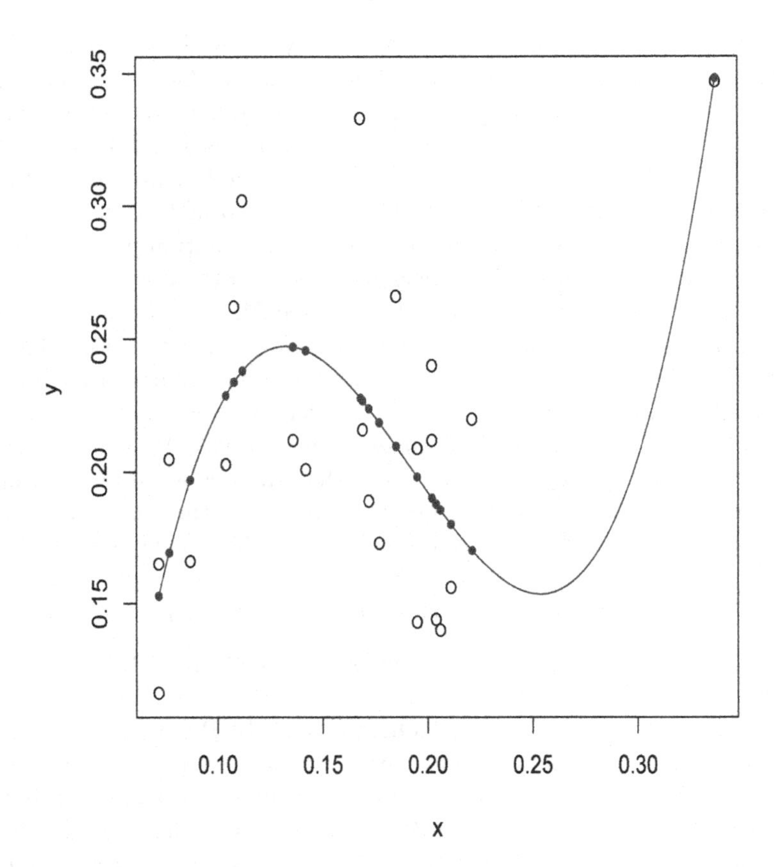

Fig. 5.4 *Data from Morris and Christiansen (1995): Circles — raw data; smooth curve & dots — cubic fit from Jiang* et al. *(2010).*

usually the case, that the underlying mean function is (still) misspecified, to some extend. For example, a plot (see Figure 5.5) suggests that the fit (by either method) is relatively poor for the middle part of the x range, where a quadratic mean function is fitted.

Also note that the Prasad-Rao MSPE estimates are almost constant across the areas, with the exception of area #5 that corresponds to the outlying case. As for the OBP, areas with the largest estimated MSPEs correspond to those for which the quadratic curve fits the data relatively

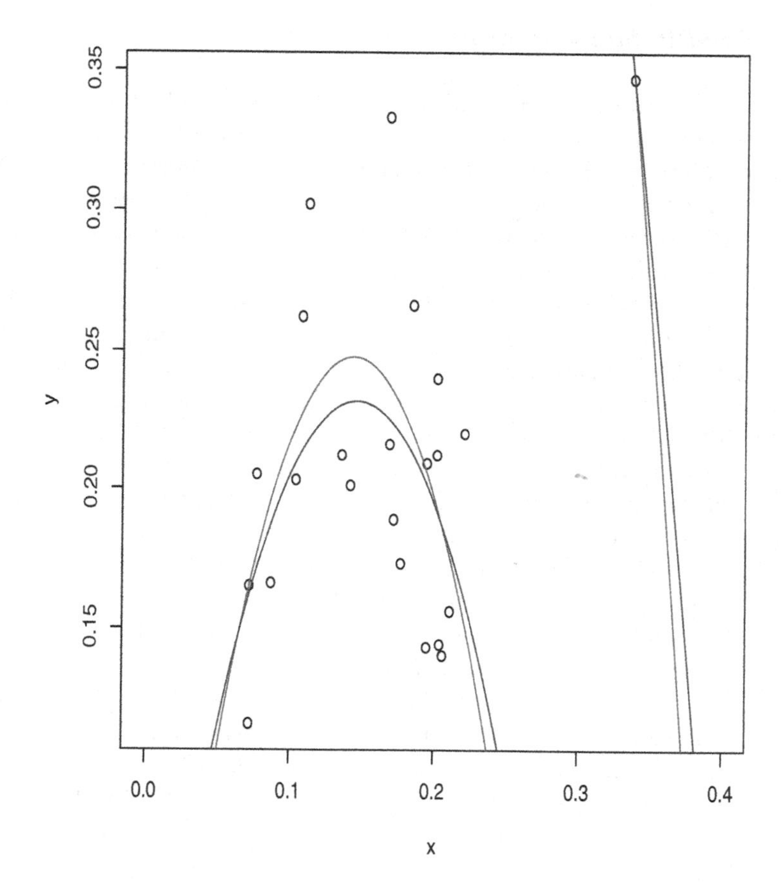

Fig. 5.5 *The Q-O mean function fits for the hospital data: Red curve (higher and slightly to the left)—OBP; blue curve (lower and slightly to the right)—EBLUP-4. The circles are the raw data.*

poorly, such as areas #1 and #4, but not area #5, for which the Q-O model fits the data well (again see Figure 5.5). This makes sense because the OBP MSPE estimates takes into account the potential bias caused by model misspecification–and this is also with respect to a specific small area.

5.9.2 *TVSFP data revisited*

The TVSFP data were used in Section 4.4 to demonstrate the L-test. Here, we use the same data to illustrate the OBP method for finite-population estimation. More specifically, we consider a problem of estimating the small area means for the difference between the immediate postintervention and pretest THKS scores. Here the "small area" is understood as a number of major characteristics (e.g., residential area, teacher/student ratio) that affect the response, but are not captured by the covariates in the model (i.e., linear combination of the CC, TV and CCTV indicators). Note that, traditionally, the words "small areas" correspond to small geographical regions or subpopulations, for which adequate samples are not available [e.g., Rao and Molina (2015)], and such information as residential characteristics or teacher/student ratios would be used as additional covariates. However, such information is not available. This is why we define these unavailable information as "area-specific", so that they can be treated as random effects. This is consistent with fundamental features of the random effects that are often used to capture unobservable effects or information [e.g., Jiang (2007)], and extends the traditional notion of SAE. Thus, a small area is the seventh graders in all of the U.S. schools that share the similar major characteristics as an LA school involved in the data over a reasonable period of time (e.g., 5 years) so that these characteristics had not change much during the time and neither had the social/educational relevance of the CC and TV programs. There are 28 LA schools in the TVSFP data that correspond to 28 sets of characteristics, so that the data are considered random samples from the 28 small areas defined as above. As such, each small area population is large enough so that $n_i/N_i \approx 0, 1 \leq i \leq 28$. Recall that the n_i's in the TVSFP sample range from 18 to 137, while the N_i's are expected to be at least tens of thousands. Note that the only place in the OBP where the knowledge of N_i is required is through the ratio n_i/N_i. The proposed NER model is (4.32) with the assumptions made below the equation. It follows that all of the auxiliary data x_i are at the area level; as a result, the value of $\bar{X}_i$ is known for every i.

As noted, the sample sizes for some small areas are quite large, but there are also areas with relatively (much) smaller sample sizes. This is quite common in real-life problems. Because the auxiliary data are at area-level, we have $\bar{X}_i'\beta = \bar{x}_i'\beta$; thus, it is easy to show that the BP (5.24) can be expressed as

$$\tilde{\theta}_i = \left\{ r_i + (1 - r_i)\frac{n_i\gamma}{1 + n_i\gamma} \right\} \bar{y}_i + \frac{1 - r_i}{1 + n_i\gamma}\bar{x}_i'\beta.$$

It is seen that, when n_i is large, the BP is approximately equal to $\bar{y}_i$, the design-based estimator, which has nothing to do with the parameter estimation. Therefore, when n_i is large, there is not much difference between the OBP and the EBLUP. On the other hand, when n_i is small or moderate, we expect some difference between the OBP and the EBLUP in terms of the MSPE. However, it is difficult to tell how much difference there is in this real data problem. The simulation results in Subsection 5.7.2 show that the difference between OBP and EBLUP in terms of the MSPE depends on to what extent the assumed model is misspecified. As noted, the response, y_{ij}, is difference in the THKS scores, and possible values of the THKS score are integers between 0 and 7. Clearly, such data are not normal. The potential impact of the nonnormality is two-fold. On the one hand, it is likely that the NER model, as proposed by Hedeker *et al.* (1994), is misspecified, in which case expression (5.24) is no longer the BP, and the Gaussian ML (REML) estimators are no longer the true ML (REML) estimators. On the other hand, even without the normality, (5.24) can still be justified as the best linear predictor [BLP; e.g., Jiang (2007), p. 75]. Furthermore, the Gaussian ML (REML) estimators are consistent and asymptotically normal even without the normality assumption [Jiang (1996)]. Other aspects of the NER model include homoscedasticity of the error variance across the small areas. Figure 5.6 shows the histogram of sample variances for the 28 small areas. The bimodal shape of the histogram suggests potential heteroscedasticity in the error variance, yet another type of possible model misspecification. Therefore, the OBP method is naturally considered.

Jiang *et al.* (2015b) carried out the OBP analysis for the 28 small areas. The results are presented in Table 5.8. The BPE of the parameters parameters are $\hat{\beta}_0 = 0.206$, $\hat{\beta}_1 = 0.687$, $\hat{\beta}_2 = 0.213$, $\hat{\beta}_3 = -0.288$, and $\hat{\gamma} = 0.003$. Although interpretation may be given for the parameter estimates, there is a concern about possible model misspecification (in which case the interpretation may not be sensible), as noted above. Regardless, our main interest is prediction, not parameter estimation; thus, we focus on the OBP. In addition to the OBPs, we also computed the corresponding $\widehat{\text{MSPE}}$, and their square roots as the measures of uncertainty. As a comparison, the EBLUPs for the small areas as well as the corresponding square roots of the MSPE estimates, $\widetilde{\text{MSPE}}$, using the Prasad-Rao method [P-R; Prasad

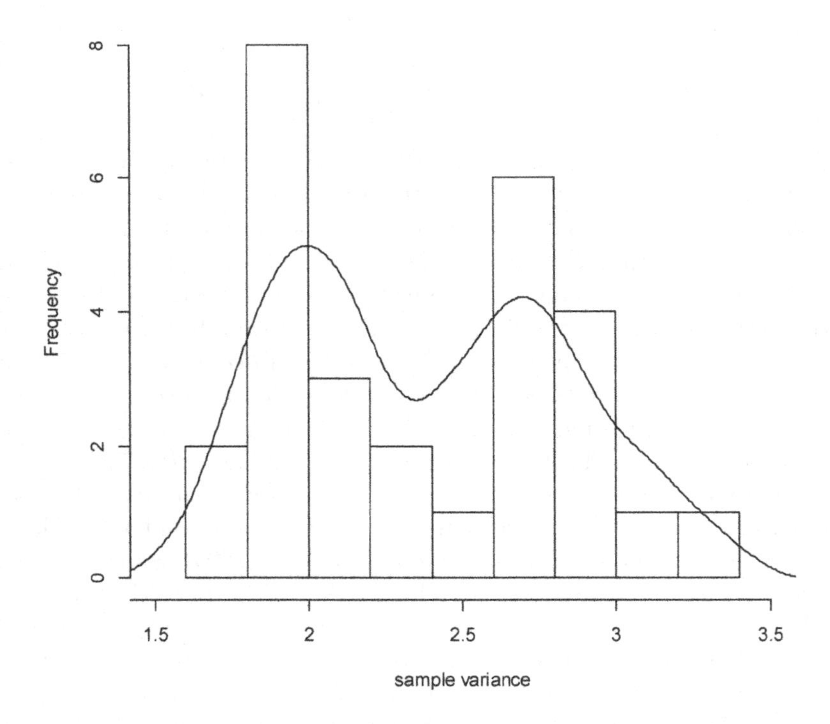

Fig. 5.6 *Histogram of sample variances; a kernel density smoother is fitted.*

and Rao (1990)] are also included. It is seen that the OBPs are all positive, even for the small areas in the control group. As for the statistical significance (here "significance" is defined as that the OBP is greater in absolute value than 2 times the corresponding square root of the MSPE estimate), the small area means are significantly positive for all of the small areas in the (1,1) group. In contrast, none of the small area mean is significantly positive for the small areas in the (0,0) group. As for the other two groups, the small area means are significantly positive for all the small areas in the (1,0) group; the small area means are significantly positive for all but two small areas in the (0,1) group. There are 7, 8, 7 and 7 small areas in the (0,0), (0,1), (1,0) and (1,1) groups, respectively.

Comparing the OBP with the EBLUP, the values of the latter are generally higher, and their corresponding MSPE estimates are mostly lower. In terms of statistical significance, the EBLUP results are significant for the (1,1), (1,0) and (0,1) groups, and insignificant for the (0,0) group. It should be noted that the P-R MSPE estimator for the EBLUP is derived under the normality assumption, while in this case the data is clearly not normal, as noted earlier. Thus, the measure of uncertainty for the EBLUP may not be accurate. In particular, just because the (square roots of the) MSPEs for the EBLUPs are lower, compared to those for the OBPs, it does not mean the corresponding true MSPEs for the EBLUPs are lower than those for the OBPs. In fact, our simulation results (see Section 5.7) have shown otherwise. It is also observed that the MSPE estimates for the EBLUPs are more homogeneous cross the small areas. This may be due to the fact that the P-R MSPE estimator for EBLUP is obtained assuming that the NER model is correct, while the proposed MSPE estimator for OBP does not use such an assumption.

In conclusion, in spite of the potential difference in the small area characteristics, the CC and TV programs appears to be successful in terms of improving the students' THKS scores (whether the improved THKS score means improved tobacco use prevention and cessation is a different matter though). It also seems apparent that the CC program was relatively more effective than the TV program. Without the intervention of any of these programs, the THKS score did not seem to improve in terms of the small area means. In terms of the statistically significant results, when $\mathrm{CC}=0$ and $\mathrm{TV}=0$, the THKS score did not seem to improve; when $\mathrm{CC}=1$, the THKS score seemed to improve; and, when $\mathrm{CC}=0$ and $\mathrm{TV}=1$, the improvement of the THKS score was not so convincing.

5.9.3 *Minnesota county data*

Finally, we present an example regarding the OBP method for count data. Torabi (2014) reported data from 87 counties in Minnesota, USA that were first reported by Jin *et al.* (2005). Both references used the data to demonstrate spatial Poisson regression with multivariate CAR errors in the log-linear model for the Poisson mean. Chen *et al.* (2015) used part of the data, namely, data regarding esophagus cancer, to illustrate the OBP method and compare it with the EBP. The data are provided in Table 5.9 for convenience, where y_i, x_i correspond to the observed and expected numbers of deaths due to esophagus cancer for county i, $i = 1, \dots, 87$.

Table 5.8 **OBP, EBLUP, Measures of Uncertainty for TVSFP Data**

ID	CC	TV	OBP	$\sqrt{\widehat{\text{MSPE}}}$	EBLUP	$\sqrt{\widetilde{\text{MSPE}}}$
403	1	0	.886	.171	.913	.121
404	1	1	.844	.296	.856	.121
193	0	0	.215	.207	.217	.120
194	0	0	.221	.137	.221	.134
196	1	0	.878	.171	.907	.124
197	0	0	.225	.158	.223	.126
198	1	1	.771	.220	.807	.131
199	0	1	.426	.142	.453	.130
401	1	1	.826	.133	.844	.127
402	0	0	.188	.171	.199	.123
405	0	1	.394	.147	.432	.129
407	0	1	.508	.300	.508	.133
408	1	0	.871	.240	.903	.123
409	0	0	.230	.125	.227	.136
410	1	1	.778	.304	.813	.124
411	0	1	.409	.195	.444	.115
412	1	0	.913	.219	.930	.126
414	1	0	.929	.257	.941	.127
415	1	1	.869	.199	.872	.135
505	1	1	.790	.154	.818	.136
506	0	1	.389	.169	.428	.134
507	0	1	.426	.148	.452	.135
508	0	1	.411	.108	.442	.136
509	1	0	.915	.097	.929	.143
510	1	0	.880	.119	.905	.143
513	0	0	.185	.215	.197	.123
514	1	1	.866	.144	.870	.140
515	0	0	.180	.102	.192	.143

We consider a GM model with $x_i'\beta = x_i$ in (5.29), or an LN model with $x_i'\beta = \log(x_i)$ in (5.28). In other words, the slope for the covariate is 1 and there is no intercept, under both models. Thus, the only unknown parameters are ϕ under the GM model, and σ under the LN model. Similar models were also proposed by Jin *et al.* (2005) and Torabi (2014). While these models may not hold exactly, a main point that we intend to make in this section is that the OBP method is more robust to potential model misspecifications.

The BPE of ϕ under the GM model is given by (5.40) with $x_i'\hat{\beta}$ replaced by $x_i, 1 \le i \le m = 87$. The value is $\hat{\phi} = 9.13$. The OBP of μ_i, under the GM model, is given by (5.34) with $x_i'\beta = x_i$ and ϕ replaced by $\hat{\phi}$. On the other hand, the ML estimator (MLE) of ϕ is $\tilde{\phi} = 3.44$; the EBP of μ_i is obtained similarly except using $\tilde{\phi}$. It is seen that the BPE and MLE are

Table 5.9 **Esophagus Cancer Data.** Observed number of deaths due to esophagus cancer for 87 Minnesota counties. Source: Torabi (2014).

County		Data									
1–10	y_i	96	673	142	126	114	41	169	100	160	124
	x_i	84	582	131	126	108	37	189	131	131	152
11–20	y_i	147	67	137	161	52	15	51	239	725	48
	x_i	129	72	124	181	42	21	73	234	695	62
21–30	y_i	116	73	79	151	149	28	3797	80	105	90
	x_i	147	96	111	166	185	39	3528	84	80	92
31–40	y_i	232	45	56	127	41	86	45	68	24	68
	x_i	192	63	57	169	32	70	51	55	21	103
41–50	y_i	53	88	86	23	35	94	66	88	113	187
	x_i	44	106	136	25	53	120	100	94	131	205
51–60	y_i	34	76	73	48	296	226	52	136	54	172
	x_i	54	95	101	46	346	280	61	102	57	156
61–70	y_i	63	1945	20	76	74	158	52	62	1000	145
	x_i	63	1741	22	92	94	176	52	60	921	162
71–80	y_i	128	54	350	89	35	53	113	21	93	53
	x_i	117	72	394	126	46	61	110	28	90	68
81–87	y_i	67	450	65	27	193	211	62			
	x_i	78	402	60	36	185	225	66			

very different. Note that the BP (5.40) can now be expressed as a weighted average of y_i and x_i with the weights being $\phi/(\phi+1)$ for y_i and $1/(\phi+1)$ for x_i. Thus, the OBP is giving relatively more weight to y_i than the EBP. This is not unreasonable. Inspection of the raw data (Exercise 5.11) suggests that there is little difference between the observed and expected; in other words, the variation is small, conditioning on the mean. Therefore, naturally, OBP is showing its faith in the data, y_i. However, because the x_i and y_i are so close, in the end, there is not much difference between the OBP and EBP anyway, as shown in Figure 5.7 and Table 5.10, where the margin of error is obtained as predicted value plus/minus the square root of the estimated MSPE ($\sqrt{\text{MSPE}}$), and the MSPE of EBP is obtained the same way as that of the OBP using the bootstrap method introduced in Section 5.8 [see (5.58)]. Due to the large range of the predicted values, it is difficult to make a good display with all of the cases in one figure. As a compromise solution, only the cases with $y_i \leq 200$ are shown in Figure 5.7, while the cases with $y_i > 200$ are reported in Table 5.10.

As for the analysis under the LN model, because the y_i are fairly large (some are very large) in this case, Laplace approximation is used [see (5.33) for the OBP; a similar approximation is used for computing the MLE for the EBP]. The BPE and MLE of σ are $\hat{\sigma} = 0.157$ and $\tilde{\sigma} = 0.158$, respectively.

Table 5.10 **Analysis of Esophagus Cancer Data.** Reported are OBP and EBP, with square–roots of corresponding estimated MSPEs in the parentheses, for counties with $y_i > 200$.

County	OBP $\left(\sqrt{\text{MSPE}}\right)$	EBP $\left(\sqrt{\text{MSPE}}\right)$
2	664.0 (25.6)	652.5 (28.8)
18	238.5 (15.1)	237.9 (14.2)
19	722.0 (24.7)	718.2 (21.5)
27	3770.5 (61.3)	3736.4 (66.2)
31	228.1 (14.1)	223.0 (14.0)
55	300.9 (13.8)	307.3 (15.5)
56	231.3 (15.7)	238.2 (15.2)
62	1924.9 (44.5)	1899.1 (57.9)
69	992.2 (32.9)	982.2 (30.4)
73	354.3 (16.3)	359.9 (18.3)
82	445.3 (18.0)	439.2 (18.4)
86	212.4 (13.8)	214.2 (11.9)

This time the BPE and MLE are very close; as a result, the OBP and EBP are also very close. The details are omitted.

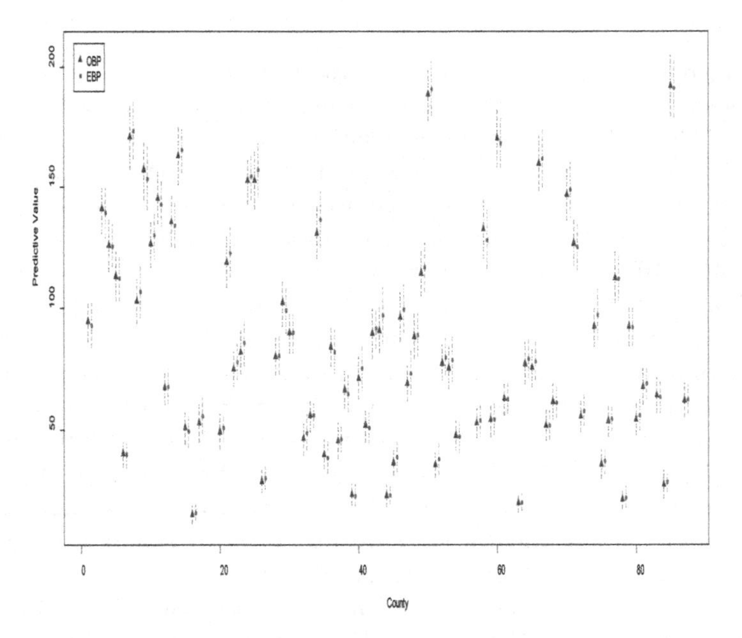

Fig. 5.7 *Analysis of Esophagus Cancer Data: OBP (blue triangle, on the left) and EBP (red square, on the right) for counties with $y_i \leq 200$; dash lines are margins of errors.*

5.10 Exercises

5.1. Show that, in Example 5.1, the ith component of the BP of ζ is given by (5.13), where β, A are the true parameters. The BPE of β is given by (5.12), and the MLE of β is given by (5.14).

5.2. This exercise has several parts.

a. Derive the expressions (5.17) and (5.18).

b. Show that, if $c \neq d$, then $g_1 \leq g_2$ and $h_1 \geq h_2$ with equality holding in both cases if and only if (5.19) holds.

c. Suppose that $A/(c-d)^2 \approx 0$, $A/b \approx 0$ and $b/a \approx 0$. Show that, in this case, the value of g_1/g_2 is approximately 0.5.

5.3. Verify the identity (5.20) and the expression (5.22).

5.4. Show that, under the finite-population setting described in Subsection 5.4.2, the BP of the small area mean, θ_i, has the expression (5.24).

5.5. Show that the design-based MSPE (5.26) can be expressed as $\text{MSPE}(\tilde{\theta}) = \text{E}\{Q(\psi) + \cdots\}$, where $\cdots$ does not depend on ψ, and $Q(\psi)$ is given by (5.26).

5.6. Show that (5.27) is a design-unbiased estimator of $\bar{Y}_i^2$, assuming $n_i > 1$.

5.7. Using Laplace approximation [e.g., Jiang (2007), §3.5.1] to derive (5.33).

5.8. Derive the BP formula (5.34) under the GM model.

5.9. Verify the expressions of BPE under the GM model, that is, (5.39) and (5.40).

5.10. Show that, in Example 5.1 (continued) in Section 5.6, the minimizer of $M(\psi)$ is given by (5.50).

5.11. Explore visually the esophagus cancer data of Table 5.9.

Chapter 6

Model Selection

Statistical models are key to potential gains that Statistics could make over a "simple-minded" approach. For example, a variance component model allows one to obtain a more efficient estimator of the fixed effects (regression coefficients) than the LS estimator; a SAE model makes it possible for one to "borrowing strengths", and thus do better than direct estimators such as the (area-level) sample mean. On the other hand, as is always said, "there is no free lunch", in the sense that such gains via statistical modeling is not risk-free for model failure, and there will be a consequence in the latter case. Therefore, the importance of careful model selection, in particular for mixed effects models, cannot be overstated.

In a way, robustness of a model-based method has to do with the model choice. For example, in some cases the linear model assumption is violated by some outliers but a higher-order model may fit the data well. In this case, it is possible that the linear model is not a good choice in the first place. For example, Recall the hospital data of §5.9.1. Ganesh (2009) proposed a Fay-Herriot model for the graft failure rates. The model is the same as that in Example 5.1 with $x_i'\beta = \beta_0 + \beta_1 x_i$. However, inspections of the raw data suggest one potential outlier (at the upper right corner) when the linear is fitted; see Figure 5.4. Jiang *et al.* (2009) used a model selection procedure, called the fence methods (see below), to identify the optimal model in this case, which led to a cubic model corresponding to the smooth curve in the figure. It is apparent that, with respect to the cubic model, there is no presence of outliers.

We shall discuss three main approaches in mixed model selection. The first is generalized information criteria; the second is the fence methods; the third is shrinkage mixed model selection. A special topic regarding nonparametric mixed model selection is deferred to the next chapter.

6.1 Generalized information criteria

The information criteria, first proposed by Akaike [Akaike's information criterion (AIC); Akaike (1973)], have had a profound impact in statistical model selection and related fields. See, for example, de Leeuw (1992), for a review. A number of similar criteria have since been proposed, including the Bayesian information criterion [BIC; Schwarz (1978)], a criterion due to Hannan and Quinn [HQ; Hannan and Quinn (1979)], and the generalized information criterion [GIC; Nishii (1984), Shibata (1984)].All of these criteria can be expressed as

$$\mathrm{GIC}(M) = -2\hat{l}(M) + \lambda_n|M|, \tag{6.1}$$

where $\hat{l}(M)$ is the maximized log-likelihood function under model M, λ_n is a penalty for complexity of the model, which may depend on the effective sample size, n, and $|M|$ is the dimension of M defined as the number of free parameters under M.

Although the information criteria are broadly used, difficulties are often encountered, especially in some non-conventional situations. One of the difficulties is that, in many cases, the distribution of the data is not fully specified (up to a number of unknown parameters); as a result, the likelihood function is not available. For example, suppose that normality is not assumed in a LMM for the random effects and errors, such as in the non-Gaussian LMM of Chapter 3, and that one wishes to select the fixed covariates using AIC, BIC, or HQ. It is not clear how to do this because the likelihood is unknown under the assumed model. Of course, one could blindly use those criteria, pretending that the data are normal, but the criteria are no longer what they mean to be. For example, Akaike's bias approximation that led to the AIC is no longer valid [e.g., Jiang and Nguyen (2015), sec. 1.1].

Nevertheless, it is of interest, from both theoretical and practical points of view, to know about the behavior of the information criteria when the assumed distribution does not hold. In fact, it is not uncommon, at all, that a practitioner uses an information criterion without making sure that the distributional assumption holds. Then what?

There are many aspects of an assumed model. Quite often in practice, only one, or some, of these aspects are of direct interest. For example, in the example mentioned above, one is only concerned about the fixed covariates that are involved in the model. So, even if the normality assumption does not hold, as long as the information criteria can correctly identify the model in terms of the fixed covariates, it is good enough for the problem of direct interest. Such an idea has been explored in the context of LMM selection.

Jiang and Rao (2003) proposed two procedures of GIC type for LMM selection. The authors considered a general LMM that can be expressed as (5.2) with the same assumptions underneath the equation. The problems of interest are regarding selection of the columns of X, which correspond to the fixed covariates, and the factors of random effects, which correspond to subvectors of α.

6.1.1 *Selecting the fixed covariates only*

The first procedure of Jiang and Rao (2003) is for the case of selecting the fixed covariates when the random-effect factors are not subject to selection. This problem is closely related to a regression model selection problem with correlated errors. Thus, consider the following general linear model:

$$y = X\beta + \zeta\,, \tag{6.2}$$

where ζ is a vector of correlated errors, and everything else is as in (5.2). It is assumes that there are a number of candidate vectors of covariates, $X_1, \dots, X_q$, from which the columns of X are to be selected. Let $K = \{1, \dots, q\}$. It is assumed that there exist a subset of $X_1, \dots, X_q$ such that, with X being the matrix corresponding to the subset, model (6.2) holds with all of the components of β nonzero. Such a subset is called optimal model. The set of all possible models can be expressed as $\mathcal{B} = \{M : M \subseteq K\}$, and there are 2^q possible models. Let $\mathcal{A}$ be a subset of $\mathcal{B}$ that is known to contain the optimal model, so the selection will be within $\mathcal{A}$. In an extreme case, $\mathcal{A}$ may be $\mathcal{B}$ itself. For any matrix A, let $\mathcal{L}(A)$ be the linear space spanned by the columns of A; P_A the projection onto $\mathcal{L}(A)$: $P_A = A(A'A)^- A'$, where B^- denotes the Moore-Penrose inverse; and $P_A^\perp$ the orthogonal projection: $P_A^\perp = I - P_A$. For any $M \in \mathcal{B}$, let $X(M)$ be the matrix whose columns are X_j, $j \in M$, if $M \neq \emptyset$; and $X(M) = 0$ if $M = \emptyset$. Consider the following model selection criterion:

$$\begin{aligned} C_N(M) &= |y - X(M)\hat{\beta}(M)|^2 + \lambda_N |M| \\ &= |P_{X(M)}^\perp y|^2 + \lambda_N |M|\,, \end{aligned} \tag{6.3}$$

$M \in \mathcal{A}$, where $|M|$ represents the cardinality of M; $\hat{\beta}(M)$ is the ordinary least squares (OLS) estimator of $\beta(M)$ in the model $y = X(M)\beta(M) + \zeta$, which can be expressed as

$$\hat{\beta}(M) = [X(M)'X(M)]^- X(M)'y,$$

and λ_N is a positive number satisfying certain conditions specified below. Note that $P_{X(M)}$ is understood as 0 if $M = \emptyset$. Denote the optimal model by M_0. If $M_0 \neq \emptyset$, denote the corresponding X and β by X and $\beta = (\beta_j)_{1 \le j \le p}$ $(p = |M_0|)$. Note that $\beta_j \neq 0$, $1 \le j \le p$ by the definition of the true model. (This is reasonable because, otherwise, the optimal model can be further simplified.) If $M_0 = \emptyset$, X, β and p are understood as 0. For $1 \le j \le q$, Let $\{j\}^c$ represent the set $K \setminus \{j\}$. Define the following sequences: $\omega_N = \min_{1 \le j \le q} |P^{\perp}_{X(\{j\}^c)} X_j|^2$, $\nu_N = \max_{1 \le j \le q} |X_j|^2$, and $\rho_N = \lambda_{\max}(ZGZ') + \lambda_{\max}(R)$, where $\lambda_{\max}$ denotes the largest eigenvalue. Let $\hat{M}$ be the minimizer of (6.4) over $M \in \mathcal{A}$, which is the selected model. The following theorem, proved in Jiang and Rao (2003), provides sufficient conditions under which the model selection is consistent in the sense that

$$P(\hat{M} \neq M_0) \longrightarrow 0 \,. \tag{6.4}$$

Theorem 6.1. Suppose that $\nu_N > 0$ for large N,

$$\rho_N/\nu_N \longrightarrow 0 \,, \quad \text{while} \quad \liminf(\omega_N/\nu_N) > 0. \tag{6.5}$$

Then, (6.4) holds for any λ_N such that

$$\lambda_N/\nu_N \longrightarrow 0 \ \text{ and } \ \rho_N/\lambda_N \longrightarrow 0. \tag{6.6}$$

Note 1. If the first part of (6.5) holds, there always exists λ_N that satisfies (6.6). For example, take $\lambda_N = \sqrt{\rho_N \nu_N}$.

Note 2. Typically, one has $\nu_N \sim N$. To see what the order of ρ_N may turn out to be, consider a special but important case of LMM. Suppose that $Z = (Z_1, \dots, Z_s)$, where each Z_r is a standard design matrix in the sense that it consists only of 0's and 1's, there is exactly one 1 in each row, and at least one 1 in each column.Likewise, $\alpha = (\alpha_1', \dots, \alpha_s')'$ such that $Z\alpha = Z_1\alpha_1 + \cdots + Z_s\alpha_s$ [see (1.1)], where α_r is a m_r-dimensional vector of uncorrelated random effects with mean 0 and variance σ_r^2. Furthermore, ϵ is a vector of uncorrelated errors with mean 0 and variance σ_0^2. Finally, $\alpha_1, \dots, \alpha_s, \epsilon$ are uncorrelated. Let n_{rk} be the number of 1's in the kth column of Z_r. Note that n_{rk} is the number of appearance of the kth component of α_r, and $Z_r'Z_r = \text{diag}(n_{rk}, 1 \le k \le m_r)$. Thus, we have

$$\lambda_{\max}(ZGZ') \le \sum_{r=1}^{s} \sigma_r^2 \lambda_{\max}(Z_r Z_r') = \sum_{r=1}^{s} \sigma_r^2 \max_{1 \le k \le m_r} n_{rk}.$$

Also, one has $\lambda_{\max}(R) = \sigma_0^2$. It follows that

$$\rho_N = O\left(\max_{1\le r\le s}\max_{1\le k\le m_r} n_{rk}\right).$$

Therefore, (6.6) is satisfied provided that $\lambda_N/N \to 0$ and

$$\max_{1\le r\le s}\max_{1\le k\le m_r}\left(\frac{n_{rk}}{\lambda_N}\right) \longrightarrow 0.$$

It is seen that, so far, only weak assumptions are made regarding the random effects and errors. Namely, neither the normality assumption, nor even the independence (or i.i.d.) assumptions, such as those under the non-Gaussian (see Chapter 3), are made. For example, in the special case noted above, it is only assumed that the components of the random effect and error vectors are uncorrelated, and the vectors of random effects and errors are uncorrelated. Because of the weak distributional assumption, the consistency result of Theorem 6.1 holds without the normality or independence assumptions. We illustrate with a specific example.

Example 6.1. Consider the following simple LMM:

$$y_{ij} = \beta_0 + \beta_1 x_{ij} + \alpha_i + \epsilon_{ij}\ , \tag{6.7}$$

$i = 1, \dots, m$, $j = 1, \dots, n$, where β_0, β_1 are unknown coefficients (the fixed effects). It is assume that the random effects $\alpha_1, \dots, \alpha_m$ are uncorrelated with mean 0 and variance σ^2. Furthermore, assume that the errors ϵ_{ij}'s have the following exchangeable correlation structure: Let $\epsilon_i = (\epsilon_{ij})_{1\le j\le n}$. Then, we have $\text{Cov}(\epsilon_i, \epsilon_{i'}) = 0$ if $i \ne i'$, and $\text{Var}(\epsilon_i) = \tau^2\{(1-\rho)I + \rho J\}$, where I is the identity matrix and J matrix of 1's. Also assume that the α's are uncorrelated with the ϵ's. Suppose that $m \to \infty$,

$$\liminf \left[\frac{1}{mn}\sum_{i=1}^{m}\sum_{j=1}^{n}(x_{ij} - \bar{x}_{\cdot\cdot})^2\right] > 0,$$

$$\text{and} \quad \limsup \left[\frac{1}{mn}\sum_{i=1}^{m}\sum_{j=1}^{n} x_{ij}^2\right] < \infty, \tag{6.8}$$

where $\bar{x}_{\cdot\cdot} = (mn)^{-1}\sum_{i=1}^m\sum_{j=1}^n x_{ij}$. Then, it is easy to show that the conditions of Theorem 6.1 are satisfied. In fact, in this case, $\rho_N \sim n$, while $\nu_N \sim \omega_N \sim mn$ (Exercise 6.1).

The above procedure requires selecting $\hat{M}$ from the entire $\mathcal{A}$. Note that $\mathcal{A}$ may contain up to 2^q subsets, if $\mathcal{A} = \mathcal{B}$. When q is relatively large, alternative procedures have been proposed in the (fixed effects) linear model context, which require less computation [e.g., Zheng and Loh (1995)]. Jiang and Rao (2003) consider an approach that is similar, in spirit, Rao and Wu (1989). First, note that one can always express $X\beta$ in (6.2) as

$$X\beta = \sum_{j=1}^{q} \beta_j X_j \tag{6.9}$$

with the understanding that some of the coefficients β_j may be zero. It follows that $M_0 = \{1 \leq j \leq q : \beta_j \neq 0\}$. Let $X_{-j} = (X_u)_{1\leq u\leq q, u\neq j}$, $1 \leq j \leq q$, $\eta_N = \min_{1\leq j\leq q} |P^{\perp}_{X_{-j}} X_j|^2$, and δ_N be a sequence of positive numbers satisfying conditions specified in Theorem 6.2 below. Let $\hat{M}$ be the subset of K such that $j \in \hat{M}$ iff

$$\frac{|P^{\perp}_{X_{-j}} y|^2 - |P^{\perp}_{X} y|^2}{|P^{\perp}_{X_{-j}} X_j|^2 \delta_N} > 1. \tag{6.10}$$

The following theorem, proved in Jiang and Rao (2003), states that, under suitable conditions, $\hat{M}$ is a consistent model selection. Recall that ρ_N is defined above Theorem 6.1.

Theorem 6.2. Suppose that $\eta_N > 0$ for large N, and

$$\rho_N/\eta_N \longrightarrow 0\ . \tag{6.11}$$

Then, (6.4) holds for any δ_N such that

$$\delta_N \longrightarrow 0 \ \text{ and } \ \rho_N/(\eta_N \delta_N) \longrightarrow 0. \tag{6.12}$$

Example 6.1 (continued). It is easy to show that, under exactly the same conditions [i.e., $m \to \infty$ and (6.8)], one has $\eta_N \sim mn$. Recall that $\rho_N \sim n$. Thus, the conditions of Theorem 6.2 are satisfied (Exercise 6.2).

6.1.2 *Selecting fixed covariates and random effect factors*

There are some major differences between selecting the fixed covariates, X_j, as considered in the previous subsection, and selecting the random effect factors. One difference is that, in selecting the latter case, one is going to determine whether the vector, α_r, as opposed to any component of α_r, should be included in the model. In other words, the components of α_r are either all "in" or all "out". Another difference is that, unlike selecting the fixed covariates, where it is reasonable assume that the X_j's are linearly independent, in a LMM, it is possible to have $r \neq r'$ but $\mathcal{L}(Z_r) \subset \mathcal{L}(Z_{r'})$. For example, see Example 6.2 below. Due to these features, the selection of random effect factors cannot be handled the same way as before.

In this section, we assume that $Z\alpha$ can be expressed as

$$Z\alpha = \sum_{r=1}^{s} Z_r \alpha_r, \tag{6.13}$$

where $Z_1, \dots, Z_s$ are known matrices; each α_r is a vector of independent random effects with mean 0 and variance σ_r^2, which is unknown, $1 \le r \le s$. Furthermore, we assume that ϵ is a vector of independent errors with mean 0 and variance $\tau^2 > 0$, and $\alpha_1, \dots, \alpha_s, \epsilon$ are independent. Such assumptions are customary in the mixed effects model context [e.g., Jiang (2007)]; therefore, (6.13) represents a fairly general class of LMMs. If $\sigma_r^2 > 0$, α_r is in the model; otherwise, it is not. It follows that selection of the random effect factors is equivalent to simultaneously determining which of the variance components $\sigma_1^2, \dots, \sigma_s^2$ are positive, and which of them are zero. The true model can be expressed as

$$y = X\beta + \sum_{r \in l_0} Z_r \alpha_r + \epsilon\ , \tag{6.14}$$

where $X = (X_j)_{j \in k_0}$ and $k_0 \subseteq K$ (see Subsection 6.1.1); $l_0 \subseteq L = \{1, \dots, s\}$ such that $\sigma_r^2 > 0$, $r \in l_0$, and $\sigma_r^2 = 0$, $r \in L \setminus l_0$.

It should noted that there is a hypothesis testing approach to selecting the random effect factors by testing the hypothesis that some of the σ_r^2 are zero (see the next subsection). However, if the null hypothesis is rejected, one knows that at least one of these variance components is non-zero, but one still does not know which one(s) are zero and which one(s) are non-zero. On the other hand, if the null hypothesis is accepted, one concludes that these variance components are zero, but it is not immediately clear what happens to other variance components not involved in the null hypothesis. Thus, the result of hypothesis testing may be inconclusive for model selection purposes. Furthermore, a hypothesis testing approach to selecting the random effect factors often rely on the normality assumption (but, see the next subsection), but no such an assumption is made here. Due to such considerations, Jiang and Rao (2003) took a different approach.

Recall that, in the previous subsection we had a procedure to select the fixed covariates of the model, which leads to $\hat{M}$ that satisfies (6.4). In fact, the only place that the determination of $\hat{M}$ might use knowledge about Z, hence l_0, is through λ_N, which depends on the order of $\lambda_{\max}(ZGZ')$. However, under (6.13), we have

$$\lambda_{\max}(ZGZ') \le \sum_{r=1}^{s} \sigma_r^2 \|Z_r\|^2,$$

where for any matrix A, $\|A\| = [\lambda_{\max}(A'A)]^{1/2}$. Thus, an upper bound for the order of $\lambda_{\max}(ZGZ')$ is $\max_{1\leq r\leq s}\|Z_r\|^2$, which does not depend on l_0. Therefore, $\hat{M}$ could be determined without knowing l_0. In any case, one may write $\hat{M} = \hat{M}(l_0)$, be it dependent on l_0 or not. Now, suppose that a selection for the random effect factors, that is, a determination of l_0, is $\hat{l}$. We then define $\hat{M} = \hat{M}(\hat{l})$. It is easy to establish the following theorem, which states that the combined procedure, which determines both the fixed covariates and the random effect factors, is consistent (Exercise 6.3).

Theorem 6.3. Suppose that $\mathrm{P}(\hat{l} \neq l_0) \to 0$ and $\mathrm{P}(\hat{M}(l_0) \neq M_0) \to 0$. Then, we have $\mathrm{P}(\hat{M} = M_0 \text{ and } \hat{l} = l_0) \to 1$.

Theorem 6.3 allows us to focus on how to select the random effect factors, because, once we have a consistent procedure of selecting the random effect factors, a procedure of jointly selecting the fixed covariates and random effect factors is readily in place.

We now describe how to obtain $\hat{l}$. First divide the vectors $\alpha_1, \dots, \alpha_s$, or, equivalently, the matrices $Z_1, \dots, Z_s$ into several groups. The first group is called "largest random factors". Roughly speaking, those are Z_r, $r \in L_1 \subseteq L$ such that $\text{rank}(Z_r)$ is of the same order as N, the sample size. We assume that $\mathcal{L}(X, Z_u, u \in L \setminus \{r\}) \neq \mathcal{L}(X, Z_u, u \in L)$, $r \in L_1$, where $\mathcal{L}(A_1, \dots, A_t)$ represents the linear space spanned by the columns of the matrices $A_1, \dots, A_t$. Such an assumption is reasonable because Z_r is supposed to be largest, hence must have contribution to the linear space. The second group consists of Z_r, $r \in L_2 \subseteq L$ such that $\mathcal{L}(X, Z_u, u \in L \setminus L_1 \setminus \{r\}) \neq \mathcal{L}(X, Z_u, u \in L \setminus L_1)$, $r \in L_2$. The ranks of the matrices in this group are of lower order of N. Similarly, the third group consists of Z_r, $r \in L_3 \subseteq L$ such that $\mathcal{L}(X, Z_u, u \in L \setminus L_1 \setminus L_2 \setminus \{r\}) \neq \mathcal{L}(X, Z_u, u \in L \setminus L_1 \setminus L_2)$, and so on. Note that if the first group, i.e., the largest random factors, does not exist, the second group becomes the first, and the other groups also move on. Intuitively, a selection procedure will not work if there is linear dependence among the candidate design matrices, because of an identifiability problem. To consider a rather extreme example, suppose that Z_1 is a design matrix consist of 0's and 1's such that there is exactly one 1 in each row, and $Z_2 = 2Z_1$. Then, to have $Z_1\alpha_1$ in the model means that there is a term α_{1i}; to have $Z_2\alpha_2 = 2Z_1\alpha_2$ in the model means that there is a corresponding term, $2\alpha_{2i}$. However, it makes no difference in terms of a LMM, because both α_{1i} and α_{2i} are random effects with mean 0 and certain variances. However, by grouping the random effect factors we have divided the Z_r's into several groups such that there is linear independence within each group. This is the motivation behind the grouping strategy described above. To illustrate the procedure and also to show that such a division of groups does exist in typical situations, consider the following.

Example 6.2. Consider the following random effects model:

$$y_{ijkl} = \mu + a_i + b_j + c_k + d_{ij} + f_{ik} + g_{jk} + h_{ijk} + e_{ijkl} , \qquad (6.15)$$

$i = 1, \dots, m_1$, $j = 1, \dots, m_2$, $k = 1, \dots, m_3$, $l = 1, \dots, n$, where μ is an unknown mean; a, b, c are random main effects; d, f, g, h are (random) two- and three-way interactions; and e is error. The model can be written as

$$y = X\mu + Z_1 a + Z_2 b + Z_3 c + Z_4 d + Z_5 f + Z_6 g + Z_7 h + e ,$$

where $X = 1_N$ with $N = m_1 m_2 m_3 n$, $Z_1 = I_{m_1} \otimes 1_{m_2} \otimes 1_{m_3} \otimes 1_n, \dots, Z_4 = I_{m_1} \otimes I_{m_2} \otimes 1_{m_3} \otimes 1_n, \dots$, and $Z_7 = I_{m_1} \otimes I_{m_2} \otimes I_{m_3} \otimes 1_n$. Here I_r and 1_r represent the r-dimensional identity matrix and vector of 1's, and $\otimes$ means Kronecker product. It is easy to see that the Z_r's are not linearly independent. For example, $\mathcal{L}(Z_r) \subset \mathcal{L}(Z_4)$, $r = 1, 2$, and $\mathcal{L}(Z_r) \subset \mathcal{L}(Z_7)$, $r = 1, \dots, 6$. Also, $\mathcal{L}(X) \subset \mathcal{L}(Z_r)$ for any r.

Suppose that $m_r \to \infty$, $r = 1, 2, 3$, while n is bounded. Then, the first group consists of Z_7; the second group Z_4, Z_5, Z_6; and the third group Z_1, Z_2, Z_3. If n also $\to \infty$, the largest random factor does not exist. However, one still has these three groups. It is easy to see that the Z_r's within each group are linearly independent (Exercise 6.4).

In general, Suppose that the Z_r's are divided into h groups such that $L = L_1 \cup \cdots \cup L_h$. We describe a procedure that determines the indexes $r \in L_1$ for which $\sigma_r^2 > 0$; then a procedure that determines the indexes $r \in L_2$ for which $\sigma_r^2 > 0$; and so on.

Group one. Consider the first group. Write $B = \mathcal{L}(X, Z_1, \dots, Z_s)$, $B_{-r} = \mathcal{L}(X, Z_u, u \in L \setminus \{r\})$, $r \in L_1$; $d = N - \text{rank}(B)$, $d_r = \text{rank}(B) - \text{rank}(B_{-r})$; $D = |P_B^{\perp} y|^2$, $D_r = |(P_B - P_{B_{-r}})y|^2$. If A is a matrix, define $\|A\|_2 = [\text{tr}(A'A)]^{1/2}$. For any $1 < \rho < 2$, let $\hat{l}_1$ be the set of indexes $r \in L_1$ such that

$$(d/D)(D_r/d_r) > 1 + d^{(\rho/2)-1} + d_r^{(\rho/2)-1}. \qquad (6.16)$$

Let $l_{01} = \{r \in L_1 : \sigma_r^2 > 0\}$.

Lemma 6.1. Suppose that $d \to \infty$; and $d_r \to \infty$, $\liminf(\|P_{B_{-r}}^{\perp} Z_r\|_2^2 / d_r) > 0$, $\|P_{B_{-r}}^{\perp} Z_r\|_2^2 / d_r^2 \to 0$, and $\|Z_r' P_{B_{-r}}^{\perp} Z_r\|_2^2 / d_r^2 \to 0$, $r \in L_1$. Then, we have $\text{P}(\hat{l}_1 = l_{01}) \to 1$.

Example 6.2 (continued). Suppose that $m_t \to \infty$, $t = 1, 2, 3$, while n is bounded but $n \geq 2$. Then, group one corresponds to a single index

$r = 7$. Furthermore, it is easy to see that $d = m_1m_2m_3(n-1)$, $d_7 \geq g = m_1m_2m_3 - m_1m_2 - m_2m_3 - m_3m_1$, $ng \leq \|P_{B_{-7}}^{\perp} Z_7\|_2^2 \leq N = m_1m_2m_3n$, and $\|Z_7' P_{B_{-7}}^{\perp} Z_7\|_2^2 \leq nN$. It follows that all of the conditions of Lemma 6.1 are satisfied.

Group two. Now consider the second group. Let $B_1 = \mathcal{L}(X, Z_u, u \in L \setminus L_1)$, $B_2 = \mathcal{L}(X, Z_u, u \in L \setminus L_1 \setminus L_2)$, $B_{1,-r} = \mathcal{L}(X, Z_u, u \in L \setminus L_1 \setminus \{r\})$, $r \in L_2$, and $B_1(l_2) = \mathcal{L}(X, Z_u, u \in (L \setminus L_1 \setminus L_2) \cup l_2)$, $l_2 \subseteq L_2$. Consider

$$C_{1,N}(l_2) = |P_{B_1(l_2)}^{\perp} y|^2 + \lambda_{1,N}|l_2|, \tag{6.17}$$

$l_2 \subseteq L_2$, where $\lambda_{1,N}$ is a positive number satisfying conditions specified below. Let $\hat{l}_2$ be the minimizer of $C_{1,N}$ over $l_2 \subseteq L_2$, and $l_{02} = \{r \in L_2 : \sigma_r^2 > 0\}$. Let $L_0 = \{0\}$, and $Z_0 = I$, the identity matrix; $\rho_{1,N} = \max_{r \in L_0 \cup L_1} \|P_{B_1} Z_r\|_2^2$, $\nu_{1,N} = \min_{r \in L_2} \|P_{B_{1,-r}}^{\perp} Z_r\|_2^2$, and $\gamma_{1,N} = \max_{r \in L_2}(\|P_{B_2}^{\perp} Z_r\|^2 / \|P_{B_{1,-r}}^{\perp} Z_r\|_2^2)$.

Lemma 6.2. Suppose that $\nu_{1,N} > 0$ for large N,

$$\rho_{1,N}/\nu_{1,N} \longrightarrow 0, \quad \text{and} \quad \gamma_{1,N} \longrightarrow 0. \tag{6.18}$$

Then, $\mathrm{P}(\hat{l}_2 \neq l_{02}) \to 0$ for any $\lambda_{1,N}$ such that

$$\lambda_{1,N}/\nu_{1,N} \longrightarrow 0 \;\text{ and }\; \rho_{1,N}/\lambda_{1,N} \longrightarrow 0. \tag{6.19}$$

Example 6.2 (continued). Group two corresponds to three indexes: $r = 4, 5, 6$. It is easy to see that $B_1 = \mathcal{L}(Z_4, Z_5, Z_6)$, $B_{1,-4} = \mathcal{L}(Z_5, Z_6)$, etc. Thus, $\|P_{B_1}\|_2^2 = \mathrm{rank}(B_1) \leq f = m_1m_2 + m_2m_3 + m_3m_1$, $\|P_{B_1} Z_7\|_2^2 \leq nf$, and $\|P_{B_2}^{\perp} Z_4\|^2 \leq \|Z_4\|^2 = m_3 n$, etc. Finally, it is easy to verify that $P_{Z_5}P_{Z_6} = P_{Z_6}P_{Z_5}$. Thus, by an inequality in Exercise 6.5, we have $P_{B_{1,-4}} = P_{(Z_5\ Z_6)} \leq P_{Z_5} + P_{Z_6}$. It follows that

$$\begin{aligned} \mathrm{tr}(Z_4' P_{B_{1,-4}} Z_4) &\leq \mathrm{tr}(Z_4' P_{Z_5} Z_4) + \mathrm{tr}(Z_4' P_{Z_6} Z_4) \\ &= n(m_2m_3 + m_3m_1); \end{aligned}$$

hence, we have

$$\begin{aligned} \|P_{B_{1,-4}}^{\perp} Z_4\|_2^2 &= \mathrm{tr}(Z_4' Z_4) - \mathrm{tr}(Z_4' P_{B_{1,-4}} Z_4) \\ &\geq n(m_1m_2m_3 - m_2m_3 - m_3m_1), \end{aligned}$$

etc. Thus, all of the conditions of Lemma 6.2 are satisfied.

General. The above procedure can be extended to the remaining groups. In general, let $B_t = \mathcal{L}(X, Z_u, u \in L \setminus L_1 \setminus \cdots \setminus L_t)$, $1 \leq t \leq h$; $B_{t,-r} = \mathcal{L}(X, Z_u, u \in (L \setminus L_1 \setminus \cdots \setminus L_t \setminus \{r\})$, $r \in L_{t+1}$, and $B_t(l_{t+1}) = \mathcal{L}(X, Z_u, u \in (L \setminus L_1 \setminus \cdots \setminus L_{t+1}) \cup l_{t+1})$, $l_{t+1} \subseteq L_{t+1}$, $1 \leq t \leq h-1$. Define

$$C_{t,N}(l_{t+1}) = |P_{B_t(l_{t+1})}^{\perp} y|^2 + \lambda_{t,N}|l_{t+1}|, \quad l_{t+1} \subseteq L_{t+1}; \tag{6.20}$$

where $\lambda_{t,N}$ is a positive number satisfying the conditions specified below. Let $\hat{l}_{t+1}$ be the minimizer of $C_{t,N}$ over $l_{t+1} \subseteq L_{t+1}$, and $l_{0\underline{t+1}} = \{r \in L_{t+1} : \sigma_r^2 > 0\}$. By exactly the same proof as that of Lemma 6.2 [see Jiang and Rao (2003)], one can establish the consistency of $\hat{l}_{t+1}$, $2 \leq t \leq h-1$. Let $\rho_{t,N} = \max_{r \in L_0 \cup \cdots \cup L_t} \|P_{B_t} Z_r\|_2^2$, $\nu_{t,N} = \min_{r \in L_{t+1}} \|P_{B_{t,-r}}^{\perp} Z_r\|_2^2$, and $\gamma_{t,N} = \max_{r \in L_{t+1}} (\|P_{B_{t+1}}^{\perp} Z_r\|^2 / \|P_{B_{t,-r}}^{\perp} Z_r\|_2^2)$. Then, we have the following theorem for determining the combination of $l_{01}, \ldots, l_{0h}$.

Theorem 6.4. Suppose that the conditions of Lemma 6.1 are satisfied,

$$\rho_{t,N}/\nu_{t,N} \longrightarrow 0, \text{ and } \gamma_{t,N} \longrightarrow 0, \quad 1 \leq t \leq h-1. \tag{6.21}$$

Then, for any $\lambda_{t,N}$ such that

$$\lambda_{t,N}/\nu_{t,N} \longrightarrow 0 \text{ and } \rho_{t,N}/\lambda_{t,N} \longrightarrow 0, \ 1 \leq t \leq h-1, \tag{6.22}$$

we have $\mathrm{P}(\hat{l}_1 = l_{01}, \ldots, \hat{l}_h = l_{0h}) \to 1$.

Note. Unlike $\hat{M}$ in Subsection 6.1.1 (see discussion above Theorem 6.3), here $\hat{l}_t$ does not depend on $\hat{l}_{t'}$, $t' < t$. In fact, $\hat{l}_1, \ldots, \hat{l}_h$ can be obtained simultaneously. Let $\hat{l} = \cup_{t=1}^h \hat{l}_t$. Then, $\hat{l}$ satisfies the requirement of Theorem 6.3.

Example 6.2 (continued). Consider $t = h = 3$, which is the final group. It is easy to see that $B_2 = \mathcal{L}(Z_1, Z_2, Z_3)$, $B_{2,-1} = \mathcal{L}(Z_2, Z_3)$, etc.; $\|P_{B_2}\|_2^2 = \mathrm{rank}(B_2) \leq h = m_1 + m_2 + m_3$, and $\|P_{B_2} Z_7\|_2^2 \leq nh$. Furthermore, by Exercise 6.5, it can be shown that $\|P_{B_2} Z_4\|_2^2 \leq (m_1+m_2+1)m_3 n$; $\mathrm{tr}(Z_1' P_{B_{2,-1}} Z_1) \leq 2m_2 m_3 n$, hence $\|P_{B_{2,-1}}^{\perp} Z_1\|_2^2 \geq (m_1 - 2)m_2 m_3 n$, etc. Finally, we have $\|P_{B_3}^{\perp} Z_1\|^2 \leq \|Z_1\|^2 = m_2 m_3 n$, etc. Thus, all of the conditions of Theorem 6.4 are satisfied.

Jiang and Rao (2003) showed via simulation studies that the finite-sample performance of the proposed model selection procedures agrees with what the theory predicts in terms of consistency.

6.1.3 *A robust approximation to BIC*

The BIC [Schwarz (1978)] is a special case of the GIC, (6.1), with $\lambda_n = \log(n)$, where n is supposed to be the effective sample size. In many applications, one is dealing with a finite population. So, an information criterion, such as the BIC, derived under the assumption of an infinite population, such as normality, would not be valid.

Fabrizi and Lahiri (2013) proposed a design-based approximation to the BIC under finite population sampling. Here, the word "design-based" refers to the fact that one is considering sampling from the finite population, which is where the randomness comes from. For example, the randomness of y_{ij} in (3.28) is not due to that v_i and e_{ij} are random variables from normal distributions, but rather that y_{ij} is randomly sampled, without replacement, from the ith finite subpopulation. Before we discuss the design-based approximation, let us first review a connection between the Bayes factor (BF) and BIC.

Let $y_{\rm s} = (y_i)_{1\leq i\leq n}$ denote a vector of i.i.d. samples, whose distribution belong to a family of probability distributions parameterized by $\psi = (\phi', \theta')'$ with $\dim(\psi) = m$ and $\dim(\phi) = m_0$. Consider testing the hypothesis:

$$M_0 : \theta = \theta_0 \quad \text{versus} \quad M_a : \theta \in R^{m-m_0}. \tag{6.23}$$

The BF is defined as the ratio of the *a posterior* and *a prior* odds in favor of a larger model, M:

$$BF = \frac{\mathrm{P}(M|y_{\rm s})}{\mathrm{P}(M_0|y_{\rm s})}\left\{\frac{\mathrm{P}(M)}{\mathrm{P}(M_0)}\right\}^{-1} = \frac{\int p(y_{\rm s}|\phi,\theta)\pi(\phi,\theta)d\phi d\theta}{\int p(y_{\rm s}|\phi,\theta_0)\pi_0(\phi)d\phi}, \tag{6.24}$$

where $\pi(\phi, \theta)$ and $\pi_0(\phi)$ are the joint prior for ϕ and θ and marginal prior for ϕ, respectively. The BF has been used in hypothesis testing and model selection. However, the calculation of the BF requires a full specification of the prior distributions under both M_0 and M. Alternatively, one may use a suitable approximation to the logarithm of the BF, and one popular approximation is the BIC [Schwarz (1978)], given by

$$S = \lambda - \frac{m - m_0}{2}\log(n), \tag{6.25}$$

where $\lambda = l(\hat{\phi}, \hat{\theta}) - l_0(\hat{\phi}_0, \theta_0)$ is the logarithm of the likelihood ratio; namely, $l(\hat{\phi}, \hat{\theta})$ is the log-likelihood evaluated at the MLE, $\hat{\psi} = (\hat{\phi}', \hat{\theta}')'$ of ψ, and $l_0(\hat{\phi}_0, \theta_0)$ is the log-likelihood evaluated at the MLE of ϕ under M_0, $\hat{\phi}_0$. Approximation (6.25) is based on Laplace approximation [e.g., Jiang (2007), sec. 3.5.1]. For a suitable choice of the prior, it can be shown [e.g., Kass and Wassermann (1995)] that

$$S = \log(BF) + O_{\rm P}(n^{-1/2}). \tag{6.26}$$

Although the derivation of S is from a hypothesis testing point of view, it can be viewed as, at least, a simple problem of model selection, that is, the choice between two models, M and M_0. In the regard, the BIC criterion, which is equivalent to minimizing $-S$, is consistent in the sense that S goes to ∞ ($-\infty$) if M (M_0) is true.

On the other hand, the BIC criterion in the above is derived under the infinite-population assumption. Fabrizi and Lahiri (2013) proposed two approaches to adapt the BIC to the finite-population setting. The first approach is an estimator of the finite-population BIC. If all of the units of the finite population were observed, the authors showed that the BIC based on all of the observations in the population is given by

$$S_{\text{pop}}(y_{\text{U}}) = \frac{N}{2}\bar{y}_{\text{U}}^2 - \frac{1}{2}\log(N), \tag{6.27}$$

where N is the population size, and $y_{\text{U}} = (y_1, \dots, y_N)$ is the population. (6.27) is called the finite population BIC Of course, one cannot compute $S_{\text{pop}}(y_{\text{U}})$ because $\bar{y}_{\text{U}}$ is unknown. Let $\widehat{\bar{y}_{\text{U}}}$ be a design-consistent estimator of $\bar{y}_{\text{U}}$. This means that, as the sample size, n, goes to ∞, $\widehat{\bar{y}_{\text{U}}}$ converges to $\bar{y}_{\text{U}}$ in probability that is induced by the sampling design. It should be noted that, here, the sampling is without replacement.Then, because $n \leq N$, N must also go to ∞ as $n \to \infty$. Replacing $\bar{y}_{\text{U}}$ in (6.27) by $\widehat{\bar{y}_{\text{U}}}$, one obtains a naive model selection criterion:

$$S_{\text{plugin}}(y_s) = \frac{N}{2}(\widehat{\bar{y}_{\text{U}}})^2 - \frac{1}{2}\log(N), \tag{6.28}$$

where s represents the sampled indexes and $y_s = (y_i)_{i \in s}$. (6.28) is called naive because it may not work even under the simple random sampling with replacement. To see this, note that, if n is large, one would expect that (6.28) to be very close to the BIC obtained under i.i.d. sampling from a normal population,

$$S_{\text{iid}}(y_s) = \frac{n}{2}\bar{y}_s^2 - \frac{1}{2}\log(n), \tag{6.29}$$

where $\bar{y}_s$ is the sample mean. It is known that $\bar{y}_s$ is a design-consistent estimator of $\bar{y}_{\text{U}}$. However, if we let $\widehat{\bar{y}_{\text{U}}} = \bar{y}_s$ in (6.28), we have

$$S_{\text{plugin}}(y_s) - S_{\text{iid}}(y_s) = \frac{N-n}{2}\bar{y}_s^2 - \frac{1}{2}\log\left(\frac{N}{n}\right). \tag{6.30}$$

Now suppose that $n = \rho N$, where $\rho \in (0, 1)$. Then, the right side of (6.30) goes to ∞ as N (and n) goes to ∞. This implies that (6.28) provides stronger evidence against M_0 than (6.29), which is known to be a consistent model selection criterion. The failure of (6.28) is due to the fact that (6.28) approximates (6.27) if all of the units in the finite population were observed, which corresponds to N, not n, in the expression. This makes the disagreement between the data and the null hypothesis look more than it actually is.

Because, here, we are dealing with a finite population, it is more reasonable to use the BIC based on the exact likelihood for the sample, or sample likelihood. The latter can be derived using a super-population model (see

Subsection 5.4.2) for the finite population and the underlying sampling design. However, survey populations usually have complex structures; as a result, misspecification of the assumed model is quite likely [e.g., Kott (1991)]. The following simple example shows that, when a model misspecification takes place, there could be serious consequence to model selection.

Example 6.3. Consider the one-way random effects model of Example 3.1. Suppose that the super-population satisfies the one-way random effects model. For simplicity, suppose that $\mathrm{var}(y_{ij}) = \sigma^2 + \tau^2 = 1$; thus, we have $\tau^2 = 1 - \sigma^2$. Suppose that the finite population has size N, and is divided into M clusters of size N_c. A sample of m clusters is selected by simple random sampling (without replacement), and all of the units in the selected clusters are selected. Thus, the total sample size is $n = mN_c$. It can be shown that the MLE of μ, based on the sample likelihood, is $\bar{y}_s$, the sample mean. Furthermore, the BIC based on the sample likelihood is given by

$$S_{\mathrm{s}}(y_s) = \frac{n}{2} \cdot \frac{\bar{y}_s^2}{1 + (N_c - 1)\sigma^2} - \frac{1}{2}\log(n). \tag{6.31}$$

It is clear that, when $N_c = 1$, (6.31) reduces to (6.30). In fact, the latter is the appropriate BIC when there is no clustering, that is, the cluster size $N_c = 1$; or when there is no cluster effect, that is, $\sigma^2 = 0$. However, when $N_c > 1$ and $\sigma^2 > 0$, the difference between the two BICs is

$$S_{\mathrm{iid}}(y_s) - S_{\mathrm{s}}(y_s) = \frac{n}{2} \cdot \frac{(N_c - 1)\sigma^2}{1 + (N_c - 1)\sigma^2}\bar{y}_s^2,$$

which goes to ∞ when $m \to \infty$ (hence $N \to \infty$ while $N_c > 1$. This means that, if one neglects the clustering in the population, resulting a model misspecification, one is likely to reject the null hypothesis more often than what should be done correctly.

In order to make the BIC more robust to model misspecification, Fabrizi and Lahiri (2013) proposed a robust design-based approximation to the BIC. The procedure was presented in a simple case described below.

Let y_U be a realization from an underlying super-population distribution, ξ, characterized by a parameter, θ. We are interested in testing

$$M_0: \quad \theta = \theta_0 \quad \text{versus} \quad M_a: \quad \theta \neq \theta_0. \tag{6.32}$$

In this special case, by (6.25), the BIC is given by $S = \lambda - (1/2)\log(n)$, where $\lambda = l(\hat{\theta}_{\mathrm{ML}}) - l(\theta_0)$ is the logarithm of the likelihood ratio. The parameter θ is estimated by solving the likelihood equation,

$$f(y_U, \theta) = 0. \tag{6.33}$$

If y_U were observed, the solution to (6.33), denoted by $\hat{\theta}_{\mathrm{ML}} = T(y_U)$, would be an estimator of θ, known as the corresponding descriptive population quantity (CDPQ) of θ. Because y_U are not entirely observed, we estimate $T(y_U)$ by a design-based estimator, $\hat{\theta} = \hat{T}(z_s)$. For example, $\hat{\theta}$ may be obtained using a quasi-likelihood approach as discussed in Section 3.2. We then consider the following approximation to S:

$$S_{\mathrm{DB}} = \frac{1}{2}W_{\mathrm{DB}} - \frac{1}{2}\log(n), \tag{6.34}$$

where $W_{\mathrm{DB}} = (\hat{\theta} - \theta_0)^2/\hat{V}_{\mathrm{D}}(\hat{\theta})$ and $\hat{V}_{\mathrm{D}}(\hat{\theta})$ is a consistent estimator of $V_{\mathrm{D}}(\hat{\theta})$, the variance of $\hat{\theta}$ under the randomization distribution. Fabrizi and Lahiri (2013) noted that the n in (6.34) is supposed to be the "effective" sample size, so some adjustment may be needed. But, typically, such an adjustment leads to a difference that is of lower order than the leading term in (6.34). So, asymptotically, (6.34) still provides the right approximation. For example, if the effective sample size is $n^* = Cn$, where C is any constant, then $\log(n^*) = \log(n) + \log(C)$, where $\log(C)$ is another constant. To further justify the approximation, we have the following theorem. The proof can be found in Fabrizi and Lahiri (2013).

Theorem 6.5. Suppose that the following regularity conditions hold:
i) $\hat{\theta}_{\mathrm{ML}} - \theta_0 = O_\xi(n^{-1/2})$ under model M_a, where O_ξ means O_{P} (e.g., Jiang 2010, sec. 3.4) with respect to the super-population distribution ξ;
ii) $l(\cdot)$ is twice differentiable with $-l''(\hat{\theta}_{\mathrm{ML}}) = I(\theta_0) + O_\xi(n^{-1/2})$, where

$$I(\theta_0) = -\mathrm{E}\left(\left.\frac{\partial^2 l}{\partial \theta^2}\right|_{\theta_0} \right)$$

is the Fisher information evaluated at θ_0;
iii) $\hat{\theta} = \hat{\theta}_{\mathrm{ML}} + o_{\xi D}(n^{-1/2})$, where $o_{\xi D}$ denotes o_{P} (e.g., Jiang 2010, sec. 3.4) with respect to the combined model/randomization distribution;
i) $\hat{V}_D(\hat{\theta}) = I^{-1}(\theta_0) + o_{\xi D}(n^{-1})$.
Then, we have $S - S_{\mathrm{DB}} = o_{\xi D}(n^{-1/2})$.

Example 6.4. Fabrizi and Lahiri (2013) carried out a simulation study under a setting in which $K = 200$ clusters, each of size $N_c = 10$, were generated under the following model:

$$y_{ij} \overset{\mathrm{ind}}{\sim} \mathrm{Bernoulli}(\pi_i),$$
$$\pi_i \overset{\mathrm{ind}}{\sim} \mathrm{Beta}\left(\frac{\mu}{\gamma}, \frac{1-\mu}{\gamma}\right),$$

$i = 1, \ldots, M$, $j = 1, \ldots, N_c$. The above model implies that the marginal proportion of the observation is μ, and the intra-cluster correlation is

$\gamma/(1+\gamma)$ (Exercise 6.6). Simple random sampling (without replacement) of 3 clusters (case I) or 6 clusters (case II) is used in selecting the clusters; once the cluster is chosen, all of the units in the selected clusters are sampled. The authors compared three BICs in testing the hypothesis $H_0 : \mu = 0.25$ versus $H_1 : \mu \neq 0.25$. These are the exact BIC, S_E, based on the exact sample likelihood; a naïve BIC, S_N, that ignores the sampling design; and the design-based BIC, S_{DB}, introduced above. In addition, the authors considered two different values of γ: $\gamma = 0.3$ and $\gamma = 1.0$. These values correspond to intra-cluster correlation coefficients of 0.25 and 0.5, respectively. The results, based on 1000 simulated runs, showed that the performance of S_{DB} is much closer to that of S_E, which is considered the gold standard, than that of S_N in terms of both the size and powers of the test. Here, the powers were considered under the alternatives $\mu = 0.5$, 0.6, 0.75 and 0.9, respectively. See Fabrizi and Lahiri (2013) for the details.

6.1.4 *A robust conditional AIC for LMM*

Consider a hybrid of the non-Gaussian mixed ANOVA model, introduced in Subsection 3.1.1, and longitudinal model, introduced in Subsection 3.1.2. The model can be expressed as (3.1) with

$$Z_i\alpha_i = \sum_{r=1}^{s} Z_{ir}\alpha_{ir}, \tag{6.35}$$

where Z_{ir} is an $n_i \times q_r$ known design matrix, with $n_i = \dim(y_i)$, and α_{ir} is a $q_r \times 1$ vector of random effects, $1 \leq r \leq s$. It is assumed that $\alpha_{ir}, 1 \leq i \leq m, 1 \leq r \leq s$ are independent with $\mathrm{E}(\alpha_{ir}) = 0$ and $\mathrm{Var}(\alpha_{ir}) = \sigma_r^2 I_{q_r}$, $1 \leq r \leq s$. Furthermore, it is assumed that $\epsilon_1, \dots, \epsilon_m$ are independent with $\mathrm{E}(\epsilon_i) = 0$, $\mathrm{Var}(\epsilon_i) = \tau^2 I_{n_i}$, $1 \leq i \leq m$, and are independent with the α's. Let $\psi = (\tau^2, \sigma_1^2, \dots, \sigma_s^2)'$.

Under the normality assumption, let $f(y|\beta, \psi)$ denote the marginal likelihood function. Then, the marginal AIC is given by

$$\mathrm{mAIC} = -2\log\{f(y|\hat{\beta}, \hat{\psi})\} + 2(p+s+1), \tag{6.36}$$

where $\hat{\beta}, \hat{\psi}$ are the MLEs of β, ψ, respectively, and $p = \dim(\beta)$. Vaida and Blanchard (2005) noted that the marginal AIC is inappropriate when selection of the random effects is of interest. They proposed a conditional AIC in the form of

$$\mathrm{cAIC} = -2\log\{f_{y|\alpha}(y|\hat{\beta}, \hat{\alpha}, \hat{\psi})\} + 2\{\mathrm{tr}(H) + 1\}, \tag{6.37}$$

where $\hat{\beta}, \hat{\psi}$ are the REML estimators of β, ψ, respectively, and $\hat{\alpha}$ is the EBLUP based on the REML estimators. Furthermore, H corresponds to a matrix that maps the observed vector, y, into the fitted vector $\hat{y} = X\hat{\beta} + Z\hat{\alpha}$, that is, $\hat{y} = Hy$. It should be noted that (6.37) is approximate version of the cAIC proposed by Vaida and Blanchard (2005); the authors also derived an exact version, which is an unbiased estimator of a conditional Akaike information, defined by the authors, as an extension of the original Akaike information [Akaike (1973)]. Consider, for example, a linear mixed model that can be expressed as (3.1). First assume that $G_i = \text{Var}(\alpha_i)$ is known, and $R_i = \text{Var}(\epsilon_i) = \tau^2 I_{n_i}$, where τ^2 is an unknown variance. If the REML method is used to estimate the model parameters, the exact version of cAIC is given by (6.37) with $\text{tr}(H) + 1$ replaced by

$$K_{\text{REML}} = \frac{(N-p-1)(\rho+1)+p+1}{N-p-2}, \tag{6.38}$$

where N is the total sample size, $p = \text{rank}(X)$ with $X = (X_i)_{1 \le i \le m}$, and $\rho = \text{tr}(H)$. If ML method is used to estimate the parameters, the expression of the exact cAIC is the same except replacing $K_{\text{ML}} = \{N/(N-p)\}K_{\text{REML}}$.

Note (6.37) is in the general form of (6.1), where the second term corresponds to a penalty for model complexity. To see the difference between cAIC and AIC [Akaike (1973)] in terms of the penalty, note that the corresponding terms to K_{REML} and K_{ML} in AIC are the same, which is $p + 1$, assuming, again, that G_i is known and $R_i = \tau^2 I_{n_i}$. There is also a finite-sample version of AIC [Vaida and Blanchard (2005)], in which case the penalty terms are given by $(N-p)(N-p-2)^{-1}(p+1)$ for REML, and $N(N-p-2)^{-1}(p+1)$ for ML.

When the covariance matrices of the random effects involve additional unknown parameters, the form of cAIC becomes more complicated. We use an example to illustrate.

Example 6.5. Consider a special case of Example 3.1 with $\mu = 0$, $a = m$, and $b_i = n, 1 \le i \le m$. In this case, the cAIC is given by (6.37) with $\text{tr}(H) + 1$ replaced by an estimator of a bias-correction term, BC. Furthermore, BC which has an asymptotic expansion:

$$\text{BC} = \rho + \frac{2}{1+n\gamma} + o(n^{-1}),$$

where $\gamma = \sigma^2/\tau^2$. Thus, the estimator of BC is given by $\widehat{\text{BC}} = \hat{\rho} + 2/(1+n\hat{\gamma})$, where $\hat{\rho}, \hat{\gamma}$ are the ML estimator of ρ, γ, respectively.

6.2 The fence methods

Although the information criteria are broadly used, difficulties are often encountered, especially in some non-conventional situations. We discuss a number of such cases below.

1. The effective sample size. In many cases, the effective sample size, n, is not the same as the number of data points. This often happens when the data are correlated. Take a look at two extreme cases. In the first case, the observations are independent; therefore, the effective sample size should be the same as the number of observations. In the second case, the data are so much correlated that all of the data points are identical. In this case, the effective sample size is 1, regardless of the number of data points. A practical situation may be somewhere between these two extreme cases, such as cases of mixed effects models, which makes the effective sample size difficult to determine.

2. The dimension of a model. Not only the effective sample, the dimension of a model, $|M|$, can also cause difficulties. In some cases, such as the ordinary linear regression, this is simply the number of parameters under M, but in other situations where nonlinear, adaptive models are fitted, this can be substantially different. Ye (1998) developed the concept of generalized degrees of freedom (gdf) to track model complexity. For example, in the case of multivariate adaptive regression splines [Friedman (1991)], k nonlinear terms can have an effect of approximately $3k$ degrees of freedom. While a general algorithm in its essence, the gdf approach requires significant computations. It is not at all clear how a plug-in of gdf for $|M|$ in (6.1) affects the selection performance of the criterion.

3. Unknown distribution. In many cases, the distribution of the data is not fully specified (up to a number of unknown parameters); as a result, the likelihood function is not available. For example, suppose that normality is not assumed under a linear mixed model (e.g., Chapter 3). Then, the likelihood function is typically not available. Now suppose that one wishes to select the fixed covariates using AIC, BIC, or HQ. It is not clear how to do this because the first term on the right side of (6.1) is not available. Of course, one could still blindly use those criteria, pretending that the data are normal, but criteria are no longer what they mean to be. For example, Akaike's bias approximation that led to the AIC [Akaike (1973)] is no longer valid. Furthermore, it is not clear how robust the performance of these criteria is to misspecification of the distribution (e.g., normal).

4. Finite-sample performance, and the effect of a constant. Even in the conventional situation, there are still practical issues regarding the use of these criteria. For example, the BIC is known to have the tendency of overly penalizing "bigger" models. In other words, the penalizer, $\lambda_n = \log n$, may be a little too much in some cases. In such a case, one may wish to replace the penalizer by $c\log(n)$, where c is a constant less than one. Question is: What c? Asymptotically, the choice of c does not make a difference in terms of consistency of model selection, so long as $c > 0$. However, practically, it does. As another example, comparing BIC with HQ, the penalizer in HQ is lighter in its order, that is, $\log n$ for BIC and $c\log\log n$ for HQ, where $c > 2$ is a constant. However, if $n = 100$, we have $\log n \approx 4.6$ and $\log\log n \approx 1.5$; hence, if c is chosen as 3, BIC and HQ are almost the same.

In fact, there have been a number of modifications of the BIC aiming at improving the finite-sample performance. For example, Broman and Speed (2002) proposed a δ-BIC method by replacing the $\lambda_n = \log n$ in BIC by $\delta\log n$, where δ is a constant carefully chosen to optimize the finite-sample performance. However, the choice of δ relies on extensive Monte-Carlo simulations, is case-by-case, and, in particular, depends on the sample size. Therefore, it is not easy to generalize the δ-BIC method.

5. Criterion of optimality. Strictly speaking, model selection is hardly a purely statistical problem–it is usually associated with a problem of practical interest. Therefore, it seems a bit unnatural to let the criterion of optimality in model selection be determined by purely statistical considerations, such as the likelihood and K-L information. Other considerations, such as scientific and economic concerns, need to be taken into account. For example, what if the optimal model selected by the AIC is not to the best interest of a practitioner, say, an economist? In the latter case, can the economist change one of the selected variables, and do so "legitimately"? Furthermore, the dimension of a model, $|M|$, is used to balance the model complexity through (6.1). However, the minimum-dimension criterion, also known as *parsimony*, is not always as important. For example, the criterion of optimality may be quite different if prediction is of main interest.

These concerns, such as the above, led to the development of a new class of strategies for model selection, known as the *fence* methods, first introduced in Jiang *et al.* (2008). Also see Jiang *et al.* (2009). The idea consists of a procedure to isolate a subgroup of what are known as correct models (those within the fence) via the inequality

$$Q(M) - Q(\tilde{M}) \leq c, \tag{6.39}$$

where $Q(M)$ is the measure of lack-of-fit for model M, $\tilde{M}$ is a "baseline model" that has the minimum Q, and c is a cut-off. The optimal model is then selected from the models within the fence according to a criterion of optimality that can be flexible; in particular, the criterion can incorporate the problem of practical interest. Furthermore, the choice of the measure of lack-of-fit, $\hat{Q}$, is also flexible and can incorporate problem of interest.

To see how the fence help to resolve the difficulties of the information criteria, note that the (effective) sample size is not used in the fence procedure, although the cut-off c, when chosen adaptively (see below), may implicitly depend on the effective sample size. Depending on the criterion of optimality for selecting the optimal model within the fence, the dimension of the model may be involved, but the criterion does allow flexibility. Also, the measure of lack-of-fit, Q, does not have to be the negative log-likelihood, as in the information criteria. For example, the residual sum of squares (RSS) is often used as Q, which does not require complete specification of the distribution. Furthermore, a date-driven approach is introduced in Section 6.2.1 for choosing the cut-off or tuning constant, c, that optimizes the finite-sample performance. Finally, the criterion of optimality for selecting the model within the fence can incorporate practical interests.

It should be noted that, as far as consistency is concerned, which is, by far, the most important theoretical property for a model selection procedure, the basic underlying assumptions for the fence are the same as those for the traditional model selection approaches, such as the information criteria and cross-validation [CV; e.g., Shao (1993)]. For the most part, it is assumed that the space of candidate models is finite, which contains a true model, and that the sample size goes to infinity while the model space remains the same.

There is a simple numerical procedure, known as the fence algorithm, which applies when model simplicity is used as the criterion to select the model within the fence. Given the cut-off c in (6.39), the algorithm may be described as follows:

> Check the candidate models, from the simplest to the most complex. Once one has discovered a model that falls within the fence and checked all the other models of the same simplicity (for membership within the fence), one stops.

One immediate implication of the fence algorithm is that one does not need to evaluate all the candidate models in order to identify the optimal one. This leads to potentially computational savings. A software package, *The Fence Package*, is available at https://cran.r-project.org/package=fence

Several variations of the fence will be discussed below. We refer to a monograph, Jiang and Nguyen (2015), for further details.

6.2.1 *Adaptive fence*

Finite-sample performance of the fence depends heavily on the choice of the cut-off, or tuning parameter, c in (6.39). In a way, this is similar to one of the difficulties with the information criteria noted in the previous section. Jiang *et al.* (2008) came up with an idea, known as *adaptive fence* (AF), to let the data "speak" on how to choose this cut-off. Let $\mathcal{M}$ denote the set of candidate models. To be more specific, assume that the minimum-dimension criterion is used in selecting the models within the fence. Furthermore, assume that there is a correct model in $\mathcal{M}$ as well as a full model, $M_{\rm f}$, so that every model in $\mathcal{M}$ is a submodel of $M_{\rm f}$. It follows that $\tilde{M} = M_{\rm f}$ in (6.39). First note that, ideally, one wishes to select c that maximizes the probability of choosing the optimal model, here defined as a correct model that has the minimum dimension among all of the correct models. This means that one wishes to choose c that maximizes

$$P = {\rm P}(M_c = M_{\rm opt}), \tag{6.40}$$

where $M_{\rm opt}$ represents the optimal model, and M_c is the model selected by the fence (6.39) with the given c. However, two things are unknown in (6.40): (i) under what distribution should the probability P be computed? and (ii) what is $M_{\rm opt}$?

To solve problem (i), note that the assumptions above on $\mathcal{M}$ imply that $M_{\rm f}$ is a correct model. Therefore, it is possible to bootstrap under $M_{\rm f}$. For example, one may estimate the parameters under $M_{\rm f}$, then use a model-based (or parametric) bootstrap to draw samples under $M_{\rm f}$. This allows us to approximate the probability P on the right side of (6.40).

To solve problem (ii), we use the idea of maximum likelihood. Namely, let $p^*(M) = {\rm P}^*(M_c = M)$, where $M \in \mathcal{M}$ and ${\rm P}^*$ denotes the empirical probability obtained by the bootstrapping. In other words, $p^*(M)$ is the sample proportion of times out of the total number of bootstrap samples that model M is selected by the fence with the given c. Let $p^* = \max_{M\in\mathcal{M}} p^*(M)$. Note that p^* depends on c. The idea is to choose c that maximizes p^*. It should be kept in mind that the maximization is not without restriction. To see this, let M_* denote a model in $\mathcal{M}$ that has the minimum dimension. Note that if $c = 0$, then $p^* = 1$ because the procedure always chooses $M_{\rm f}$. Similarly, $p^* = 1$ for very large c, if M_* is unique (because, when c is large enough, every $M \in \mathcal{M}$ is in the fence; hence, the

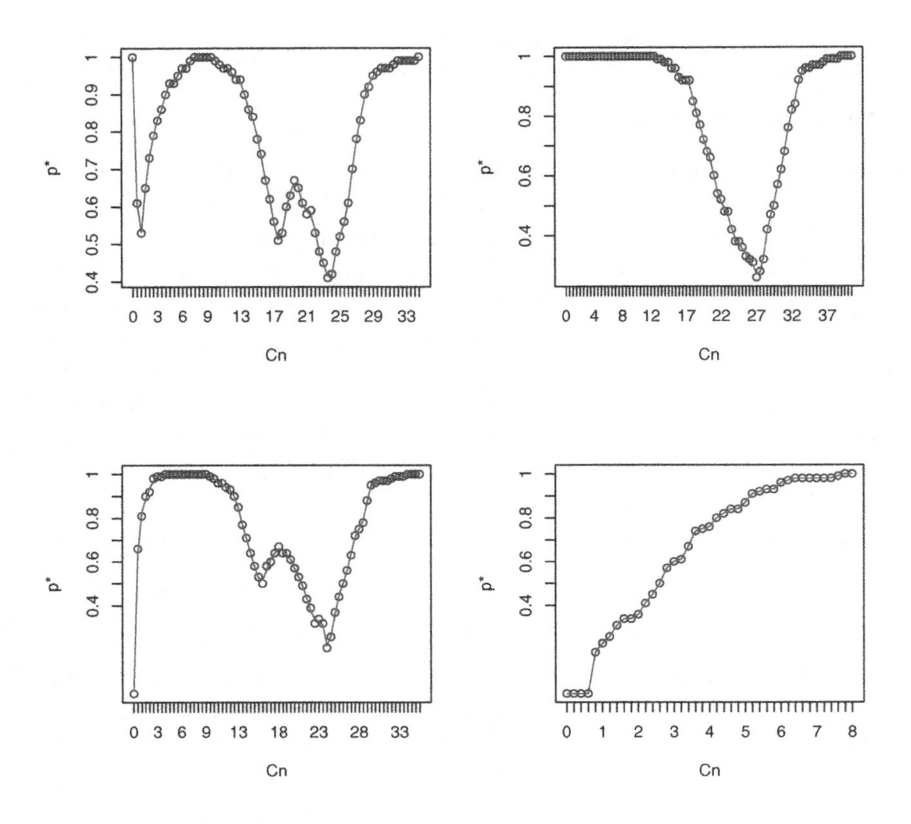

Fig. 6.1 *A few plots of* p^* *against* $c_n = c$

procedure always chooses M_*). Therefore, what one looks for is "a peak in the middle" of the plot of p^* against c. See the upper-left plot of Figure 6.1 for an illustration, where $c = c_n$. The highest peak in the middle of the plot (corresponding to approximately $c = 9$) gives the optimal choice. Here is another look at the AF. Typically, the optimal model is the model from which the data is generated, then this model should be the most likely given the data. Thus, given c, one is looking for the model (using the fence) that is most supported by the data or, in other words, one that has the highest posterior probability. The latter is estimated by bootstrapping. One then pulls off the c that maximizes the posterior probability.

Note: In the original paper of Jiang *et al.* (2008), the fence inequality (6.39) was presented with the c on the right side replaced by $c\hat{\sigma}_{M,\tilde{M}}$, where $\hat{\sigma}_{M,\tilde{M}}$ is an estimated standard deviation of the left side. Although, in some special cases, such as when Q is the negative log-likelihood, $\hat{\sigma}_{M,\tilde{M}}$ is easy to obtain, the computation of $\hat{\sigma}_{M,\tilde{M}}$, in general, can be time-consuming. This is especially the case for the AF, which calls for repeated computation of the fence under the bootstrap samples. Jiang *et al.* (2009) proposed to merge the factor $\hat{\sigma}_{M,\tilde{M}}$ with the tuning constant c, which leads to (6.39), and use the AF idea to choose the tuning constant adaptively. The latter authors called this modification *simplified adaptive fence*, and showed that it enjoys similarly impressive finite-sample performance as the original AF.

Recall the AF looks to find a peak in the middle of the plot of p^* against c. Sometimes, there may not be a peak in the middle, or there may be multiple peaks. Figure 6.1 shows a few other possible patterns. See, for example, Jiang *et al.* (2015b) for discussion on how to deal with such cases.

6.2.2 *Invisible fence*

Another variation of the fence that is intended for high-dimensional selection problems is *invisible fence* [Jiang *et al.* (2011b)]. A critical assumption in Jiang *et al.* (2008) is that there exists a correct model among the candidate models. Although the assumption is necessary in establishing consistency of the fence, it limits the scope of applications because, in practice, a correct model simply may not exist, or exist but not among the candidate models. We can extend the fence by dropping this assumption.

Note that the measure of lack-of-fit, Q in (6.39), typically has the expression, $Q(M) = \inf_{\theta_M \in \Theta_M} Q(M, \theta_M; y)$ for some measure Q. A vector $\theta^*_M \in \Theta_M$ is called an optimal parameter vector under M with respect to Q if it minimizes $\mathrm{E}\{Q(M, \theta_M; Y)\}$, that is,

$$\mathrm{E}\{Q(M, \theta^*_M; Y)\} = \inf_{\theta_M \in \Theta_M} \mathrm{E}\{Q(M, \theta_M; Y)\} \equiv Q(M), \qquad (6.41)$$

where the expectation is with respect to the true distribution of Y (which may be unknown, but not model-dependent). A correct model with respect to Q is a model $M \in \mathcal{M}$ such that

$$Q(M) = \inf_{M' \in \mathcal{M}} Q(M'). \qquad (6.42)$$

When M is a correct model with respect to Q, the corresponding θ^*_M is called a true parameter vector under M with respect to Q. Note that here a correct model is defined as a model that provides the best approximation, or best fit to the data (note that Q typically corresponds to a measure of lack-of-fit), which is not necessarily a correct model in the traditional sense. However, the above definitions are extensions of the traditional concepts in model selection [e.g., Jiang *et al.* (2008)]. The main difference is that, in the latter reference, the measure Q must satisfy a minimum requirement that $\mathrm{E}\{Q(M, \theta_M; y)\}$ is minimized when M is a correct model, and θ_M a true parameter vector under M. With the extended definition, the minimum condition is no longer needed, because it is automatically satisfied.

Let us now take another look at the fence. To be specific, assume that the minimum-dimension criterion is used to select the optimal model within the fence. In case there are ties (i.e., two models within the fence, both with the minimum dimension), the model with the minimum dimension and minimum $Q(M)$ will be chosen. Recall the cut-off c in (6.39). As it turns out, whatever one does in choosing the cut-off (adaptively or otherwise), only a fixed small subset of models have nonzero chance to be selected; in other words, the majority of the candidate models do not even have a chance. We illustrate this with an example. Suppose that the maximum dimension of the candidate models is 3. Let $M_j^\dagger$ be the model with dimension j such that $c_j = Q(M_j^\dagger)$ minimizes $\hat{Q}(M)$ among all models with dimension j, $j = 0, 1, 2, 3$. Note that $c_3 \leq c_2 \leq c_1 \leq c_0$; assume no equality holds for simplicity. The point is that any $c \geq c_0$ does not make a difference in terms of the final model selected by the fence, which is $M_0^\dagger$. Similarly, any $c_1 \leq c < c_0$ will lead to the selection of $M_1^\dagger$; any $c_2 \leq c < c_1$ leads to the selection of $M_2^\dagger$; any $c_3 \leq c < c_2$ leads to the selection of $M_3^\dagger$; and any $c < c_3$ will lead to non-selection, because no model is in the fence. In conclusion, any fence methods, adaptive or otherwise, will eventually select a model from one of the four: $M_j^\dagger$, $j = 0, 1, 2, 3$. Question is: Which one?

To solve this problem we use the AF idea by drawing bootstrap samples, say, under the full model. The idea is to select the model that has the highest empirical probability to best fits the data when controlling the dimension of the model. More specifically, for each bootstrap sample, we find the best-fitting model at each dimension, that is, $M_j^{*\dagger}$, such that $Q^*(M_j^{*\dagger})$ minimizes $Q^*(M)$ for all models with dimension j, where Q^* represents Q computed under the bootstrap sample. We then compute the relative frequency, out of the bootstrap samples, for different models selected, and the maximum relative frequency, say p_j^*, at each dimension j. Let $M_{j^*}^{*\dagger}$ be the model that corresponds to the maximum p_j^* (over different j's) and this is the model we select. In other words, if at a certain dimension we find a model that has the highest empirical probability to best fit the data, this is the model we select. Although the new procedure might look quite different from the fence, it actually uses implicitly the principle of the AF. For such a reason, the procedure is called invisible fence, or IF.

Computation is a major concern in high dimension problems. For example, for the 522 gene pathways developed by Subramanian *et al.* (2005) at dimension $k = 2$ there are 135,981 different $Q(M)$'s to be evaluated; at dimension $k = 3$ there are 23,570,040 different $\hat{Q}(M)$'s to be evaluated, ... If one has to consider all possible k's, the total number of evaluations is 2^{522}, an astronomical number. Jiang *et al.* (2011b) proposed the following strategy, called the *fast algorithm*, to meet the computational challenge. Consider the situation where there are a (large) number of candidate elements (e.g., gene-sets, variables), denoted by $1, \dots, m$, such that each candidate model corresponds to a subset of the candidate elements. A measure Q is said to be *subtractive* if it can be expressed as

$$Q(M) = s - \sum_{i \in M} s_i, \tag{6.43}$$

where s_i, $i = 1, \dots, m$ are some nonnegative quantities computed from the data, M is a subset of $1, \dots, m$, and s is some quantity computed from the data that does not depend on M. Typically we have $s = \sum_{i=1}^m s_i$, but the definition does not impose such a restriction. For example, in gene-set analysis [GSA; e.g., Efron and Tibshirani (2007), s_i corresponds to the gene-set score for the ith gene-set. As another example, Mou (2012) considered $s_i = |\hat{\beta}_i|$, where $\hat{\beta}_i$ is the estimate of the coefficient for the ith candidate variable under the full model, in selecting the fixed covariates.

For a subtractive measure, the models that minimize $Q(M)$ at different dimensions are found almost immediately. Let $r_1, r_2, \dots, r_m$ be the ranking of the candidate elements in terms of decreasing s_i. Then, the model that minimizes $Q(M)$ at dimension one is r_1; the model that minimizes $Q(M)$ at dimension two is $\{r_1, r_2\}$; the model that minimizes $Q(M)$ at dimension three is $\{r_1, r_2, r_3\}$, and so on (Exercise 6.8).

Jiang *et al.* (2011b) implemented the IF with the fast algorithm, for which a natural choice of s_i is the gene-set score, as noted, and showed that IF significantly outperforms GSA in simulation studies. The authors also showed that IF has a nice theoretical property, called *signal-consistency*, which GSA does not have. Consider, for example, the problem of selecting the fixed covariates in a linear mixed model. Instead of letting the sample size increase (in a certain way), one may consider letting the absolute values of the fixed effects, called the signals, increase. A model selection procedure is called signal-consistent if the probability of selecting the optimal model goes to one as the signals go to infinity. In a way, signal-consistency is equivalent to consistency in the traditional sense.

6.2.3 *Model selection with incomplete data*

The missing-data problem has a long history [e.g., Little and Rubin (2002)]. While there is an extensive literature on statistical analysis with missing or incomplete data, the literature on model selection in the presence of missing data is relatively sparse. See, Jiang *et al.* (2015a) for a review of literature on model selection with incomplete data. Existing model selection procedures face special challenges when confronted with missing or incomplete data. Obviously, the naive complete-data-only strategy is inefficient, sometimes even unacceptable by the practitioners due to the overwhelmingly wasted information. For example, in a study of backcross experiments [e.g., Lander and Botstein (1989)], a data set was obtained by researchers at UC-Riverside (personal communications; see Zhan *et al.* (2011)). Out of the 150 or so subjects, only 4 have complete data record. Situations like are, unfortunately, the reality that we often have to deal with.

Verbeke *et al.* (2008) offered a review of formal and informal model selection strategies with incomplete data, but the focus is on model comparison, instead of model selection. As noted by Ibrahim *et al.* (2008), while model comparisons "demonstrate the effect of assumptions on estimates and tests, they do not indicate which modeling strategy is best, nor do they specifically address model selection for a given class of models". The latter authors further proposed a class of model selection criteria based on the output of the E-M algorithm. Jiang *et al.* (2015a) points out a potential drawback of the E-M approach of Ibrahim *et al.* (2008) in that the conditional expectation in the E-step is taken under the assumed (candidate) model, rather than an objective (true) model. Note that the complete-data log-likelihood is also based on the assumed model. Thus, by taking the conditional expectation, again, under the assumed model, it may bring false supporting evidence for an incorrect model. Similar problems have been noted in the literature, which are sometimes referred to as "double-dipping" [e.g., Copas and Eguchi (2005)].

On the other hand, the AF idea works naturally with the incomplete data. For the simplicity of illustration, let us assume, for now, that the candidate models include a correct model as well as a full model. It then follows that the full model is, at least, a correct model, even though it may not be the most efficient one. Thus, in the presence of missing data, we can run the E-M [Dempster *et al.* (1977)] to obtain the MLE of the parameters, under the full model. Note that, here, we do not have the double-dipping problem, because the conditional expectation (under the full model) is "objective". Once the parameter estimates are obtained, we can use the model-based (or parametric) bootstrap to draw samples, under the full model, as in the AF. The best part of this strategy is that, when one draws the bootstrap samples, one draws samples of complete data, rather than data with the missing values. Therefore, one can apply any existing model selection procedure that is built for the complete-data situation, such as the fence, to the bootstrap sample. Suppose that B bootstrap samples are drawn. The model with the highest frequency of being selected (as the optimal model), out of the bootstrap samples, is the (final) optimal model. We call this procedure the EMAF algorithm due to its similarity to the AF idea. One can extend the EMAF idea to situations where a correct model may not exist, or exists but not among the candidates. In such a case, the bootstrap samples may be drawn under a model $\tilde{M}$, which is the model with the minimum Q in the sense of the second paragraph of §6.2.2.

6.2.4 *Examples*

1. Fay-Herriot model. The Fay-Herriot model was introduced in Example 5.1. Jiang *et al.* (2008) reported results of a simulation study, in which AF was compared with several other non-adaptive choices of the cut-off, c in (6.39). The measure Q was taken as the negative log-likelihood, and the sample size was $n = m = 30$. The candidate predictors are $x_1, \dots, x_5$, generated from the $N(0,1)$ distribution, and then fixed throughout the simulations. The candidate models included all possible models with at least an intercept. Five cases were considered, in which the data y were generated under the model $y = \sum_{j=1}^{5} \beta_j x_j + v + e$, where $\beta' = (\beta_1, \dots, \beta_5) = (1,0,0,0,0)$, $(1,2,0,0,0)$, $(1,2,3,0,0)$, $(1,2,3,2,0)$ and $(1,2,3,2,3)$, denoted by Model 1, 2, 3, 4, 5, respectively. The authors considered the simple case $D_i = 1, 1 \le i \le n$. The true value of A is 1 in all cases. The number of bootstrap samples for the evaluation of the p^*'s is 100. In addition to the AF, five different non-adaptive choice of $c = c_n$ were considered, which satisfy the consistency requirements given in Jiang *et al.* (2008), namely, that $c_n \to \infty$ and $c_n/n \to 0$ in this case. The results are presented in Table 6.1, which were the percentage of times, out of the 100 simulations, that the optimal model was selected by each

Table 6.1 **Fence with different choice of c**

Optimal Model	1	2	3	4	5
Adaptive c	100	100	100	99	100
$c = \log\log(n)$	52	63	70	83	100
$c = \log(n)$	96	98	99	96	100
$c = \sqrt{n}$	100	100	100	100	100
$c = n/\log(n)$	100	91	95	90	100
$c = n/\log\log(n)$	100	0	0	0	6

method. It is seen that the performances of the fence with $c = \log(n)$, $\sqrt{n}$ or $n/\log(n)$ are fairly close to that of the AF. Of course, in any particular case one might get lucky to find a good c value, but one cannot be lucky all the time. Regardless, the AF always seems to pick up the optimal value, or something close to the optimal value of c in terms of the finite-sample performance.

2. Gene set analysis. Efron and Tibshirani (2007) carried out an empirical study, in which the authors simulated 1000 genes and 50 samples in each of 2 classes, control and treatment. The genes were evenly divided into 50 gene-sets, with 20 genes in each gene-set. The data matrix was originally generated independently from the $N(0, 1)$ distribution, then the treatment effect was added according to one of the following five scenarios:

1. All 20 genes of gene-set 1 are 0.2 units higher in class 2.
2. The first 15 genes of gene-set 1 are 0.3 units higher in class 2.
3. The first 10 genes of gene-set 1 are 0.4 units higher in class 2.
4. The first 5 genes of gene-set 1 are 0.6 units higher in class 2.
5. The first 10 genes of gene-set 1 are 0.4 units higher in class 2, and the second 10 genes of gene-set 1 are 0.4 units lower in class 2.

Jiang *et al.* (2011b) considered the same five scenarios in a simulation study. In Efron and Tibshirani's study only the first gene-set is of potential interest. Jiang *et al.* (2011b) expanded the one-gene-set case to a two-gene-set case, in which they duplicated the five scenarios to the second gene-set.

Also, in Efron and Tibshirani's study the genes were simulated independently. Jiang *et al.* (2011b) considered, in addition to the independent case ($\rho = 0$), a case where the genes are correlated with equal correlation coefficient $\rho = 0.3$. The correlation is generated by associating with each microarray a random effect. The genes on the same microarray are then correlated for sharing the same random effect. Let x_{ij} be the (i, j) element of the data matrix, X, where i represents the gene and j the microarray, $i = 1, \dots, 1000$, $j = 1, \dots, 100$. Here $j = 1, \dots, 50$ correspond to the controls and $j = 51, \dots, 100$ the treatments. Then, we have

$$x_{ij} = \alpha_j + \epsilon_{ij}, \tag{6.44}$$

where the α_j's and ϵ_{ij}'s are independent random effects and errors that are distributed as $N(0, \rho)$ and $N(0, 1-\rho)$, respectively. It follows that each x_{ij} is distributed as $N(0, 1)$, and $\text{cor}(x_{ij}, x_{i'j}) = \rho$, $i \neq i'$. The treatment effects are then added to the right side of (6.44) for $j = 51, \ldots, 100$ and genes i in the given gene-set(s), as above.

Compare the performance of IF with GSA in gene-set identification. In Efron and Tibshirani's simulation study, the authors showed that the maxmean has the best overall performance as compared with other methods, including the mean, the absolute mean, GSEA (Gene Set Enrichment Analysis; Subramanian *et al.* (2005) and GSEA version of the absolute mean. Therefore, the comparisons focus on the best performer of GSA, that is, the maxmean. In addition to the one-gene-set and two-gene-set cases, each with the five scenarios listed above, the simulation comparisons also include the case where no gene-set is potentially interesting, that is, no treatment effect is added to any gene-set. This is what we call the null scenario. For GSA one needs to choose the FDR as well as the number of permutation samples for the test of significance of gene-sets. For IF, on the other hand, one also needs to specify the level of significance as well as the number of permutation samples for the test for no gene-set. The FDR and level of significance are both chosen as $\alpha = 0.05$. The number of permutations for both GSA and IF is 200 [which is the number that Efron and Tibshirani (2007) used for their simulations].

The first comparison is on the probability of correct identification, or true-positive (TP). For IF this means that the gene-sets selected match exactly those to which the treatment effects are added, which we call true gene-sets; similarly, for GSA this means that the gene-sets that are found significant are exactly those true gene-sets. Table 6.2 reports the empirical probability of TP based on 100 simulation runs. For example, for the Null Scenario, One-Gene-Set case, with $\rho = 0$, the numbers mean that for 95 out of the 100 simulation runs, IF selected no (0) gene-sets; while for 59 of the 100 simulation runs, GSA found no (0) gene-sets. As another example, for Scenario 2, Two-Gene-Set case, with $\rho = 0.3$, IF selected the exact two gene-sets, to which the treatment effects are added, for 97 out of the 100 simulation runs; while GSA found the exact two gene-sets for 66 out of the 100 simulation runs. Note that these are results of same-data comparisons, that is, for each simulation run, the results for both methods are based on the same simulated data. Also reported (in the parentheses) are empirical probabilities of overfit (OF, in the sense that the identified gene-sets include all the true gene-sets plus some false discoveries) and underfit (UF, in the

Table 6.2 **IF vs GSA - Empirical Probabilities (in %) of TP (OF, UF)**

		$\rho = 0$		$\rho = 0.3$	
Scenario	Method	1-Gene-Set	2-Gene-Set	1-Gene-Set	2-Gene-Set
Null	IF	95 (5,0)	95 (5,0)	64 (36,0)	64 (36,0)
	GSA	59 (41,0)	59 (41,0)	52 (48,0)	52 (48,0)
1	IF	80 (6,14)	68 (1,31)	80 (16,4)	88 (4,8)
	GSA	53 (37,10)	53 (25,22)	61 (36,3)	62 (30,8)
2	IF	88 (5,7)	88 (0,12)	88 (12,0)	97 (2,1)
	GSA	67 (32,1)	65 (26,9)	65 (35,0)	66 (32,2)
3	IF	87 (5,8)	84 (0,16)	83 (14,3)	96 (2,2)
	GSA	66 (31,3)	68 (24,8)	66 (33,1)	69 (27,4)
4	IF	73 (6,21)	63 (2,35)	75 (15,10)	80 (7,13)
	GSA	64 (28,8)	57 (19,24)	66 (29,5)	63 (21,16)
5	IF	87 (6, 7)	84 (0,16)	91 (9,0)	99 (0,1)
	GSA	70 (30,0)	76 (17,7)	82 (18,0)	86 (12,2)

sense that at least one of the true gene-sets is not discovered). It appears that IF has better performance than GSA in terms of TP uniformly across all the cases and scenarios. While most of the losses for IF are due to UF, OF appears to be the major problem for GSA. Furthermore, both methods appear to be fairly robust against correlations between genes.

The next comparison focuses on the signal-consistency properties of both methods (see the last paragraph of Subsection 6.2.2). As noted, traditionally, consistency in model identification (including parameter estimation and model selection) involves sample size going to infinity. Such an assumption, however, is not very realistic in gene-set analysis, because the sample size n usually is much smaller than the number of genes under consideration. Therefore, signal-consistency is considered. A gene-set identification procedure is signal-consistent if its probability of TP goes to one as the treatment effects, or signals, increase to infinity. Of course, one may not be able to increase the signals in real-life, but the point is to see if a procedure works perfectly well in the "ideal situation", just like consistency in the traditional sense. To investigate signal-consistency property of IF and GSA, one of the cases, namely, the two-gene-set case of Scenario 5, was expanded by increasing the treatment effects in two different ways. First, one increases the signals in a balanced manner, that is, the signals increase at the same pace for both gene-sets. Next, one lets the signals increase in an unbalanced manner, so that the pace is much faster for the first gene-set than for the second one. Table 6.3 reports the empirical probabilities of TP based on 100 simulation runs. Here the signals are expressed in the form of (a, b, c, d), where the values a, b, c, d are added to the right side

Table 6.3 **IF vs GSA - Empirical Probabilities (in %) of TP with Increasing Signals**

Case #	Signals	$\rho = 0$ IF	$\rho = 0$ GSA	$\rho = 0.3$ IF	$\rho = 0.3$ GSA
1	(0.4,-0.4,0.4,-0.4)	84	76	99	86
2	(0.5,-0.5,0.5,-0.5)	100	88	100	97
3	(1.0,-1.0,1.0,-1.0)	100	100	100	100
4	(1.0,-1.0,0.5,-0.5)	100	97	100	99
5	(1.5,-1.5,0.5,-0.5)	100	88	100	88
6	(2.0,-2.0,0.5,-0.5)	100	64	100	56
7	(2.5,-2.5,0.5,-0.5)	100	26	100	23
8	(3.0,-3.0,0.5,-0.5)	100	10	100	3
9	(3.5,-3.5,0.5,-0.5)	100	2	100	0
10	(4.0,-4.0,0.5,-0.5)	100	0	100	0

of (6.44) for $51 \leq j \leq 100$ and 1st 10 genes of gene-set one, 2nd 10 genes of gene-set one, 1st 10 genes of gene-set two, and 2nd 10 genes of gene-set two, respectively.

Case 1 is taken from the bottom two rows of Table 6.2 (two-gene-set case), which serves as a baseline. Then, one can see what happens when the signals increase. In cases 1–3, where the signals increase in the balanced way, both IF and GSA seem to work perfectly well as both methods show signs of signal-consistency. However, in cases 1, 4–10, where the signals increase in the unbalanced way, the empirical probability drops, and eventually falls apart for GSA, even with increasing signals. On the other hand, IF still shines in this situation, having perfect empirical probabilities of TP. It is interesting to know what happens to GSA in the latest situation. Jiang *et al.* (2011b) noted that this was due to the restandardization method used in Efron and Tibshirani (2007) (Exercise 6.9).

In introducing the GSA method, Efron and Tibshirani (2007) considered a situation where the same treatment effect is added to all the gene-sets. In other words, all the gene-sets are equally d.e. The authors used this example to make the point for the need of restandardization. The claim is that, in this case, there is "nothing special about any one gene-set". While the claim is arguable from a practical point of view, it would be interesting to see how the two methods, IF and GSA, work in a situation like this. Jiang *et al.* (2011b) simulated data according to Scenario 1 above, except that the 0.2 units are added to all the gene-sets. If, as Efron and Tibshirani claimed, there is nothing special about any gene-set, one expects a procedure to identify no (zero) gene-set in this case. According to the results based on 100 simulation runs, when $\rho = 0$, the empirical probabilities of identifying zero gene-set is 50% for IF and 47% for GSA; when $\rho = 0.3$, the corresponding empirical probabilities are 58% for IF and 54% for GSA. So, in the latest comparison, the two methods performed similarly with IF doing slightly better.

3. Backcross experiment: A real-data example. Recall the data set obtained by the UC-Riverside researchers mentioned in Subsection 6.2.3. The gene expression data were originally published by Luo *et al.* (2007). The phenotypic values of eight quantitative traits of barley were published by Hayes *et al.* (1993). Detailed description of the experiment can be found in the latter reference, which involved 150 double haploid (DH) lines derived from the cross of two spring barley varieties, Morex and Steptoe. The DH lines are considered as the subjects here. In all there were 495 SNP markers on seven chromosomes that are under investigation. As mentioned, there are significant missing values in the data so that only 4 of the 150 subjects have complete genotype records. On the other hand, there are no missing values in the phenotypic data.

We use this data set to illustrate a variation of the EMAF algorithm, described in Subsection 6.2.3, with the AF replaced by IF, introduced in Subsection 6.2.2. The new algorithm is thus called EMIF.

Following Broman and Speed (2002), we have a conditional linear regression model for the phenotype variable, Y, such that, given the marker indicators, x, we have $Y_i = \sum_{k=1}^{r} \sum_{j \in M_k} \beta_{jk} x_{ijk} + \epsilon_i$, where r is the number of chromosomes, M_k is a subset of $\{1, \ldots, q\}$ and q is the number of markers on each chromosome, and ϵ_i is a normal error, with mean zero and unknown variance σ^2. The ϵ_i's are uncorrelated and also independent with the X_{ijk}'s. Furthermore, the marker indicators, X_{ijk}, are assumed to be a Markov chain within each chromosome with $\mathrm{P}(X_{i1k} = 0) = \mathrm{P}(X_{i1k} = 1) = 1/2$ (Mendel's rule) and $\mathrm{P}(X_{i,j+1,k} = 1|X_{ijk} = 0) = \mathrm{P}(X_{i,j+1,k} = 0|X_{ijk} = 1) = \theta$, where θ is the *recombination fraction.* The problem of interest is to identify the subset $M = (M_1, \ldots, M_r)$, which is viewed as a model selection problem as in Broman and Speed (2002). However, the high-dimensional nature of the data presents a problem for the direct application of the EMIF, because the total number of markers (495) is much larger than the sample size ($n = 150$). More specifically, the least squares (LS) fit is unfeasible when the number of predictors is larger than the sample size. To overcome this difficulty, Jiang *et al.* (2015a) used the following idea of *conditional modeling*, described under a more general setting.

Suppose that, conditional on $X = (x_i')_{1 \le i \le n}$, one has a linear regression $Y = X\beta + \epsilon$, where $Y = (Y_i)_{1 \le i \le n}$ are the observations, and $\epsilon = (\epsilon_i)_{1 \le i \le n}$

are the errors such that the components of ϵ are independent with mean 0, and ϵ is independent of X. Furthermore, suppose that $X = [X_{(1)}\ X_{(2)}]$ with $X_{(r)} = (X'_{ir})_{1\leq i\leq n}, r = 1, 2$ such that $X_{(1)}, X_{(2)}$ are independent [e.g., Broman and Speed (2002)]. Then, it is easy to show that $X_{(1)}$ is independent of $[X_{(2)}, \epsilon]$. Note that we can express the regression model as $Y = X_{(1)}\beta_1 + X_{(2)}\beta_2 + \epsilon$. Without loss of generality, we assume that $X_{(1)}\beta_1$ does not involve an intercept [which, if exist, belongs to $X_{(2)}\beta_2$].

Now suppose that $X_{i2}, i = 1, \dots, n$ are independent, and that $\mathrm{E}(X_{i2})$ does not depend on i. Then, $\mathrm{E}(X'_{i2}\beta_2 + \epsilon_i) = \mathrm{E}(X_{i2})'\beta_2$ is a constant, say, β_0. Let $e_i = X'_{i2}\beta_2 + \epsilon_i - \beta_0$. It is easy to show that $e_i, i = 1, \dots, n$ are independent with $\mathrm{E}(e_i) = 0$, and $Y = [1_n\ X_{(1)}](\beta_0\ \beta'_1)' + e$, e being independent of $[1_n\ X_{(1)}]$. In other words, conditional on $X_{(1)}$, we, once again, have a standard linear regression model (i.e., the errors are independent with mean zero, and independent with the predictors).

The point is that $X_{(1)}$ can be of much lower dimension than X. For the barley cross data, we can let $X_{(1)}$ correspond to markers on any particular chromosome. The number of markers on the 7 chromosomes are 60, 78, 81, 60, 93, 56 and 67, respectively, all of which are smaller than the sample size 150. Within each chromosome, we apply the EMIF in conjunction with the IF (see Subsection 6.2.2). The number of bootstrap samples is $B = 100$.

It is known that, for high-dimensional data the IF may suffer from the so-called dominant factor effect [Jiang *et al.* (2011b), sec. 3.3]. For the most part, this means that the IF frequency (i.e., the empirical probability of the most frequently selected model; see Subsection 2.2) tends to be in favor of a lower dimensional model than the true model, if the "signals" are relatively weak due due to the limited sample size. This problem is dealt with naturally by the EMIF. First we apply the IF, under the full model, that is, all the markers on a given chromosome, to obtain the IF frequencies at different dimensions, say, $p_1^*, p_2^*, \dots, p_q^*$, where p_j^* is the IF frequency at dimension j, and q is the total number of markers, for the chromosome. If the frequencies show a "peak", that is, there is a $1 < j < q$ such that $p_j^* > p_{j-1}^*$ and $p_j^* > p_{j+1}^*$, the EMIF shall continue; otherwise, we conclude that there is no more than one QTL on the chromosome. In the latter case, the highest IF frequency must take place at the boundary, that is, either at dimension one or at the highest dimension corresponding to all the markers on the chromosome. However, it is unlikely that all the markers are QTLs; therefore, dimension one is chosen, and the EMIF stops.

If the frequency plot show a "peak", and therefore the E-MS is to continue, we first look for the last peak, that is, the highest dimension that

Table 6.4 **EMIF Results for Grain Protein**

Chromosome	Marker ID#				Chromosome	Marker ID#			
1	12	13			5	280	285	332	333
2	65	66			6	379	380		
3	184	186	199	200	7	467	470		
4	176								

corresponds to a peak in order to be conservative. This is similar to the AF (Subsection 6.2.1), where the first significant peak is chosen in order to determine the cut-off for the fence [e.g., Jiang *et al.* (2009), Jiang and Nguyen (2015)]. The first peak for the AF corresponds to the last peak for the IF. The markers corresponding to the last peak are selected, the current model is updated, and the updated model is treated as the (new) full model for the next step of iteration. The procedure is repeated until either the updated model is identical to the current model, or no peak is found during the current step; in both cases, the current model is chosen as the final model. For the latter case, when no peak is found, we choose the highest dimension, instead of dimension one as above in the initial step. This is because, at this stage, we have already determined that there are more than one QTLs on the chromosome (the EMIF would not have continued otherwise); furthermore, the highest dimension possibly has been updated, so it no longer corresponds to all of the markers on the chromosome.

The results for the grain protein phenotype are presented in Table 6.4. The results show some consistency with the foundings of Zhan *et al.* (2011). For example, the latter authors found that chromosomes 2, 3, 5 "seem to control more genes than other chromosomes". According to our results, those three chromosomes contain nearly 60% of all the QTLs found. In particular, chromosomes 3 and 5 are the top two according to the number of QTLs found. It should be noted that the number of QTLs found on a chromosome is not the only thing that represents the relative importance of the chromosome; the magnitude of the QTL effect is also important. In this application, however, our focus is identification of the QTLs, rather than estimation of the QTL effects.

6.3 Shrinkage mixed model selection

There has been some recent work on joint selection of the fixed and random effects in mixed effects models. Bondell *et al.* (2010) considered such a selection problem in a certain type of linear mixed models, which can be expressed as

$$y_i = X_i\beta + Z_i\alpha_i + \epsilon_i, \quad i = 1, \ldots, m, \tag{6.45}$$

where y_i is an $n_i \times 1$ vector of responses for subject i, X_i is an $n_i \times p$ matrix of explanatory variables, β is a $p \times 1$ vector of regression coefficients (the fixed effects), Z_i is an $n_i \times q$ known design matrix, α_i is a $q \times 1$ vector of subject-specific random effects, ϵ_i is an $n_i \times 1$ vector of errors, and m is the number of subjects. It is assumed that the $\alpha_i, \epsilon_i, i = 1, \ldots, m$ are independent with $\alpha_i \sim N(0, \sigma^2\Psi)$ and $\epsilon_i \sim N(0, \sigma^2 I_{n_i})$, where Ψ is an unknown covariance matrix. The problem of interest, using the terms of shrinkage model selection, is to identify the nonzero components of β and $\alpha_i, 1 \leq i \leq m$. For example, the components of α_i may include a random intercept and some random slopes.

6.3.1 *An E-M based approach*

To take advantage of the idea of shrinkage variable selection, Bondell *et al.* (2010) adopted a modified Cholesky decomposition. Note that the covariance matrix, Ψ, can be expressed as $\Psi = D\Omega\Omega' D$, where $D = \text{diag}(d_1, \ldots, d_q)$ and $\Omega = (\omega_{kj})_{1 \leq k,j \leq q}$ is a lower triangular matrix with 1's on the diagonal. Thus, one can express (6.45) as

$$y_i = X_i\beta + Z_i D\Omega\xi_i + \epsilon_i, \quad i = 1, \ldots, m, \tag{6.46}$$

where the ξ_i's are independent $N(0, \sigma^2)$ random variables. The idea is to apply shrinkage estimation to both $\beta_j, 1 \leq j \leq p$ and $d_k, 1 \leq k \leq q$. Note that setting $d_k = 0$ is equivalent to setting all of the elements in the kth column and kth row of Ψ to zero, and thus creating a new submatrix by deleting the corresponding row and column, or the exclusion of the kth component of α_i. However, direct implementation of this idea is difficult, because the ξ_i's are unobserved, even though their distribution is much simpler. To overcome this difficulty, Bondell *et al.* (2010) used the E-M algorithm [Dempster *et al.* (1977)]. By treating the ξ_i's as observed, the complete-data log-likelihood can be expressed as (Exercise 6.10)

$$l_c = c_0 - \frac{N + mq}{2} \log \sigma^2 - \frac{1}{2\sigma^2}(|y - X\beta - Z\tilde{D}\tilde{\Omega}\xi|^2 + |\xi|^2), \tag{6.47}$$

where c_0 is a constant, $N = \sum_{i=1}^m n_i$, $X = (X_i)_{1 \leq i \leq m}$, $Z = \text{diag}(Z_1, \ldots, Z_m)$, $\tilde{D} = I_m \otimes D$, $\tilde{\Omega} = I_m \otimes \Omega$ ($\otimes$ means Kronecker product), $\xi = (\xi_i)_{1 \leq i \leq m}$, and $|\cdot|$ denotes the Euclidean norm. (6.47) leads to the shrinkage estimation by minimizing

$$P_{\rm c}(\phi|y,\xi) = |y - X\beta - Z\tilde{D}\tilde{\Omega}\xi|^2 + \lambda\left(\sum_{j=1}^{p}\frac{|\beta_j|}{|\tilde{\beta}_j|} + \sum_{k=1}^{q}\frac{|d_j|}{|\tilde{d}_j|}\right), \quad (6.48)$$

where ϕ represents all of the parameters, including the β's, the d's, and the ω's, $\tilde{\beta} = (\tilde{\beta}_j)_{1\le j\le p}$ is given by the right side of (5.4) with the variance components involved in $V = \text{Var}(y)$ replaced by their REML estimators (see Section 3.2)], $\tilde{d}_k, 1 \le k \le q$ are obtained by decomposition of the estimated Ψ via the REML, and λ is the regularization parameter. Bondell *et al.* (2010) proposed to use the BIC in choosing the regularization parameter. Here the form L^1 penalty in (6.48) is in terms of the adaptive Lasso [Zou (2006)]. To incorporate with the E-M algorithm, one replaces (6.48) by its conditional expectation given y, and the current estimate of ϕ. Note that only the first term on the right side of (6.48) involves ξ, with respect to which the conditional expectation is taken. The conditional expectation is then minimized with respect to ϕ to obtain the updated (shrinkage) estimate of ϕ. A similar approach was taken by Ibrahim *et al.* (2011) for joint selection of the fixed and random effects in GLMMs [see Section 2.4], although the performance of the proposed method was studied only for the special case of linear mixed models.

As in Subsections 6.2.1, one may use the AF idea to derive a data-driven approach to selection of the regularization parameters in shrinkage mixed model selection, such as the λ in (6.48). Below we consider selection problems from a different perspective.

6.3.2 *Predictive shrinkage selection*

Suppose that the purpose of the joint selection is for predicting some mixed effects. We can incorporate this into the model selection criterion [see discussion in the paragraph below the one containing (6.39)]. We consider a predictive measure of lack-of-fit developed in Section 5.2., and incorporate it with the shrinkage idea of Bondell *et al.* (2010) and Ibrahim *et al.* (2011). Namely, we replace the first term on the right side of (6.48) by the measure,

$$Q(\phi|y) = (y - X\beta)'\Gamma\Gamma'(y - X\beta) - 2\text{tr}(\Gamma'\Sigma), \quad (6.49)$$

which can be derived in a similar way as (5.10) [or (5.20); Exercise 6.11]. The regularization parameter, λ, may be chosen using a similar procedure as in Subsection 6.2.1. We refer to this method as *predictive shrinkage selection*, or PSS. The idea has been explored by Hu *et al.* (2015). The authors found that PSS performs better than the shrinkage selection method based on (6.48) not only in terms of the predictive performance, but also in terms of *parsimony*. The latter refers to the classical criterion of selecting a correct model with the minimum number of parameters. The authors also extended the PSS to Poisson mixed models, a special case of GLMM, and obtained similar results. There is another advantage of the PSS in terms of computation. Denote the first term on the right side of (6.48) by $Q_{\rm c}(\phi|y,\xi)$. Note that, unlike $Q_{\rm c}(\phi|y,\xi)$, the unobserve ξ is not involved in $Q(\psi|y)$ defined by (6.49), which is what the PSS is based on. This means that, unlike Bondell *et al.* (2010) and Ibrahim *et al.* (2011), PSS does not need to run the E-M algorithm, and thus is computationally (much) more efficient. We illustrate this with a real-data example.

6.3.3 *Real-data example: Analysis of high-speed network*

Efficient data access is essential for sharing massive amounts of data among geographically distributed research collaborators. The analysis of network traffic is getting more and more important today for utilizing limited resources offered by the network infrastructures and planning wisely large data transfers. The latter can be improved by learning the current conditions and accurately predicting future network performance. Short-term prediction of network traffic guides the immediate scientific data placements for network users; long-term forecast of the network traffic enables capacity-planning of the network infrastructure needs for network designers. Such predictions become non-trivial when the amount of network data grows in unprecedented speed and volumes. One such available data source is NetFlow [Cisco Systems (1966)].

The NetFlow measurements provide high volume, abundant specific information for each data flow; some sample records are shown in Table 6.5 (with IP addresses masked for privacy). For each record contains the following list of variables:

Start, End The start and end time of the recorded data transfer.

Sif, Dif The source and destination interface assigned automatically for the transfer.

SrcIPaddress, DstIPaddress The source and destination IP addresses of the transfer.

SrcP, DstP The source and destination Port chosen based on the transfer type such as email, FTP, SSH, etc.

Table 6.5 **Sample NetFlow Records.**

Start	End	Sif	SrcIPaddress(masked)	SrcP	Dif
DstIPaddress(masked)	DstP	P	Fl	Pkts	Octets
0930.23:59:37.920	0930.23:59:37.925	179	xxx.xxx.xxx.xxx	62362	175
xxx.xxx.xxx.xxx	22364	6	0	1	52
0930.23:59:38.345	0930.23:59:39.051	179	xxx.xxx.xxx.xxx	62362	175
xxx.xxx.xxx.xxx	28335	6	0	4	208
1001.00:00:00.372	1001.00:00:00.372	179	xxx.xxx.xxx.xxx	62362	175
xxx.xxx.xxx.xxx	20492	6	0	2	104
0930.23:59:59.443	0930.23:59:59.443	179	xxx.xxx.xxx.xxx	62362	175
xxx.xxx.xxx.xxx	26649	6	0	1	52
1001.00:00:00.372	1001.00:00:00.372	179	xxx.xxx.xxx.xxx	62362	175
xxx.xxx.xxx.xxx	26915	6	0	1	52
1001.00:00:00.372	1001.00:00:00.372	179	xxx.xxx.xxx.xxx	62362	175
xxx.xxx.xxx.xxx	20886	6	0	2	104

P The protocol chosen based on the general transfer type such as TCP, UDP, etc.

Fl The flags measured the transfer error caused by the congestion in the network.

Pkts The number of packets of the recorded data transfer.

Octets The Octets measures the size of the transfer in bytes.

Features of NetFlow data have led to consideration of GLMMs (see Section 2.4) for predicting the network performance. First, NetFlow record is composed of multiple time series with unevenly collected time stamps. Because of this feature, traditional time series methods such as ARIMA model, wavelet analysis, and exponential smoothing [e.g., Fan and Yao (2003), sec. 1.3.5] encounter difficulties, because these methods are mainly designed for evenly collected time stamps and dealing with a single time series. Thus, there is a need for modeling a large number of time series without constraints on even collection of time stamps. On the other hand, GLMM is able to fully utilize all of the variables involved in the data set without requiring evenly-spaced time variable.

Second, there are empirical evidences of associations, as well as heteroscedasticity, found in the NetFlow records. For example, with increasing number of packets in a data transfer, it takes longer, in general, to finish the transfer. This suggests that the number of packets may be considered as a (fixed) predictor for the duration of data transfer. Furthermore, there appear to be fluctuation among network paths in terms of slope and range in the plots of duration against the number of packets. See Fig. 1 of Hu *et al.* (2015). This suggests that the network path for data transfer may be associated with a random effect to explain the duration under varying conditions. Thus, again, a mixed effects model seems to be plausible. Moreover, GLMM is more flexible than linear mixed model in terms of the mean-variance association.

Third, NetFlow measurements are big data with millions of observation for a single router within a day and 14 variables in each record with 30s or 40s interaction terms as candidates. The large volume and complexity of the data require efficient modeling. However, this is difficult to do with fixed effects modeling. For example, traditional hierarchical modeling requires dividing the data into groups, but the grouping is not clear, and requires investigation to identify the variable that classify the observed data. Explorative data analysis shows that the grouping factor could be the path of the data transfer, the delivering time of the day, the transfer protocol used, or the combination of some or all of these. With so many uncertainties, one approach to simplifying the modeling is via the use of random effects [Jiang (2007)].

The NetFlow data used for the current analysis was provided by ESnet for the duration from May 1, 2013 to June 30, 2013. Considering the network users' interests, the established model should be able to predict the duration of a data transfer so that the users can expect how long it would take for the data transfer, given the size of their data, the start time of the transfer, selected path and protocols. Considering the network designers' interests, the established model should be able to predict the long-time usage of the network so that the designer will know which link in the network is usually congested and requires more bandwidth, or rerouting of the path.

In the following, we illustrate a model built for these interests, and compare its prediction accuracy with two traditional GLMM procedures: Backward-Forward selection (B-F) and Estimation-based Lasso (E-Lasso). The latter refers to Lasso for the penalized likelihood method, corresponding to (6.48) [Bondell *et al.* (2010)].

The full model predicts the transfer duration, assuming influences from the fixed effects including transfer start time, transfer size (Octets and Packets) and the random effects including network transfer conditions such as Flag and Protocol, source and destination Port numbers, and transfer path such as source and destination IP addresses and Interfaces. The PSS

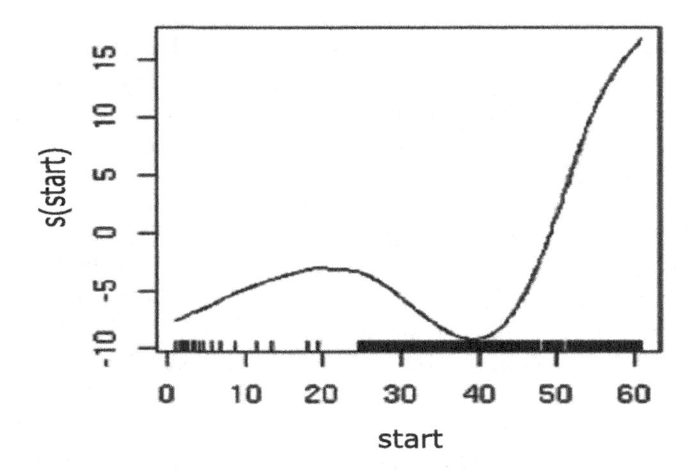

Fig. 6.2 *Fitted Smoothing Spline: Transfer Duration vs Start Time.*

procedure, with the Lasso penalty, that is, (6.48) without the denominators, $|\tilde{\beta}_j|$ and $\tilde{d}_j|$, $1 \leq j \leq p$, has selected the model

$$y = \beta_{\text{start}} s(x_{\text{start}}) + \beta_{\text{pkt}} x_{\text{pkt}} + Z_{\text{ip-path}} v_{\text{ip-path}} + e, \tag{6.50}$$

where $s(\cdot)$ is a fitted smoothing spline implemented to taken into account that the mean response is usually nonlinearly associated with the time variable, x_{start}, with the parameters of the smoothing spline chosen automatically by cross-validation. The parameter estimates and their corresponding P-values for model (6.50) are given in Table 6.6. The fitted smoothing spline is plotted in Figure 6.2, which shows how the transfer duration varies with start time.

Furthermore, in (6.50), $Z_{\text{ip-path}}$ is the design matrix whose columns correspond to the ip-paths, and $v_{\text{ip-path}}$ is a vector-valued random effect whose components correspond to the ip-paths, and e is an additional error corresponding to the background noise. The PSS has identified six paths with non-zero random-effect standard deviations, indexed as 14, 16, 38, 41, 61, and 83. Among those paths, all except path 61 have the estimated standard deviation of at least 10, while the estimated standard deviation for path 61 is almost zero. A plot is shown in Figure 6.3. The estimated standard deviation for the background noise is 11.239.

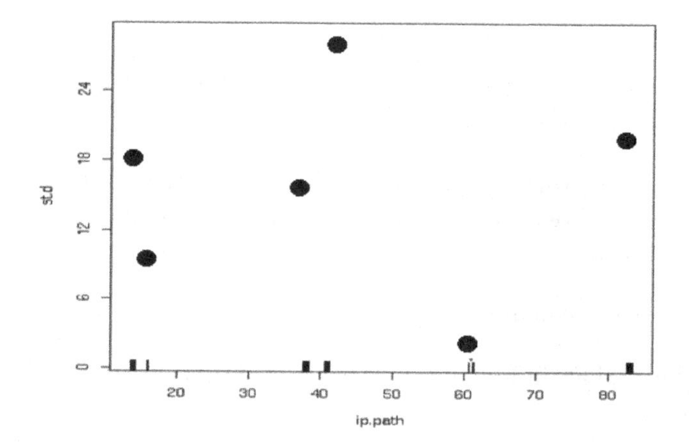

Fig. 6.3 *Estimated Non-zero Standard Deviation.*

Table 6.6 **Estimates of Non-zero Fixed Effects.**

Fixed Effects	Estimates	Standard Error	P-value
Intercept	-13.809	0.914	<2e-16
Start Time	0.574	0.0169	<2e-16
Packets	1.115	0.035	<2e-16

Table 6.7 **Comparison of PSS, B-F, E-Lasso in Term of MSPE and Computing Time.**

	E-Lasso	B-F	PSS
MSPE	2306	42230	127
Time (in seconds)	6.26×10^{7}	5.43×10^{10}	142

Regarding the comparison of PSS with B-F and E-Lasso, the overall MSPE as well as computation speed for the comparing methods are presented in Table 6.7. Note that, in this case, one actually knows the truth for the prediction, and therefore is able to compute the (exact) MSPE, and record the total computational time, of course. The results show that, in terms of prediction accuracy (MSPE), PSS is about 18 times better than E-Lasso, and 330 times better than B-F; in terms of computational time, PSS is about 4×10^5 times less than E-Lasso and 3.8×10^8 times less than B-F. In conclusion, at least for this application, PSS greatly improves the prediction accuracy that fits the interests of modeling noted earlier and, at the same time, provides efficient fast algorithm compared to the E-M based E-Lasso and regression based B-F procedure.

6.4 Exercises

6.1. Show that, in Example 6.1, the conditions of Theorem 6.1 are satisfied with $\rho_N \sim n$ and $\nu_N \sim \omega_N \sim mn$.

6.2. Continue with the previous exercise. Show that, in this case, $\eta_N \sim mn$, where η_N is defined above (6.10). Thus, the conditions of Theorem 6.2 are satisfied.

6.3. Prove Theorem 6.3.

6.4. Show that, in Example 6.2, the Z_r's within each group are linearly independent.

6.5. Let A and B be matrices with the same number of rows such that $P_A P_B = P_B P_A$. Show that $P_{(A\ B)} \leq P_A + P_B$.

6.6. In Example 6.4, it is understood that $y_{ij}, j = 1, \ldots, N_c$ are conditionally independent given π_i. Show that
a. $\mathrm{E}(y_{ij}) = \mu$;
b. $\mathrm{var}(y_{ij}) = \mu(1-\mu)$; and
c. $\mathrm{cor}(y_{ij}, y_{ik}) = \gamma/(1+\gamma)$, $1 \leq j \neq k \leq N_c$.

6.7. In Subsection 6.2.2, the paragraph below the one containing (6.42), an example is given on how many different models can be possibly selected by the fence. Extend the example to a general rule, assuming that the maximum dimension of the candidate models is k.

6.8. In Subsection 6.2.2, a statement is made in the paragraph below the one that contains (6.43) regarding the fast algorithm under a subtractive measure. Prove the statement.

6.9. Consider the GSA method mentioned in Example 2 of Subsection 6.2.4. Let s_i denote the gene set score for gene set i, $1 \leq i \leq n$. Consider a simple case where each gene set consists of a single gene. Then, by restandardization, one subtracts the mean of the s_i's from each s_i, then divides the difference by the standard deviation of the s_i's. Can you explain what would happen to the performance of GSA when one gene set is so dominant that all but one gene set scores are below the overall mean?

6.10. Show that, by treating the ξ_i's in Section 6.3 as observed, the complete-data log-likelihood can be expressed as (6.47).

6.11. Derive the predictive measure of lack-of-fit (6.49) [Hint: this is similar to the derivation of (5.10)].

Chapter 7

Other Topics

There are other types, or aspects, of robust mixed model analysis that, so far, have not been systematically covered. However, these topics are relatively unrelated, or have some unique features of their own. Also, the literature covering each one of these topics is not as extensive as the ones covered in the previous chapters. Due to such considerations, we have put together this final chapter, with a collection of several topics. These include mixed model diagnostics, nonparametric and semiparametric methods, Bayesian analysis, outliers, benchmarking, and more about prediction. It should be noted that some of these topics may be associated, in a certain way, with topics covered in the previous chapters; yet, there are obvious reasons that justify their inclusion in the current chapter. As we have been doing all along, the focus is on robustness, or implication to robustness.

7.1 Mixed model diagnostics

In a way, the first topic is closely related to the previous chapter. Model checking, or model diagnostics, is sometimes the first step of a modeling process in practice; some other times, it is the last step. In fact, quite often the process involves back-and-forth practices of model selection and diagnostics. At each step, a model is selected from amongst a set of candidate models; the selected model is then checked using some model diagnostic techniques (see below). If the diagnostics does not find any problem with the selected model, the latter will be used, and the process ends; otherwise, some further consideration will be implemented, either to the candidate models or to the selection procedure, and the cycle is repeated.

McCullagh and Nelder (1989) (ch. 12) categorizes model diagnostic tools as informal and formal. Informal checking is mainly based on diagnostic plots, while formal checking mostly rely on goodness-of-fit tests. Since the latter have been discussed earlier in Subsection 4.4, the present section will focus on informal checking.

An important feature of mixed effects models is the presence of random effects; however, the latter are unobservable. Thus, in addition to the standard tools for informal checking [e.g., McCullagh and Nelder (1989), ch. 12], special attentions have been paid on checking the assumptions about the random effects. Several authors have used the idea of EBLUP or empirical Bayes estimators (EB), discussed in the previous chapter, for diagnosing distributional assumptions regarding the random effects [e.g., Dempster and Ryan (1985); Calvin and Sedransk (1991)]. The approach is reasonable because the EBLUP or EB are natural estimators of the random effects. In the following we first describe a method proposed by Lange and Ryan (1989) based on a similar idea.

A commonly used assumption regarding the random effects, and errors, is that they are normally distributed. If such an assumption holds, one has a case of Gaussian mixed models; otherwise, one is dealing with a non-Gaussian linear mixed model (see Chapter 3). Lange and Ryan considered the longitudinal model (see Subsection 3.1.2), assuming that $G_i = G$, $R_i = \tau^2 I_{k_i}$, $i = 1, \dots, m$, and developed a weighted normal plot for assessing normality of the random effects in a such a model. First, under the model (3.1), and normality, one can derive the BPs, or Bayes estimators, of the random effects α_i $i = 1, \dots, m$ (see Section 5.1), assuming that β and γ, the vector of variance components, are known. This is given by

$$\tilde{\alpha}_i = \mathrm{E}(\alpha_i|y_i) = GZ_i'V_i^{-1}(y_i - X_i\beta),$$

where $V_i = \mathrm{Var}(y_i) = \tau^2 I_{k_i} + Z_iGZ_i'$. Furthermore, the covariance matrix of $\tilde{\alpha}_i$ is given by $\mathrm{Var}(\tilde{\alpha}_i) = GZ_i'V_i^{-1}Z_iG$. Lange and Ryan proposed to examine a Q–Q plot of the standardized linear combinations

$$z_i = \frac{c'\tilde{\alpha}_i}{\{c'\mathrm{Var}(\tilde{\alpha}_i)c\}^{1/2}}, \quad i = 1, \dots, m, \tag{7.1}$$

where c is a known vector. They argued that, through appropriate choices of c, the plot can be made sensitive to different types of model departures. For example, for a model with two random effects factors, a random intercept and a random slope, one may choose $c_1 = (1, 0)'$ and $c_2 = (0, 1)'$ and produce two Q–Q plots. On the other hand, such plots may not reveal possible nonzero correlations between the (random) slope and intercept. Thus, Lange and Ryan suggested producing a set of plots ranging from one marginal to the other by letting $c = (1-u, u)'$ for some moderate number of values $0 \le u \le 1$.

Dempster and Ryan (1985) suggested that the normal plot should be weighted to reflect the differing sampling variances of $\tilde{\alpha}_i$. Following the same idea, Lange and Ryan proposed a generalized weighted normal plot. They suggested plotting z_i against $\Phi^{-1}\{F^*(z_i)\}$, where F^* is the weighted empirical cdf defined by $F^*(x) = \sum_{i=1}^m w_i 1_{(z_i \le x)} / \sum_{i=1}^m w_i$, with $w_i = c'\text{Var}(\tilde{\alpha}_i)c = c'GZ_i'V_i^{-1}Z_iGc$.

In practice, however, β and γ are unknown. In such cases, Lange and Ryan suggested using the ML or REML estimators in place of these parameters. They argued that, under suitable conditions, the limiting distribution of $\sqrt{n}\{\hat{F}^*(x) - \Phi(x)\}$ is normal with mean zero and variance equal to the variance of $\sqrt{n}\{F^*(x) - \Phi(x)\}$ minus an adjustment, where $\hat{F}^*(x)$ is $F^*(x)$ with the unknown parameters replaced by their ML or REML estimators. See Lange and Ryan (1989) for details. This suggests that, in the case of unknown parameters, the Q–Q plot will be $\hat{z}_i$ against $\Phi^{-1}\{\hat{F}^*(\hat{z}_i)\}$, where $\hat{z}_i$ is z_i with the unknown parameters replaced by their ML (REML) estimates. However, the (asymptotic) variance of $\hat{F}^*(x)$ is different from that of $F^*(x)$, as indicated above. Therefore, if one wishes to include, for example, a ± 1 SD bound in the plot, the adjustment for estimation of parameters must be taken into account. We consider an example.

Example 7.1. Consider, again, the one-way random effects model of Example 3.1 with normality assumption. Note that $a = m$ and $b_i = k_i$ under the new notation. Because α_i is real-valued, $c = 1$ in (7.1). If μ, σ^2, τ^2 are known, the EB estimator of α_i is given by

$$\hat{\alpha}_i = \frac{k_i\sigma^2}{\tau^2 + k_i\sigma^2}(\bar{y}_{i\cdot} - \mu),$$

where $\bar{y}_{i\cdot} = k_i^{-1}\sum_{j=1}^{k_i} y_{ij}$, with

$$w_i = \text{var}(\hat{\alpha}_i) = \frac{k_i\sigma^4}{\tau^2 + k_i\sigma^2}.$$

Therefore, in this case, we have

$$z_i = \frac{\hat{\alpha}_i}{\text{sd}(\hat{\alpha}_i)} = \frac{\bar{y}_{i\cdot} - \mu}{\sqrt{\sigma^2 + \tau^2/k_i}},$$

$i = 1, \dots, m$ and

$$F^*(x) = \left(\sum_{i=1}^m \frac{k_i\sigma^4}{\tau^2 + k_i\sigma^2}\right)^{-1} \sum_{i=1}^n \frac{k_i\sigma^4}{\tau^2 + k_i\sigma^2} 1_{(z_i \le x)}.$$

In practice, μ, σ^2, and τ^2 are unknown and therefore replaced by their REML (ML) estimators when making a Q–Q plot (Exercise 7.1).

7.2 Nonparametric/semiparametric methods

7.2.1 *A P-spline nonparametric model*

One way to achieve robustness is to make the underlying model less restrictive. For example, instead of assuming a linear model, one may include higher order terms, such as quadratic or cubic functions of the covariates. The higher-order model includes the linear model as a special case (when the coefficients of the higher-order terms are zero), and thus is less restrictive than the linear model in that, even if the linear model fails, the higher-order model may still be valid. More generally, one may model the mean function non-parametrically, or semi-parametrically. Let us begin with the Fay-Herriot model of Example 5.1. An extension of the may be written as

$$y_i = f(x_i) + v_i + e_i, \quad i = 1, \dots, m, \tag{7.2}$$

where the assumptions about v_i and e_i are the same as in the Fay-Herriot model (see Subsection 2.1), but $f(\cdot)$ is an unknown function. In order to make inference about $f(\cdot)$ trackable, Opsomer *et al.* (2008) used a P-spline approximation to $f(\cdot)$ in the form of

$$\begin{aligned} \tilde{f}(x) &= \beta_0 + \beta_1 x + \cdots + \beta_p x^p \\ &\quad + \gamma_1 (x - \kappa_1)_+^p + \cdots + \gamma_q (x - \kappa_q)_+^p, \end{aligned} \tag{7.3}$$

where p is the degree of the spline, q is the number of knots, κ_j, $1 \leq j \leq q$ are the knots, and $x_+ = x 1_{(x>0)}$. Graphically, a P-spline is pieces of (pth degree) polynomials smoothly connected at the knots. The P-spline model, which is (7.2) with f replaced by $\tilde{f}$, is fitted by *penalized least squares*—that is, by minimizing

$$|y - X\beta - Z\gamma|^2 + \lambda |\gamma|^2, \tag{7.4}$$

with respect to β and γ, where $y = (y_i)_{1 \leq i \leq n}$, the ith row of X is $(1, x_i, \dots, x_i^p)$, the ith row of Z is $[(x_i - \kappa_1)_+^p, \dots, (x_i - \kappa_q)_+^p]$, $1 \leq i \leq n$, and λ is a penalty, or smoothing, parameter. To determine λ, Wand (2003) used the following interesting connection to a linear mixed model (see the previous chapter). Suppose that the ϵ_i's are distributed as $N(0, \tau^2)$. Then if the γ's are treated as independent random effects with the distribution $N(0, \sigma^2)$, the minimizer of (7.4) is the same as the best linear unbiased estimator (BLUE) for β and the best linear unbiased predictor (BLUP) for γ, provided that λ is identical to the ratio τ^2/σ^2 (Exercise 7.2). Thus, the P-spline model is fitted the same way as the linear mixed model

$$y = X\beta + Z\gamma + \epsilon \tag{7.5}$$

[e.g., Jiang (2007), §2.3.1].

Although the above P-spline–linear mixed model connection is convenient for computational purpose, Jiang (2010) (p. 453) noted that the connection is asymptotically valid only if the true underlying function f is not a P-spline. Nevertheless, in most applications of P-splines, the unknown function f is unlikely to be a P-spline, so, in a way, (7.3) is only used as an approximation, with the approximation error vanishing as $q \to \infty$.

Opsomer *et al.* (2008) incorporated the spline model with random effects. By making use of the P-spline–mixed-model connection, their approximating model has the term $U\alpha$ added to the right side of (7.5):

$$y = X\beta + Z\gamma + U\alpha + \epsilon, \tag{7.6}$$

where α is the vector of random effects and U is a known matrix. It is assumed that γ, α, and ϵ are uncorrelated with means 0 and covariance matrices Σ_γ, Σ_α, and Σ_ϵ, respectively. The BLUE and BLUP are given by

$$\begin{aligned}\tilde{\beta} &= (X'V^{-1}X)^{-1}X'V^{-1}y,\\ \tilde{\gamma} &= \Sigma_\gamma Z'V^{-1}(y - X\tilde{\beta}),\\ \tilde{\alpha} &= \Sigma_\alpha U'V^{-1}(y - X\tilde{\beta}),\end{aligned}$$

where $V = \mathrm{Var}(y) = Z\Sigma_\gamma Z' + U\Sigma_\alpha U' + \Sigma_\epsilon$ [see (5.4), (5.5)]. The EBLUE and EBLUP, denoted by $\hat{\beta}$, $\hat{\gamma}$, and $\hat{\alpha}$, are obtained by replacing V by $\hat{V}$ in the corresponding expressions. Here, we assume that $V = V(\varphi)$, where φ is a vector of variance components, so that $\hat{V} = V(\hat{\varphi})$. The REML estimator is used for $\hat{\varphi}$, as suggested by Opsomer *et al.* (2008).

7.2.2 *Nonparametric model selection*

Note that a P-spline is characterized by p, q, and also the location of the knots. Note that, however, given p, q, the location of the knots can be selected by the space-filling algorithm implemented in R [**cover.design()**]. But the question how to choose p and q remains. The general "rule-of-thumb" is that p is typically between 1 and 3, and q proportional to the sample size, n, with 4 or 5 observations per knot. But there may still be a lot of choices given the rule-of-thumb. For example, if $n = 200$, the possible choices for q range from 40 to 50, which, combined with the range of 1 to 3 for p, gives a total of 33 choices for the P-spline.

Jiang *et al.* (2010) proposed to use a version of the fence methods to choose the degree of the spline, p, the number of knots, q, and the smoothing parameter, λ at the same time. Note that, here again, the true underlying model is not among the class of candidate models, that is, the approximating splines (7.3). This is similar to what was discussed in the beginning of Subsection 6.2.2. Furthermore, the role of λ in the model should be made clear: λ controls the degree of smoothness of the underlying model. A natural measure of lack-of-fit is $Q(M, \theta_M; y) = |y - X\beta - Z\gamma|^2$. However, $Q(M)$ is not obtained by minimizing $Q(M, \theta_M; y)$ over β and γ without constraint. Instead, we have $Q(M) = |y - X\hat{\beta} - Z\hat{\gamma}|^2$, where $\hat{\beta}$ and $\hat{\gamma}$ are the minimizer of (7.4), and hence depends on λ. Note that, instead of using the P-spline-linear mixed model connection, Jiang *et al.* (2010) suggests to solve the minimization problem directly.

Another difference is that there may not be a full model among the candidate models. Therefore, the fence inequality (6.39) is replaced by

$$Q(M) - Q(\tilde{M}) \leq c, \tag{7.7}$$

where $\tilde{M}$ is the candidate model that has the minimum $Q(M)$. We use the following criterion of optimality within the fence which combines model simplicity and smoothness. For the models within the fence, choose the one with the smallest q; if there are more than one such models, choose the model with the smallest p. This gives the best choice of p and q. Once p, q are chosen, we choose the model within the fence with the largest λ. The tuning constant c is chosen adaptively using the AF idea (see Subsection 6.2.1), where parametric bootstrap is used for computing p^*. More specifically, $\tilde{M}$ and ML estimators under $\tilde{M}$ are used for the bootstrapping.

The following theorem is proved in Jiang *et al.* (2010). For simplicity, assume that the matrix $W = (X\ Z)$ is of full rank. Let $P_{W^\perp} = I_n - P_W$, where $n = \sum_{i=1}^m n_i$ and $P_W = W(W'W)^{-1}W'$.

Theorem 7.1. Computationally, the above fence procedure is equivalent to the following: (i) first use the AF to select p and q using (7.7) with $\lambda = 0$ and $Q(M) = y'P_{W^\perp}y$ (see Lemma 7.1 below), and same criterion as above for choosing p, q within the fence; (ii) let M_0^* denotes the model corresponding to the selected p and q, find the maximum λ such that

$$Q(M_0^*, \lambda) - Q(\tilde{M}) \leq c^*, \tag{7.8}$$

where for any model M with the corresponding X and Z, we have

$$\begin{aligned}
Q(M, \lambda) &= |y - X\hat{\beta}_\lambda - Z\hat{\gamma}_\lambda|^2, \\
\hat{\beta}_\lambda &= (X'V_\lambda^{-1}X)^{-1}X'V_\lambda^{-1}y, \\
\hat{\gamma}_\lambda &= \lambda^{-1}(I_q + \lambda^{-1}Z'Z)^{-1}Z'(y - X\hat{\beta}_\lambda), \\
X'V_\lambda^{-1}X &= X'X - \lambda^{-1}X'Z(I_q + \lambda^{-1}Z'Z)^{-1}Z'X, \\
X'V_\lambda^{-1}y &= X'y - \lambda^{-1}X'Z(I_q + \lambda^{-1}Z'Z)^{-1}Z'y,
\end{aligned}$$

and c^* is chosen by the AF (V_λ is defined below but not directly needed here for the computation because of the last two equations).

Note that in step (i) of the Theorem one does not need to deal with λ. The motivation for (7.8) is that this inequality is satisfied when $\lambda = 0$, so one would like to see how far λ can go. In fact, the maximum λ is a solution to the equation $Q(M_0^*, \lambda) - Q(\tilde{M}) = c^*$. The purpose of the last two equations is to avoid direct inversion of $V_\lambda = I_n + \lambda^{-1}ZZ'$, whose dimension is equal to n, the total sample size. Note that V_λ does not have a block diagonal structure because of ZZ', so if n is large direct inversion of V_λ may be computationally burdensome.

The proof of the Theorem requires the following lemma, whose proof is left as an exercise (Exercise 7.3).

Lemma 7.1. For any M and y, $Q(M, \lambda)$ is an increasing function of λ with $\inf_{\lambda>0} Q(M, \lambda) = Q(M)$.

7.2.3 *Examples*

1. A simulation study. Jiang *et al.* (2010) carried out a simulation study designed to evaluate performance of the proposed fence method. Consider (7.2) with $D_i = D$, $1 \le i \le m$. The model can be written as

$$y_i = f(x_i) + \epsilon_i, \quad i = 1, \ldots, m, \tag{7.9}$$

where $\epsilon_i \sim N(0, \sigma^2)$ with $\sigma^2 = A + D$, which is unknown. Thus, the model is the same as a nonparametric regression model.

Consider three different cases that cover various situations and aspects. In the first case, Case 1, the true underlying function is a linear function, $f(x) = 1 - x$, $0 \le x \le 1$, hence the model reduces to the traditional Fay-Herriot model. The goal is to find out if the fence can validate the traditional Fay-Herriot model in the case that it is valid.

In the second case, Case 2, the true underlying function is the quadratic spline with two knots, given by

$$f(x) = 1 - x + x^2 - 2(x-1)_+^2 + 2(x-2)_+^2, \quad 0 \le x \le 3 \tag{7.10}$$

(the shape is half circle between 0 and 1 facing up, half circle between 1 and 2 facing down, and half circle between 2 and 3 facing up). Note that this function is smooth in that it has a continuous derivative (Exercise 7.4). Here the intention is to investigate whether the fence can identify the true function in the "perfect" situation, that is, when $f(x)$ itself is a spline.

The last case, Case 3, is perhaps the most practical situation, in which no spline can provide a perfect approximation to $f(x)$. In other words, the

Table 7.1 **Nonparametric Model Selection - Case 1 & Case 2.** *Reported are empirical probabilities, in terms of percentage, based on 100 simulations that the optimal model is selected.*

	Case 1			Case 2		
Sample size	$m = 10$	$m = 15$	$m = 20$	$m = 30$	$m = 40$	$m = 50$
Highest Peak	62	91	97	71	83	97
Confidence L.B.	73	90	97	73	80	96

true underlying function is not among the candidates. In this case $f(x)$ is chosen as $0.5\sin(2\pi x)$, $0 \le x \le 1$, which is one of the functions considered by Kauermann (2005).

Consider situations of small or median sample size, namely, $m = 10$, 15 or 20 for Case I, $m = 30$, 40 or 50 for Case 2, and $m = 10$, 30 or 50 for Case 3. The covariate x_i are generated from the Uniform$[0, 1]$ distribution in Case 1, and from Uniform$[0, 3]$ in Case 2; then fixed throughout the simulations. Following Kauermann (2005), we let x_i be equidistant in Case 3. The error standard deviation σ in (7.9) is chosen as 0.2 in Case 1 and Case 2. This value is chosen such that the signal standard deviation in each case is about the same as the error standard deviation. As for Case 3, we consider three different values for σ, 0.2, 0.5 and 1.0. These values are also of the same order as the signal standard deviation in this case.

The candidate approximating splines for Case 1 and Case 2 are the following: $p = 0, 1, 2, 3$, $q = 0$ and $p = 1, 2, 3$, $q = 2, 5$ (so there are a total of 10 candidates). As for Case 3, following Kauermann (2005), we consider only linear splines (i.e., $p = 1$); furthermore, we consider the number of knots in the range of the "rule-of-thumb" (i.e., roughly 4 or 5 observations per knot; see Section 7.2.2.), plus the intercept model ($p = q = 0$) and the linear model ($p = 1$, $q = 0$). Thus, for $m = 10$, $q = 0, 2, 3$; for $m = 30$, $q = 0, 6, 7, 8$; and for $m = 50$, $q = 0, 10, 11, 12, 13$.

Table 7.1 shows the results based on 100 simulations under Case 1 and Case 2. As in Jiang *et al.* (2009), we consider both the highest peak, that is, choosing c with the highest p^*, and 95% lower bound, that is, choosing a smaller c corresponding to a peak of p^* in order to be conservative, if the corresponding p^* is greater than the 95% lower bound of the p^* for any larger c that corresponds to a peak of p^*. It is seen that performance of the AF is satisfactory even with the small sample size. Also, it appears that the confidence lower bound method works better in smaller sample, but makes almost no difference in larger sample. These are consistent with the findings of Jiang *et al.* (2009).

Table 7.2 shows the results for Case 3. Note that, unlike Case 1 and

Table 7.2 **Nonparametric Model Selection - Case 3.** *Reported are empirical distributions, in terms of percentage, of the selected models.*

	Sample Size	$m = 10$		$m = 30$		$m = 50$	
	# of Knots	0,2,3		0,6,7,8		0,10,11,12,13	
		(p,q)	%	(p,q)	%	(p,q)	%
$\sigma = .2$	Highest Peak	(0,0)	1	(1,0)	9	(1,10)	100
		(1,0)	31	(1,6)	91		
		(1,2)	68				
	Confidence L.B.	(1,0)	24	(1,0)	9	(1,10)	100
		(1,2)	76	(1,6)	91		
$\sigma = .5$	Highest Peak	(0,0)	14	(1,0)	21	(1,0)	13
		(1,0)	27	(1,6)	77	(1,10)	84
		(1,2)	56	(1,7)	2	(1,11)	2
		(1,3)	3			(1,12)	1
	Confidence L.B.	(0,0)	8	(1,0)	8	(1,0)	2
		(1,0)	23	(1,6)	89	(1,10)	94
		(1,2)	65	(1,7)	3	(1,11)	2
		(1,3)	4			(1,12)	2
$\sigma = 1$	Highest Peak	(0,0)	27	(0,0)	15	(0,0)	10
		(1,0)	20	(1,0)	18	(1,0)	26
		(1,2)	49	(1,6)	63	(1,10)	60
		(1,3)	4	(1,7)	4	(1,11)	2
						(1,12)	2
	Confidence L.B.	(0,0)	20	(0,0)	1	(0,0)	2
		(1,0)	13	(1,0)	13	(1,0)	13
		(1,2)	59	(1,6)	82	(1,10)	80
		(1,3)	8	(1,7)	4	(1,11)	2
						(1,12)	3

Case 2, here there is no optimal model (an optimal model must be a true model, according to our definition, which does not exist). So, instead of giving the empirical probabilities of selecting the optimal model, we give the empirical distribution of the selected models in each case. It is apparent that, as σ increases, the distribution of the models selected becomes more spread out. A reverse pattern is observed as m increases. The confidence lower bound method appears to perform better in picking up a model with splines. Within the models with splines, fence seems to overwhelmingly prefer fewer knots than more knots.

Note that the fence procedure allows one to choose not only p and q but also λ. In each simulation we compute $\hat{\beta} = \hat{\beta}_\lambda$ and $\hat{\gamma} = \hat{\gamma}_\lambda$, given below (7.8), based on the λ chosen by the adaptive fence. The fitted values are calculated by (7.3) with β and γ replaced by $\hat{\beta}$ and $\hat{\gamma}$, respectively. We then average the fitted values over the 100 simulations. Figure 7.1 shows the average fitted values for the three cases ($m = 10, 30, 50$) with $\sigma = 0.2$

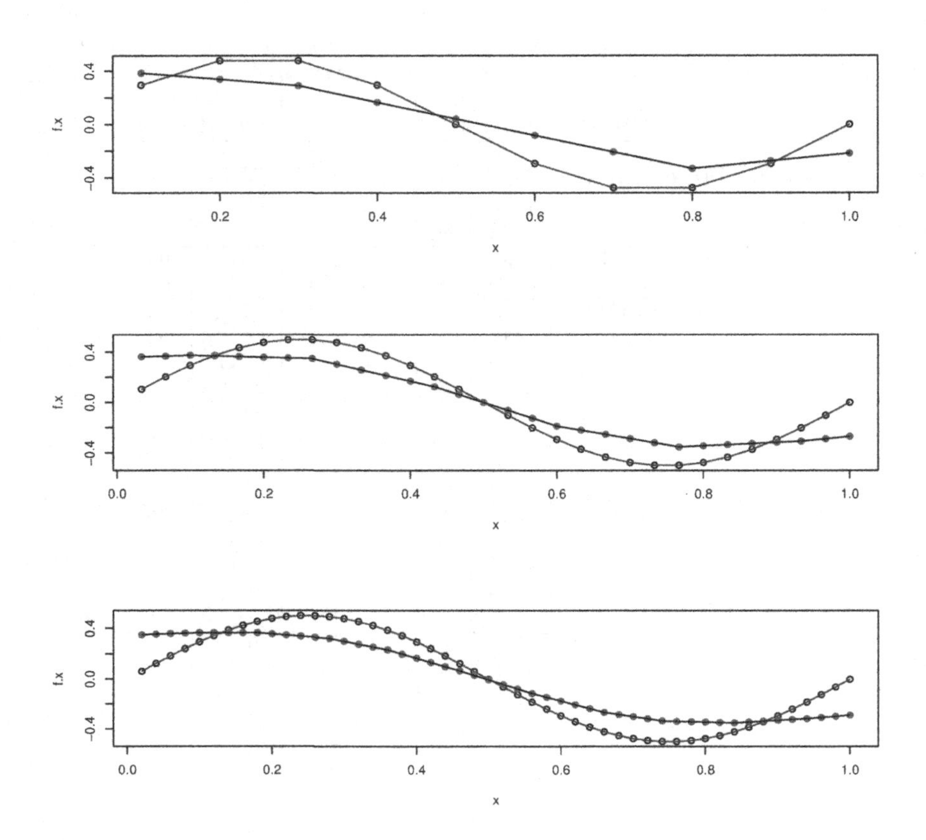

Fig. 7.1 *Case 3 Simulation. Top figure: Average fitted values for $m = 10$. Middle figure: Average fitted values for $m = 30$. Bottom figure: Average fitted values for $m = 50$. In all cases, the red dots represent the fitted values, while the blue circles correspond to the true underlying function.*

under Case 3. The true underlying function values, $f(x_i) = 0.5\sin(2\pi x_i)$, $i = 1, \ldots, m$ are also plotted for comparison.

2. Hospital data revisited. Recall the hospital data of Subsection 5.9.1. Ganesh (2009) proposed a Fay-Herriot model for the graft failure rates as follows: $y_i = \beta_0 + \beta_1 x_i + v_i + e_i$, where the v_i's are hospital-specific random effects and e_i's are sampling errors. It is assumed that v_i, e_i are independent with $v_i \sim N(0, A)$ and $e_i \sim N(0, D_i)$. Here the variance A is unknown. Based on the model, Ganesh obtained credible intervals for selected contrasts. However, inspections of the raw data suggest some nonlinear trends, which raises the question on whether the fixed effects part of the model can be made more flexible in its functional form.

To answer this question, we consider the Fay-Herriot model as a special member of a class of approximating spline models discussed in this section. More specifically, we assume the model (7.2) with x_i being the severity index. We then consider the approximating spline (7.3) with $p = 0, 1, 2, 3$ and $q = 0, 1, \dots, 6$ ($p = 0$ is only for $q = 0$). Here the upper bound 6 is chosen according to the "rule-of-thumb" (because $m = 23$, so $m/4 = 5.75$). Note that the Fay-Herriot model corresponds to the case $p = 1$ and $q = 0$. The question is then to find the optimal model, in terms of p and q, from this class.

We apply the AF method described in Subsection 7.2.2 to this case. Here to obtain the bootstrap samples needed for obtaining c^*, we first compute the ML estimator under the model $\tilde{M}$, which minimizes $Q(M) = y'P_{W^\perp}y$ among the candidate models [i.e., (6.2); see Theorem 6.1], then draw parametric bootstrap samples under $\tilde{M}$ with the ML estimators treated as the true parameters. This is reasonable because $\tilde{M}$ is the best approximating model in terms of the fit, even though under model (7.2) there may not be a true model among the candidate models. The bootstrap sample size is chosen as 100.

The fence method has selected the model $p = 3$ and $q = 0$, that is, a cubic function with no knots, as the optimal model. We repeated the analysis 100 times, each time using different bootstrap samples. All results led to the same model: a cubic function with no knots. The left figure of Figure 7.2 shows the plot of p^* against $c = c_n$ in the AF model selection.

A few comparisons are always helpful. Our first comparison is to fence itself but with a more restricted space of candidate models. More specifically, we consider (7.3) with the restriction to linear splines only, that is, $p = 1$, and knots in the range of the "rule-of-thumb", namely, $q = 4, 5, 6$, plus the intercept model ($p = q = 0$) and the linear model ($p = 1$, $q = 0$). In this case, the fence method selected a linear spline with four knots (i.e., $p = 1$, $q = 4$) as the optimal model. The value of λ corresponding to this model is approximately equal to 0.001. The plot of p^* against c_n for this model selection is very similar to the left figure of Figure 7.2, and therefore omitted. In addition, the right figure of Figure 7.2 shows the fitted values and curves under the two models selected by the fence from within the different model spaces as well as the original data points.

A further comparison can be made by treating (7.2) as a generalized additive model [GAM; e.g., Hastie and Tibshirani (1990)] with heteroscedastic errors. A weighted fit can be obtained with the amount of smoothing optimized by using a generalized cross-validation (GCV) criterion. Here the weights used are $w_i = 1/(A + D_i)$ where the ML estimate for A is used as a plug-in estimate. Recall that the D_i's are known. This fitted function is also overlayed in the right plot of Figure 7.2. Notice how closely this fitted function resembles the restricted space fence fit.

To expand the class of models under consideration by GCV-based smoothing, we used the BRUTO procedure [Hastie and Tibshirani (1990)] which augments the class of models to look at a null fit and a linear fit for the spline function; and embeds the resulting model selection (i.e., null, linear or smooth fits) into a weighted backfitting algorithm using GCV for computational efficiency. Interestingly here, BRUTO finds simply an overall linear fit for the fixed effects functional form. While certainly an interesting comparison, BRUTO's theoretical properties for models like (7.2) have not really been studied in depth.

Finally, as mentioned in Subsection 7.2.1, by using the connection between P-spline and linear mixed model one can formulate (6.2) as a linear mixed model, where the spline coefficients are treated as random effects. The problem then becomes a (parametric) mixed model selection problem, hence the method of Section 6.2 can be applied. In fact, this was our initial approach to this dataset, and the model we found was the same as the one by BRUTO. However, there is some reservation about this approach, as explained in Subsection 7.2.1.

7.2.4 *Functional and semiparametric mixed models*

There have been studies in functional mixed effects models. These include models for functional data, in which the random effects are Euclidean (i.e., not functional), and models in which both the data and the random effects are functional. Guo Guo (2002) noted that data in many cases arise as curves, such as growth curves, bormone profiles, and biomarkers measured over time. The author proposed a functional mixed effects model, in which the random effects are modeled as realizations of a Gaussian process. Suppose that there is a response curve associated with each of the m subjects. Let y_{ij} be the observed value of the ith curve at time t_{ij}, $i = 1, \dots, m, j = 1, \dots, n_i$ such that $y_{ij} = x_{ij}'\beta(t_{ij}) + z_{ij}'\alpha_i(t_{ij}) + \epsilon_{ij}$, where $\beta(t) = [\beta_k(t)]_{1 \le k \le p}$ is a $p \times 1$ vector of fixed functions, $\alpha_i(t) = [\alpha_k(t)]_{1 \le k \le q}$ is a $q \times 1$ vector of random functions that are modeled as realizations of Gaussian processes, $A(t) = [\alpha_k(t)]_{1 \le k \le q}$, with zero means,

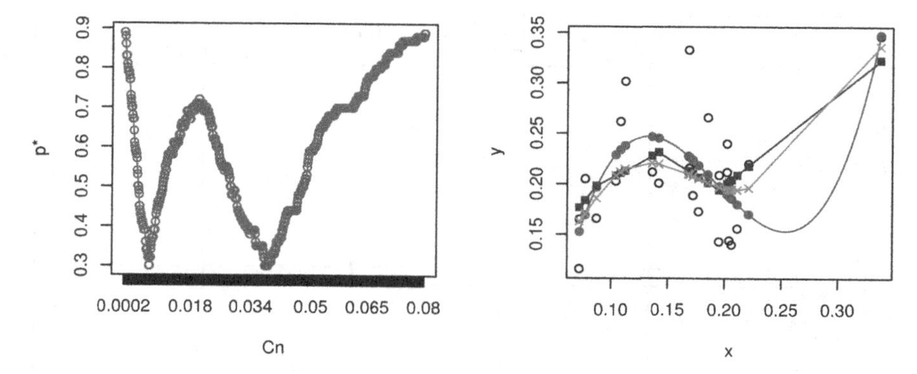

Fig. 7.2 *Left: A plot of p^* against $c = c_n$ from the search over the full model space. Right: The raw data and the fitted values and curves; red dots and curve correspond to the cubic function resulted from the full model search; blue squares and lines correspond to the linear spline with 4 knots resulted from the restricted model search; green X's and lines represent the GAM fits.*

$x_{ij} = (x_{ijk})_{1 \leq k \leq p}$ and $z_{ij} = (z_{ijk})_{1 \leq k \leq q}$ are design matrices that include covariates as well as dummy variables, and $\epsilon_{ij} \sim N(0, \sigma_\epsilon^2)$ and are independent with the α's.

In fitting the proposed model, Guo first approximated both the fixed and random functions by smoothing splines. He then used a connection built by Wahba Wahba (1978), [Wahba (1983)] to model $\beta(t)$ and $A(t)$ as

$$\beta_k(t) = b_{1k} + b_{2k}t + \lambda_{b,k}^{-1/2} \int_0^t W_{b,k}(s)ds, \quad k = 1, \ldots, p;$$

$$\alpha_k(t) = a_{1k} + a_{2k}t + \lambda_{a,k}^{-1/2} \int_0^t W_{a,k}(s)ds, \quad k = 1, \ldots, q,$$

where $(b_{1k}, b_{2k})' \sim N(0, \tau I_2)$ with $\tau \to \infty$ (diffuse prior), and $(a_{1k}, a_{2k})' \sim N(0, \sigma_k^2 D)$, D being an unknown covariance matrix, and $W_{b,k}(s), W_{a,k}(s)$ are Weiner processes. Here, a_{1k} and a_{2k} are considered random intercept and random slope, respectively.

In terms of statistical inference, the latest author was mainly interested in testing linearity of $\beta_k(t)$ and $\alpha_k(t)$, that is, $\lambda_{b,k}^{-1} = 0$, or $\lambda_{a,k}^{-1} = 0$. Using a result from [Self, S. G. and Liang, K. Y. (1987)] on likelihood ratio test for parameters on the boundary of the parameter space. More specifically, Guo Guo (2002) showed that, under the null hypothesis of $\lambda_{b,k}^{-1} = 0$, the asymptotic distribution of the likelihood ratio statistic (LRS) is a mixture of 0 and χ_1^2 with equal weight (1/2 for both). As for testing for a single component of $\beta(t)$, that is, $\beta_k(t) = 0$, it is equivalent to testing

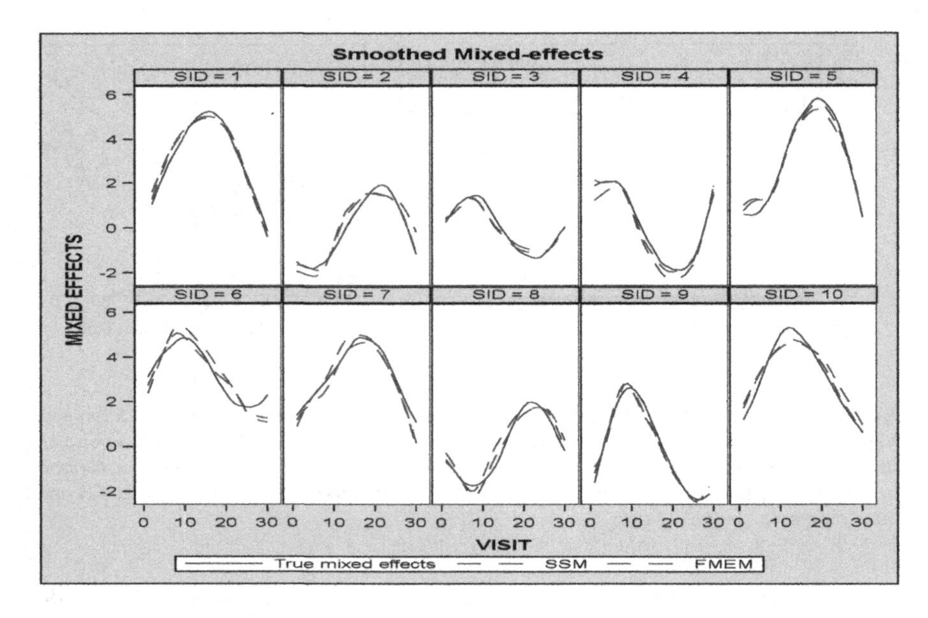

Fig. 7.3 *Fitted Curves via SSM and FMEM*

for $b_{1k} = b_{2k} = \lambda_{b,k}^{-1} = 0$. The asymptotic null distribution of the LRS is a mixture of χ_2^2 and χ_3^2 with equal weight.

Example 7.2. We illustrate the fitting of the functional mixed effects model using a simulated example. The data were generated under the model $y_i(t) = 5x_{i1}\sin(2\pi t) + 5x_{i2}\sin(\pi t) + r_i(t) + e_i$, $i = 1, \ldots, 10$, $t \in \{t_1, \ldots, t_{30}\}$. For each subject i, x_{i1} is generated from Uniform$[-0.5, 0.5]$ and x_{i2} is binary with equal number of 0s and 1s. Thus, we have $\beta_1(t) = 2\sin(2\pi t)$, $\beta_2(t) = 5\sin(\pi t)$. Furthermore, $r_i(t) \sim N(0, \sigma_1^2\Sigma(t))$ with $\sigma_1^2 = 90$ and $\Sigma(t)$ is simulated by the reproducing kernel $R^1(t_i, t_i)$ [e.g., Guo (2002)]. Finally, $e_i = (e_{ij})_{1\leq j\leq 30}$ and $e_{ij} \overset{\text{i.i.d.}}{\sim} N(0, 1)$. The generated data are subject to random missing with probability 0.1. The model were fitted using two methods proposed in Guo's paper, the state space model (SSM) and functional mixed effects model (FMEM). There are software packages in SAS that implement these methods, PROC SSM for SSM and PROC MIXED for FMEM. See Figure 7.3 for the illustration.

Krafty *et al.* (2011) considered functional mixed effects spectral analysis. In biomedical experiments, data are often collected from multiple subjects as time series. It is natural to use carry out time series analysis to study the effects of design covariates. One of the standard analysis in time series is spectral analysis (e.g., [Koopmans (1995)]). Here, a random effect is associated with a group of subjects, while multiple time series correspond to subjects within the group. The random effects are function-valued, and different groups are assumed to be independent. A novel contribution of the work is a mixed effects Cramér representation in the following form:

$$X_{jkt} = \int_{-1/2}^{1/2} A_0(\omega; U_{jk}) A_j(\omega; V_{jk}) e^{2\pi i t\omega} dZ_{jk}(\omega),$$

$j = 1, \ldots, N, k = 1, \ldots, n_j$, where N is the number of groups and n_j the number of time series within the jth group; X_{jkt} is the kth replicate time series of the jth group; U_{jk}, V_{jk} are vectors of covariates, $A_0(\cdot; U_{jk}), A_j(\cdot; V_{jk})$ are functions corresponding to the fixed and random effects, respectively. Through asymptotic analysis, the authors showed that, when the replicate-specific spectra are smooth, the log-periodograms converge to a functional mixed effects model. The authors also developed BLUP for the unit-specific random effects.

In another related work, Rady *et al.* (2015) considered estimation in mixed-effects functional ANOVA models. The authors considered the following so-called mixed functional ANOVA model:

$$y_{ij}(t) = \mu(t) + \alpha_i(t) + \beta_j(t) + \epsilon_{ij}(t), \quad i = 1, \ldots, a, j = 1, \ldots, b.$$

This is similar to a two-way random effects ANOVA model (see Example 3.2) except that everything involved is a function. However, in their analysis, the authors have focused on a fixed value of t. Such results have little interest in terms of understanding the functional relationship.

Semiparametric GLMM also has received attention in the literature. For example, Lombardía and Sperlich (2008) considered an extension of GLMM in that the conditional mean of the response given the random effects can be expressed as $\mu_{ij} = \mathrm{E}(y_{ij}|\alpha_i, T_{ij}, X_{ij}) = g\{\lambda(T_{ij}) + x'_{ij}\beta + z'_{ij}\alpha_i\}$, where α_i is a vector-valued random effect, x_{ij}, t_{ij} are observed vectors of regressors, and z_{ij} is a subvector of $(1, x'_{ij})'$. Furthermore, $\lambda(\cdot)$ is an unknown function. It is assumed that
(i) the responses y_{ij} are conditionally independent given α, T, X;
(ii) the random effects are i.i.d. with mean 0 and covariance matrix Σ; and
(iii) (T, X) are independent with α.
The authors combine likelihood approach for mixed effects models with kernel methods. In term of asymptotic properties, a point-wise asymptotic normality result regarding estimation of the function $\lambda(\cdot)$. As for measure

of uncertainty, the authors proposed a bootstrap procedure and provided a theoretical justification, also in terms of asymptotic approximation of the distribution of (y, X, T) by the bootstrap distribution. The authors discussed application of their method to SAE [e.g., Rao and Molina (2015); also see Chapter 5], currently a very active area of research and applications.

7.2.5 *Nonparametric bootstrapping*

Model-based (or parametric) bootstrap methods have been used in mixed model analysis; see, for example, Jiang (2017) (sec. 4.5.3). It is a bit tricky, though, to bootstrap nonparametrically in a mixed model situation. Clearly, Efron's original idea [Efron (1979)] of i.i.d. bootstrapping would not work due to the fact that the observations are neither independent nor identically distributed.

On the other hand, the random effects in a mixed effects model are often assumed to be i.i.d., for each random effect factor. The difficulty is that the random effects are unobserved so, once again, Efron's i.i.d. bootstrap would not work. [Hall and Maiti (2006)] proposed a method to bootstrap the random effects nonparametrically, at least in some cases of mixed effects models. Consider the NER model (e.g., Subsection 5.4.2). Suppose that the random effects are i.i.d. with an unknown distribution. The question is how to generate samples of the random effects? An answer by Hall and Maiti: Depending on what one wants to do. In many cases, the quantity of interest does not involve every piece of information about the distribution of the random effects. For example, Hall and Maiti observed that the MSPE of EBLUP involves only the second and fourth moments of the random effects and errors, up to the order of $o(m^{-1})$. This means that for random effects and errors from any distributions with the same second and fourth moments, the MSPE of EBLUP, with some suitable estimators of the variance components, are different only by a term of $o(m^{-1})$.

This observation leads to a seemingly simple strategy: First, estimate the second and fourth moments of the random effects and errors; then, draw bootstrap samples of the random effects and errors from certain distributions that match the first (which is 0) and estimated second and fourth moments; given the bootstrapped random effects and errors, use the NER model to generate the bootstrapped data. Specifically, let the NER model be expressed as

$$y_{ij} = x_{ij}'\beta + v_i + e_{ij}, \quad i = 1, \ldots, m, j = 1, \ldots, n_i. \tag{7.11}$$

Then, the bootstrapped data, y_{ij}, are generated by (7.11) with β replaced by $\hat{\beta}$, the empirical BLUE (EBLUE) of β given by (5.4) with the variance components involved in V by consistent estimators, and v_i, e_{ij} replaced by v_i^*, e_{ij}^*, respectively, where v_i^*, e_{ij}^*, $i = 1, \dots, m$, $j = 1, \dots, n_i$ are the bootstrapped random effects and errors.

Regarding the moment-matching bootstrap of the random effects and errors, clearly, the choice is not unique. A simple choice is a three-point distribution depending on a single parameter $p \in (0, 1)$, defined as $\mathrm{P}(\xi = 0) = 1 - p$ and

$$\mathrm{P}\left(\xi = \frac{1}{\sqrt{p}}\right) = \mathrm{P}\left(\xi = -\frac{1}{\sqrt{p}}\right) = \frac{p}{2}. \tag{7.12}$$

It is easy to show (Exercise 7.5) that $\mathrm{E}(\xi) = 0, \mathrm{E}(\xi^2) = 1$, and $\mathrm{E}(\xi^4) = 1/p$. Thus, if we let $p = \mu_2^2/\mu_4$, the distribution of $\sqrt{\mu_2}\xi$ has second moment μ_2 and fourth moment μ_4). Note that the inequality $\mu_2^2 \le \mu_4$ always hold with the equality holding if and only if ξ^2 is degenerate (i.e., a.s. a constant).

Another choice is the rescaled Student t-distribution whose degrees of freedom (d.f.) $\nu > 4$ and is not necessarily an integer. The distribution has first and third moments 0, $\mu_2 = 1$, and $\mu_4 = 3(\nu-2)/(\nu-4)$, which is always greater than 3, meaning that the tails are heavier than those of the normal distribution. However, the moment-matching may not always take place. To see this, let $\hat{\mu}_2$ and $\hat{\mu}_4$ be the estimated second and fourth moments of the random effect, and η have the rescaled t-distribution as above. Then, $\hat{\mu}_2^{1/2}\eta$ has second moment $\hat{\mu}_2$ and fourth moment $3(\nu - 2)\hat{\mu}_2^2/(\nu - 4)$. By letting the latter equal to $\hat{\mu}_4$, one has

$$\hat{\mu}_4/\hat{\mu}_2^2 = 3(\nu - 2)/(\nu - 4) > 3, \tag{7.13}$$

which may not hold even if the true moments satisfy $\mu_4/\mu_2^2 > 3$. When (7.16) does not hold, the matching d.f., ν, cannot be found.

Based on the moment-matching bootstrap, Hall and Maiti (2006) developed a double-bootstrap MSPE estimator and showed that it is second-order unbiased. On the other hand, the moment-matching, double-bootstrap procedure may be computationally intensive, and so far there have not been published comparisons of the method with other existing methods that also produce second-order unbiased MSPE estimators, such as the Prasad–Rao (see Subsection 5.8) and jackknife [Jiang *et al.* (2002)]. On the other hand, the idea of [Hall and Maiti (2006)] is potentially applicable to mixed models with more complicated covariance structure, such as mixed models with crossed random effects. We conclude this section with an illustrative example.

Example 7.3. Consider a two-way random effects model

$$y_{ij} = \mu + u_i + v_j + e_{ij},$$

$i = 1, \dots, m_1, j = 1, \dots, m_2$, where μ is an unknown mean; the u_i's are i.i.d. with mean 0, variance σ_1^2, and an unknown distribution F_1; the v_j's are i.i.d. with mean 0, variance σ_2^2, and an unknown distribution F_2; the e_{ij}'s are i.i.d. with mean 0, variance σ_0^2, and an unknown distribution F_0; and u, v, and e are independent. Note that the observations are *not clustered* under this model. As a result, the jackknife method of [Jiang *et al.* (2002)] may not apply in estimating the MSPE of EBLUP in this case. However, it is fairly easy to obtain estimators of the second and fourth moments of the random effects and errors. For the second moments one may use the ML or REML methods (see Section 3.2); for the fourth moments one may use the EMM method (see Subsection 3.3.1; also Theorem 4.8). It would be interesting to see if the moment-matching, double-bootstrap method can produce a second-order unbiased MSPE estimator in this case.

7.3 Bayesian analysis

7.3.1 *A robust hierarchical Bayes method*

In the context of SAE, Chakraborty *et al.* (2018) showed that a hierarchical Bayes (HB) small area predictor proposed by Datta and Ghosh (1991) is not robust to outliers in a way similar to the EBLUP. Following Sinha and Rao (2009)'s effort to robustify the EBLUP (see below), Chakraborty *et al.* (2018) proposed a robust Bayesian method based on a normal mixture idea. A similar approach was considered by Gershunskaya (2018), but under a non-Bayesian framework. Below we shall focus on describing the method of Chakraborty *et al.* (2018).

They considered a Bayesian version of the NER model [Battese *et al.* (1988), see Subsection 5.4.2] except that the distribution of the unit-level errors is assumed to follow a two-component normal mixture distribution, instead of normal distribution. Specifically, their proposed normal-mixture (NM) HB model is defined as follows:
(I) Conditional on $\beta = (\beta_j)_{1 \le j \le p}$, $v = (v_i)_{1 \le i \le m}$, z_{ij}, $1 \le i \le m$, $1 \le j \le N_i$, p_e, σ_1^2, σ_2^2, and σ_v^2, $Y_{ij}, 1 \le i \le m, 1 \le j \le N_i$ are independent with

$$Y_{ij} \sim z_{ij} N(x_{ij}'\beta + v_i, \sigma_1^2) + (1 - z_{ij}) N(x_{ij}'\beta + v_i, \sigma_2^2). \tag{7.14}$$

(II) The z_{ij}'s are i.i.d. indicators with $\mathrm{P}(z_{ij} = 1|p_e) = p_e$, and are independent of β, v, σ_1^2, σ_2^2, and σ_v^2.

(III) Conditional on β, the z_{ij}'s, p_e, σ_1^2, σ_2^2, and σ_v^2, the random effects $v_i, 1 \le i \le m$ are independent and distributed as $N(0, \sigma_v^2)$.

Note that, in (I), both components of the NM have the same mean but different variances; equivalently, this means that the unit-level error, e_{ij}, has a NM distribution, in which both components have mean zero but different variances, σ_1^2 and σ_2^2, with the larger variance corresponding to source of the outliers. The prior for the parameters was assumed to be noninformative; sufficient conditions were given to ensure propriety of the posterior. The prior was also chosen carefully so that the conditional distributions, used in the Markov chain Monte-Carlo computation are simple.

The authors studied empirical, frequentist properties of their proposed Bayesian predictors, and showed that they performed similarly to the robust EBLUP of Sinha and Rao (2009) when outliers were present, and both methods outperformed the HB method of Datta and Ghosh (1991) and the M-quantile method of Chambers and Tzavidis (2006). In the absence of outliers, the proposed method performed similarly to that of Datta and Ghosh (1991). As an application, Chakraborty *et al.* (2018) reanalyzed the Iowa crops data of Battese *et al.* (1988). We present their analysis results below as an example.

7.3.2 *Real-data example: Iowa crops data*

One of the well-known problems in SAE was discussed in Battese *et al.* (1988), in which the authors presented data from 12 Iowa counties obtained from the 1978 June Enumerative Survey of the U.S. Department of Agriculture as well as from land observatory satellites on crop areas involving corn and soybeans. The objective was to predict the mean hectares of corn and soybeans per segment for the 12 counties using the satellite information. In this paper, the authors introduced, for the first time, the NER model that has since become popular in SAE. The model for the crops data can be expressed as

$$y_{ij} = \beta_0 + \beta_1 x_{1ij} + \beta_2 x_{2ij} + v_i + e_{ij}, \tag{7.15}$$

$i = 1, \ldots, 12$, $j = 1, \ldots, n_i$, where n_i ranges from 1 to 6. Here i represents county and j segment within the county; y_{ij} is the number of hectares of corn (or soybeans); x_{1ij} and x_{2ij} are the number of pixels classified as corn and soybeans, respectively, according to the satellite data. Furthermore, v_i is a county-specific random effect, and e_{ij} is the sampling error.

Table 7.3 **Data for Hardin County [Battese *et al.* (1988)]**

Reported		# of Pixels		Mean # of Pixels	
Corn	Soybeans	Corn	Soybeans	Corn	Soybeans
88.59	102.59	220	262	325.99	177.05
88.59	29.46	340	87		
165.35	69.28	355	160		
104.00	99.15	261	221		
88.63	143.66	187	345		
153.70	94.49	350	190		

It is assumed that the random effects are independent and distributed as $N(0, \sigma_v^2)$, the sampling errors are independent and distributed as $N(0, \sigma_e^2)$, and the random effects and sampling errors are uncorrelated. The primary interest was to estimate the mean hectares of crops, which can be expressed as $\zeta_i = \beta_0 + \beta_1 \bar{x}_{1i(p)} + \beta_2 \bar{x}_{2i(p)} + v_i$, where $\bar{x}_{1i(p)}$ and $\bar{x}_{2i(p)}$ are the population mean numbers of pixels classified as corn and soybeans per segment, respectively, for the ith county, which are available.

The corn data are the focus in this analysis. Battese *et al.* (1988) suggested that the second observation from Hardin county was an outlier. An inspection of the data from that county (see Table 7.3) tells why: the second observation is exactly the same as the first one, possibly due to an recording error. Battese *et al.* (1988) recommended to simply remove this case from the analysis; also see Datta and Ghosh (1991). A concern raised by Chakraborty *et al.* (2018) is that "removing any data which may be a non-representative outlier from analysis will result in loss of valuable information about a part of the non-sampled units of the population which may contain outliers." The latter authors proposed to analyze the corn data using the robust HB method based on the NM distribution, as described above. They compared the analysis resulting using four different methods: the robust HB method, the Sinha-Rao method, the M-quantile method proposed by Chambers and Tzavidis (2006), and the Datta-Ghosh method. The first three methods were developed on the basis of robustness against outliers (see the next section for more details about the Sinha-Rao and M-quantile methods); the Datta-Ghosh method was known to be not robust against outliers [Chakraborty *et al.* (2018)].

The results show that, for the first 11 counties, which do not contain any potential outlier, there is a close agreement among the robust HB method, the Sinha-Rao method, and the Datta-Ghosh method in the estimated county mean hectares. This suggests that the robust HB method performs similarly as the well-established methods, such as Datta-Ghosh, when no outlier is present. On the other hand, for the 12th county which contains the outlier, the estimated county mean hectares based on the robust HB method and Sinha-Rao method are very similar; they are, however, very different from that based on the Datta-Ghosh method. This agrees with the expectation, in view of the non-robustness of the Datta-Ghosh method against outliers. As for the M-quantile method, it was found that for the first three counties the estimates based on the M-quantile method are widely different from those based on the robust HB and Sinha-Rao method. Chakraborty *et al.* (2018) suggests that this indicates some potential bias of the M-quantile estimates.

7.3.3 *Bayesian empirical likelihood*

Chaudhuri and Ghosh (2011) proposed a Bayesian empirical likelihood (EL) approach to semiparametric Bayesian inference. El [Owen (1988); also see Owen (2001)] is a method of estimation that relies on fewer assumptions than the traditional maximum likelihood method. To introduce the Bayesian EL method, let us begin with the hierarchical model that, given $\eta_1, \dots, \eta_m$, observations $y_1, \dots, y_m$ are independent with

$$y_i|\eta_i \sim \exp\left\{\frac{\eta_i y_i - b(\eta_i)}{\phi_i} + c(y_i, \phi_i)\right\}, \tag{7.16}$$

$$\theta_i = x_i'\beta + v_i, \ \ \mathrm{E}(v_i|\beta, A) = 0, \ \ \mathrm{var}(v_i|\beta, A) = A, \tag{7.17}$$

where $\theta_i = h(\eta_i)$ for some function $h(\cdot)$. (7.16) is in the form of an exponential family [McCullagh and Nelder (1989)]. Let $w_i, 1 \le i \le m$ denote the possible jumps in the empirical distribution of $y_i, 1 \le i \le m$. For a given $\theta = (\theta_i)_{1\le i\le m}$, the EL is defined as $L = \prod_{i=1}^m w_i$ with the constraints $w_i \ge 0, 1 \le i \le m$, $\sum_{i=1}^m w_i = 1$, and that

$$\sum_{i=1}^m w_i\{y_i - \mu(\theta_i)\} = 0, \ \ \sum_{i=1}^m w_i\left[\frac{\{\{y_i - \mu(\theta_i)\}^2}{V(\theta_i)} - 1\right] = 0, \tag{7.18}$$

for specified functions $\mu(\cdot)$ and $V(\cdot)$ (Exercise 7.6). Note that (7.18) corresponds to the mean and variance under the empirical distribution.

The next step is to specify a prior distribution for the random effects, $v = (v_i)_{1\le i\le m}$, and parameters β and A. Any proper prior for these leads to a proper prior for θ through (7.17). Chaudhuri and Ghosh showed that, if a prior for θ is proper, the corresponding posterior is also proper. The posterior is evaluated via Markov chain Monte-Carlo.

Now consider the case where the data are divided into independent clusters, $y_{ij}, j = 1, \dots, n_i$, for $i = 1, \dots, m$. The exponential family (7.16) is now replaced by

$$y_{ij}|\eta_{ij} \sim \exp\left\{\frac{\eta_{ij}y_{ij} - b(\eta_{ij})}{\phi_{ij}} + c(y_{ij}, \phi_{ij})\right\}$$

with $\theta_{ij} = h(\eta_{ij}) = x'_{ij}\beta + v_i$. For a given $\theta = (\theta_{ij})$, the EL is defined as $L = \prod_{i=1}^{m}\prod_{j=1}^{n_i} w_{ij}$ with the constraints $w_{ij} \geq 0$, $\sum_{i,j} w_{ij} = 1$, and

$$\sum_j w_{ij}\{y_{ij} - \mu(\theta_{ij})\} = 0, \quad \sum_j w_{ij}\left[\frac{\{y_{ij} - \mu(\theta_{ij})\}^2}{V(\theta_{ij})} - 1\right] = 0,$$

for each $1 \leq i \leq m$, where the functions $\mu(\cdot)$ and $V(\cdot)$.

Intuitively, because EL avoids full parametric model assumptions, the method is potentially more robust to departure from model assumptions. Somehow, the robustness feature was not fully explored in Chaudhuri and Ghosh (2011) although, through an application to median family income data, the authors showed that the EL approach produced better estimates than the Current Population Survey estimates [e.g., Ghosh *et al.* (1996)] and the standard HB estimates.

7.3.4 *Bayesian model diagnostics*

Yan and Sedransk (2007) and Yan and Sedransk (2010) discussed problems associated with Bayesian model diagnostics. The specific aspect of model inadequacy has to do with fitting a model that does not account for all of the hierarchical structures that are present. The authors proposed two testing procedures based on the predictive posterior distribution,

$$f(\tilde{y}|y) = \int f(\tilde{y}|\theta)p(\theta|y)d\theta. \tag{7.19}$$

In (7.19), y denotes the observed data and $\tilde{y}$ the future data, $f(\cdot|\theta)$ is the pdf given θ, the parameter vector, and $p(\cdot|y)$ is the posterior. It is assumed that $\tilde{y}$ and y are independent given θ. Note that (7.19) is very similar to the predictive likelihood, (4.45), except that now $\tilde{y}$ and y are different [while $\tilde{y} = y$ in (4.45)].

Yan and Sedransk (2007) developed two procedures for the diagnostics. The first is to compute the posterior predictive p-value, defined as

$$p_{ij} = \mathrm{P}(\tilde{y}_{ij} \leq y_{ij}|y), \tag{7.20}$$

where $\tilde{y}_{ij}, y_{ij}$ are the (i, j) components of $\tilde{y}, y$, respectively. The second procedure is to compute the p-value of a diagnostic statistic, $t(\cdot)$, that is

$$\mathrm{P}[t(\tilde{y} \geq t(y)|y).$$

Examples of $t(\cdot)$ include the sample mean, median, and standard deviation. We consider an example.

Example 7.4. Consider a simple model in which y_{ij}, $i = 1, \dots, m$, $j = 1, \dots, k$ are independent and distributed as $N(\mu, \tau^2)$, given μ, τ^2. It can be shown that, if the assumed model is correct, then, as $N = mk \to \infty$, the distribution of y and $\tilde{y}|y$ are the approximately the same, hence, the p-value (7.20) has a Uniform[0, 1] distribution (Exercise 7.7). This suggests a graphical way of checking the model but means of a Q-Q plot of the distribution of p_{ij} against the uniform distribution. For example, suppose that the true underlying distribution has an extra hierarchy, that is, $y_{ij}|\mu_i, \tau^2 \sim N(\mu_i, \tau^2)$ and $\mu_i|\sigma^2, \tau^2 \sim N(\mu, \sigma^2)$; the conditional independence assumption is unchanged. Yan and Sedransk (2007) showed that, in this case, the previous model is correct in terms of the mean and variance, but not in terms of the within-cluster covariance. Thus, if the intra-cluster correlation, defined as $\text{cov}(y_{ij}, y_{ik})/\text{var}(y_{ij})$ for $j \neq k$, is sufficiently high, the Q-Q plot would detect such a difference.

In a subsequent work, Yan and Sedransk (2010) proposed another method of detecting a missing hierarchical structure, which is based on the Q-Q plot of the predictive standardized residuals,

$$r_{ij} = \frac{y_{ij} - \text{E}(\tilde{y}_{ij}|y)}{\{\text{var}(\tilde{y}_{ij}|y)\}^2}.$$

The authors showed that the new procedure works under a similar way as described in Example 7.4.

7.4 More about outliers

The work of Sinha and Rao (2009) was mentioned in Subsections 7.3.1 and 7.3.2. We now describe their method in further detail. The goal was to study impact of "representative outliers" on the normality-based EBLUP (see Section 5.1). A representative outlier is defined as [Chambers (1986)] a "sample element with a value that has been correctly recorded and cannot be regarded as unique" that "there is no reason to assume that there are no more similar outliers in the nonsampled part of the population". Although the EBLUPs are efficient under the assumed Gaussian mixed model, they are sensitive to outliers that deviate from the assumed model. Such outliers exist practically, because the Gaussian model may never hold exactly. To robustify the EBLUP, Sinha and Rao focused on the likelihood equation, derived under normality. Consider the longitudinal model (see Subsection

3.1.2). The likelihood equation can be written as

$$\sum_{i=1}^{m} X_i' V_i^{-1}(y_i - X_i\beta) = 0, \tag{7.21}$$

$$\sum_{i=1}^{m}\left\{(y_i - X_i'\beta)'V_i^{-1}\frac{\partial V_i}{\partial \psi_k}V_i^{-1}(y_i - X_i\beta)\right.$$
$$\left. -\mathrm{tr}\left(V_i^{-1}\frac{\partial V_i}{\partial \psi_k}\right)\right\} = 0, \quad k = 1, \dots, q, \tag{7.22}$$

where $\psi = (\psi_k)_{1\le k\le q}$ is the vector of variance components involved in the V_is. Sinha and Rao propose to robustify the EBLUP by robustifying the likelihood equation. Write $y_i - X_i\beta$ as $U_i^{1/2}U_i^{-1/2}(y_i - X_i\beta) = U_i^{1/2}r_i$, where U_i is a diagonal matrix with diagonal elements equal to those of V_i. Replacing r_i by $\psi(r_i)$, where $\psi(\cdot)$ is the vector-valued function with dimension $n_i = \dim(y_i)$ and component function $\psi_{\rm h}(\cdot)$, and $\psi_{\rm h}$ is Huber's ψ-function [Huber (1964)], and adding a matrix K_i inside the trace in (7.22) (to be explained below), one has the robust versions of (7.21) and (7.22):

$$\sum_{i=1}^{m} X_i' V_i^{-1}U_i^{1/2}\psi(r_i) = 0, \tag{7.23}$$

$$\sum_{i=1}^{m}\left\{\psi'(r_i)U_i^{1/2}V_i^{-1}\frac{\partial V_i}{\partial \psi_k}V_i^{-1}U_i^{1/2}\psi(r_i)\right.$$
$$\left. -\mathrm{tr}\left(K_iV_i^{-1}\frac{\partial V_i}{\partial \psi_k}\right)\right\} = 0, \quad k = 1, \dots, q. \tag{7.24}$$

In (7.24), K_i is a diagonal matrix chosen as $K_i = cI_{n_i}$, where $c = \mathrm{E}\{\psi_{\rm h}(Z)\}$ with $Z \sim N(0,1)$. The robustified ML estimator of (β, ψ) is defined as the solution to (7.23) and (7.24).

Sinha and Rao established asymptotic normality of the robustified ML estimator, which may be viewed as M-estimator [Huber (1981)]. In terms of measure of uncertainty for the robust EBLUP, Sinha and Rao adopted a parametric bootstrap approach, and studied its empirical performance.

Also mentioned earlier (see Subsection 7.3.2) was the M-quantile method of Chambers and Tzavidis (2006). Quantiles have been used in statistics as alternatives to the mean as measures of location. One particular case is the median of, say, a data set, which corresponds to a point such that 50% of the data are less than it. In the context of SAE, Chambers and Tzavidis (2006) proposed a quantile-based approach. The intention was to offer an alternative was to model the between-area variation to the random effects approach. Motivated by the quantile regression [Koenker and Bassett

(1978)], the authors defined the M-quantile as a quantity, $Q = Q_q(x;\psi)$, that satisfies

$$\int \psi_q(y - Q)f(y|x)dy = 0,$$

where q is a given number in $(0, 1)$ and

$$\psi_q(r) = 2\psi^{-1}(r/s)\{q1_{(r>0)} + (1 - q)1_{(r\le 0)}\},$$

$\psi(\cdot)$ being an influence function, and s is a "suitable robust estimator of scale". Here y and x denote the response and covariate, respectively. The authors argued that the use of M-quantile instead of standard quantile regression is mainly due to practical reasons, because the M-quantile regression is easier to fit by utilizing a iteratively reweighted least squares algorithm. In applying the M-quantiles to SAE, the authors define a unit-level M-quantile coefficient, q_i, via the equation

$$Q_{q_i}(x_i;\psi) = y_i,$$

where x_i and y_i are x and y for the ith unit. The area-specific M-quantile coefficient is then defined as average of the unit-level ones.

Estimation of the area M-quantile coefficients are obtained by fitting the model with sample M-quantiles. The authors claimed that a main advantage of their M-quantile model is that it allows for outlier-robust inference using "widely available M-estimation software". The authors tried to link their area-specific M-quantile coefficient to the random effect by arguing that, for example, if all area M-quantile coefficients are equal to 0.5, one concludes "there is no between-area variation beyond that explained by the model covariates". In terms of measure of uncertainty, the authors suggested to use the mean squared error (MSE). This seems a bit contradicting as the point was thought to avoid using the mean-based approach, which leads to the best predictor under the MSE. Also, unlike the mean-based approach, such as the EBLUP, there was no optimality consideration under the M-quantile framework, whose theoretical foundation appears to be lacking.

7.5 Benchmarking

Benchmarking is a technique that is often used, especially in the context of surveys. For example, Pfeffermann (2013), wrote, in his review of "new important developments" in SAE: "Benchmarking robustifies the inference by forcing the model-based predictors to agree with a design-based estimator for an aggregate of the areas for which the design-based estimator is reliable". A benchmarking equation typically looks like the following:

$$\sum_{i=1}^{m} w_i\hat{\theta}_i = \sum_{i=1}^{m} w_i y_i, \tag{7.25}$$

where $\hat{\theta}_i$ is a model-based predictor of small area mean for the ith small area, and y_i is a design-based estimator for the same small area mean; the w_i's are known weights.

Unfortunately, the standard predictors, such as the EBLUP, do not necessarily satisfy the benchmarking equation, (7.25). In order to handle this problem, two main approaches have been proposed. The first approach is to make some adjustment of the EBLUP to force it to satisfy the benchmarking condition. A ratio-benchmarked predictor is defined as

$$\hat{\theta}_{\mathrm{rb},i} = \hat{\theta}_i \frac{\sum_{i=1}^{m} w_i y_i}{\sum_{i=1}^{m} w_i \hat{\theta}_i}. \tag{7.26}$$

A difference-benchmarked predictor is defined as

$$\hat{\theta}_{\mathrm{db},i} = \hat{\theta}_i + \sum_{i=1}^{m} w_i y_i - \sum_{i=1}^{m} w_i\hat{\theta}_i. \tag{7.27}$$

It is easy to see that both $\hat{\theta}_{\mathrm{rb},i}, 1 \le i \le m$ and $\hat{\theta}_{\mathrm{db},i}, 1 \le i \le m$ satisfy (7.25). A question of both theoretical and practical interest is regarding the "optimal" adjustment. For example, one can extend (7.27) as

$$\hat{\theta}_{i,\lambda} = \hat{\theta}_i + \lambda_i\left(\sum_{i=1}^{m} w_i y_i - \sum_{i=1}^{m} w_i\hat{\theta}_i\right) \tag{7.28}$$

for any constants $\lambda_i, 1 \le i \le m$ with the resulting predictor, $\hat{\theta}_{i,\lambda}, 1 \le i \le m$, still satisfying the benchmarking condition, (7.25), provided that

$$\sum_{i=1}^{m} \lambda_i w_i = 1. \tag{7.29}$$

The difference-benchmarked predictor is a special case of $\hat{\theta}_{i,\lambda}$ with $\lambda_i = 1, 1 \le i \le m$. In the subsequent discussion, we shall focus on the area-level model, or Fay-Herriot model, which is widely used in SAE.

Wang *et al.* (2008) considered a weighted sum of MSPE, that is,

$$S(\lambda) = \sum_{i=1}^{m} \phi_i \mathrm{E}(\hat{\theta}_{i,\lambda} - \theta_i)^2, \tag{7.30}$$

where $\phi_i, 1 \le i \le m$ are specified positive weights. The authors showed that, under suitable conditions, the optimal choice of λ, in the sense of minimizing $S(\lambda)$, is given by (Exercise 7.8)

$$\lambda_i = \frac{w_i/\phi_i}{\sum_{j=1}^m w_j^2/\phi_j}, \quad 1 \le i \le m. \tag{7.31}$$

It is also easy to see that the λ_is defined by (7.31) satisfy (7.29), therefore, the benchmarking condition (7.25) is satisfied.

Another approach to benchmarking is called *self benchmarking*. The idea is to make some modification in the estimation procedure so that the resulting predictors automatically satisfy the benchmarking condition. You and Rao (2002) proposed to modify the BLUE of β [see (5.4)] so that the resulting EBLUP is self benchmarking. Another interesting approach was proposed by Wang *et al.* (2008), in which the authors introduced the following *augmented* Fay-Herriot model:

$$y_i = x_i'\beta + \beta_a w_i D_i + v_i + e_i, \quad i = 1, \dots, m, \tag{7.32}$$

with the same assumptions as for the Fay-Herriot model, where β_a is an additional unknown parameter. It can be shown (Exercise 7.9) that the EBLUP, $\hat{\theta}_i$, of the small area mean, $\theta_i = x_i'\beta + v_i$, $1 \le i \le m$, obtained under the augmented model, (7.32), are self benchmarking. It should be noted that (7.32) is not the true underlying model—it is just a model that one fits to get the EBLUPs; the true underlying model is still the Fay-Herriot model (see Example 5.1).

Bandyopadhyay (2017) extended both approaches of benchmarking the EBLUP described above to benchmarking the OBP (see Chapter 5). We describe below the second approach based on an augmented model (extension of the first approach is relatively straightforward). The augmented model is different from (7.32), expressed as

$$y_i = x_i'\beta + \beta_a w_i (1 - B_i)^{-1} + v_i + e_i, \quad i = 1, \dots, m, \tag{7.33}$$

where $B_i = A/(A + D_i)$; other assumptions are the same as in the Fay-Herriot model. A main difference between (7.33) and (7.32) is that, in (7.33), B_i is not a known quantity, because it depends on the unknown variance A. Thus, in a way, the latter model is not a linear model, because the mean function depends on a parameter that is involved in the variance function. Nevertheless, for any fixed $A \ge 0$, one can define $X(A) = [X_1 \ X_2(A)]$, where $X_1 = (x_i')_{1 \le i \le m}$ and $X_2(A) = [w_i(1 - B_i)^{-1}]_{1 \le i \le m}$, and $\beta^* = (\beta', \beta_a)'$. Then, (7.33) can be written in a matrix form as

$$y = X(A)\beta^* + v + e, \tag{7.34}$$

where $y = (y_i)_{1 \le i \le m}$ and v, e are defined similarly. According to (5.11), if A is known, the BPE of β^* is given by

$$\tilde{\beta}^*(A) = \{X(A)'\Gamma^2(A)X(A)\}^{-1}X(A)'\Gamma^2(A)y \tag{7.35}$$

with $\Gamma = \text{diag}(1 - B_i, 1 \le i \le m)$. In case that A is unknown (the practical situation), its BPE is obtained by minimizing (5.20), that is,

$$Q^*(A) = \{y - X(A)\tilde{\beta}^*(A)\}'\Gamma^2(A)\{y - X(A)\tilde{\beta}^*(A)\} + 2A\text{tr}\{\Gamma(A)\} \tag{7.36}$$

over $A \ge 0$. Denote the minimizer of (7.36) by $\hat{A}$. Then, the BPE of β^* is given by $\hat{\beta}^* = \tilde{\beta}^*(\hat{A})$. One then define the OBP of $\theta = (\theta_i)_{1 \le i \le m}$ as

$$\hat{\theta} = \hat{A}\hat{V}^{-1}y + D\hat{V}^{-1}X(\hat{A})\hat{\beta}^*, \tag{7.37}$$

where $\hat{V} = \hat{A}I_m + D$ with $D = \text{diag}(D_i, 1 \le i \le m)$. It can be shown (Exercise 7.9) that the OBP defined by (7.37), with the ith component $\hat{\theta}_i$, $1 \le i \le m$, are self benchmarking, that is, they automatically satisfy (7.25).

To compare performance of the benchmarked OBP with those of the benchmarked EBLUP, Bandyopadhyay (2017) carried out a simulation study under a setting used by Wang *et al.* (2008). The data were simulated under the model

$$y_{ij} = x_i'\beta + v_i + e_{ij}, \quad i = 1, \dots, 50, \ j = 1, \dots, n_i, \tag{7.38}$$

where $x_i' = (1, z_i)$ with $z_i = \text{pop}_i^{0.2} - \overline{\text{pop}^{0.2}}$, with pop_i being the approximate population size of state i [see Table 1 of Wang *et al.* (2008)]; the sample sizes n_i are approximately proportional to $\sqrt{\text{pop}_i}$. One then take $y_i = n_i^{-1}\sum_{j=1}^{n_i} y_{ij}$, hence $e_i = n_i^{-1}\sum_{j=1}^{n_i} e_{ij}$, to imply a Fay-Herriot model: $y_i = x_i'\beta + v_i + e_i$.

The simulation setting is exactly as that of Wang *et al.* (2008). Namely, the v_is are generated from the $N(0, 1)$ and e_{ij}s from the $N(0, 16)$ distributions, so that $e_i \sim N(0, 16/n_i)$, thus $D_i = 16/n_i$, $1 \le i \le m$; the true β is $(6.0, 3.0)$.

(7.38) is the true underlying model; however, it is not the assumed model. The assumed model is (7.38) with $x_i = 1$ (i.e., without z_i). In other words, the assumed model is misspecified. The performance of three different types of predictors: (i) predictor without benchmarking; (ii) benchmarking via adjustment; and (iii) benchmarking via augmented model, each with two different methods: EBLUP and OBP, are compared in terms of empirical MSPE based on 10,000 simulation runs. The results are presented in Figure 7.4.

Note that the first column (from the left) of Figure 7.4 correspond to EBLUP and OBP without benchmarking; the last two columns correspond

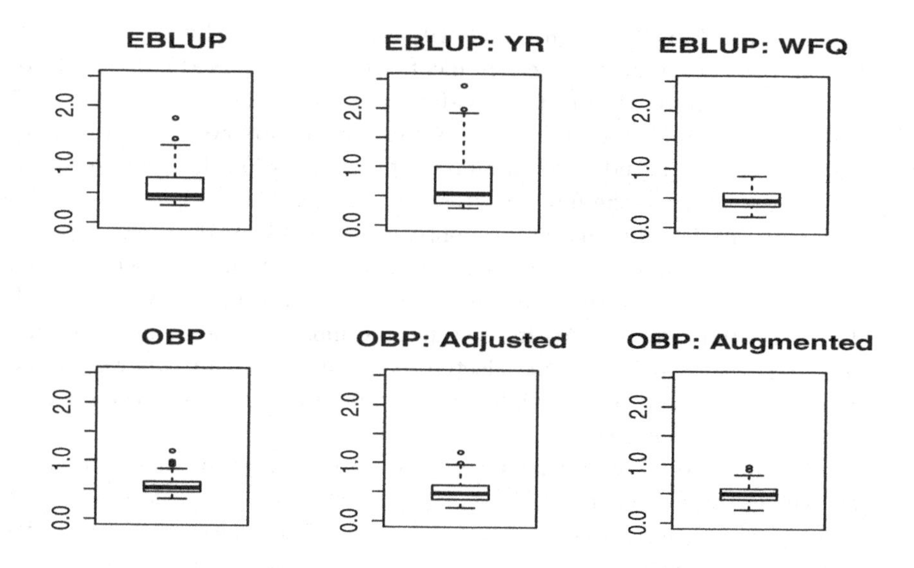

Fig. 7.4 *Empirical MSPE of Predictors*

to EBLUP and OBP with benchmarking. Overall, it appears that OBP performs better than EBLUP either without benchmarking or with benchmarking via adjustment. As for benchmarking via augmented models, OBP performs slightly better than EBLUP in terms of the interquartile range.

7.6 More about prediction

7.6.1 *Prediction of future observation*

So far as prediction is concerned, the majority of the literature on mixed effects models focuses on prediction of mixed effects. However, prediction of future observations are also of interest in practice.

Consider the problem of predicting a future observation under a non-Gaussian LMM (see Chapter 3). Because normality is not assumed, our approach is distribution-free; that is, it does not require any specific assumption about the distribution of the random effects and errors. First note that for this type of prediction, it is reasonable to assume that a future observation is independent of the current ones. We offer some examples.

Example 7.5. In longitudinal studies, one may be interested in prediction, based on repeated measurements from the observed individuals, of a future observation from an individual not previously observed. It is of less interest to predict another observation from an observed individual, because longitudinal studies often aim at applications to a larger population (e.g., drugs going to the market after clinical trials).

Example 7.6. In surveys, responses may be collected in two steps: in the first step, a number of families are randomly selected; in the second step, some family members (e.g., all family members) are interviewed for each of the selected families. Again, one may be more interested in predicting what happens to a family not selected, because one already knows enough about selected families (especially when all family members in the selected families are interviewed).

Therefore, we assume that a future observation, y_*, is independent of the current ones. Then, we have $E(y_*|y) = E(y_*) = x_*^t\beta$, so the best predictor is $x_*^t\beta$, if β is known; otherwise, an empirical best predictor (EBP) is obtained by replacing β by an estimator. So the point prediction is fairly straightforward. A question that is often of practical interest but has been so far neglected, for the most part, is that of prediction intervals.

A prediction interval for a single future observation is an interval that has a specified coverage probability for the future observation. In model-based statistical inference, it is assumed that the future observation has a certain distribution. Sometimes, the distribution is specified up to a finite number of unknown parameters, for example, mean and variance of the normal distribution. Then, a prediction interval may be obtained, if the parameters are adequately estimated, and the uncertainty in the parameter estimations is suitably assessed. Clearly, such a procedure is dependent on the underlying distribution in that, if the distributional assumption fails, the prediction interval may be seriously inaccurate, that is, either wider than necessary, or not having the claimed coverage probability. An alternative to the parametric method is a distribution-free one, in which one does not assume that the form of the distribution is known. For a review of literature on prediction interval for a future observation using either distribution-dependent or distribution-free approaches, see, for example, sec. 2.3.2 of Jiang (2007).

Note that, even if β is unknown, it is still fairly easy to obtain a prediction interval for y_* if one is willing to make the assumption that the distributions of the random effects and errors are known up to a vector of variance components. To see this, consider (7.11), where the random effect

v_i and error e_{ij} are independent such that $v_i \sim N(0, \sigma^2)$ and $e_{ij} \sim N(0, \tau^2)$. It follows that the distribution of y_{ij} is $N(x'_{ij}\beta, \sigma^2 + \tau^2)$. Because methods are well developed for estimating fixed parameters such as β, σ^2, and τ^2 (see Chapter 5), a prediction interval with asymptotic coverage probability $1 - \rho$ is easy to obtain. However, it is much more difficult if one does not know the forms of the distributions of the random effects and errors, or one is not willing to make specified distributional assumptions about the random effects and errors due to robustness concerns. This is the case that we consider below.

According to Chapter 3, for consistency of REML or ML estimators of the fixed effects and variance components, one does not need to assume that the random effects and errors are normally distributed. A LMM is said to be standard if it can be expressed as (1.1), where each Z_r $(1 \leq r \leq s)$ consists only of 0s and 1s, there is exactly one 1 in each row and at least one 1 in each column. The methods described below were developed by Jiang and Zhang (2002), which treat the standard and non-standard cases quite differently.

For standard LMM, the method is surprisingly simple. First, one throws away the middle terms in (1.1) that involve the random effects, and pretends that it is a linear regression model, with i.i.d. errors, $y = X\beta + \epsilon$. Next, one computes the ordinary least squares (OLS) estimator, $\hat{\beta} = (X'X)^{-1}X'y$, and the residuals, $\hat{\epsilon} = y - X\hat{\beta}$. Let $\hat{a}$ and $\hat{b}$ be the $\rho/2$ and $1 - \rho/2$ quantiles of the residuals. Then, a prediction interval for y_* with asymptotic coverage probability $1 - \rho$

$$[\hat{y}_* + \hat{a}, \hat{y}_* + \hat{b}], \tag{7.39}$$

where $\hat{y}_* = x'_*\hat{\beta}$. Note that, although the method sounds almost the same as the residual method in linear regression, its justification is not so obvious because, unlike linear regression, the observations under a (standard) LMM are not independent. The method may be improved if one uses more efficient estimators such as the EBLUE, instead of the OLS estimator.

To see how (7.39) is derived, let y_* be a future observation that one wishes to predict. Suppose that y_* satisfies the same standard LMM. Then, y_* can be expressed as

$$y_* = x'_*\beta + \alpha_{*1} + \cdots + \alpha_{*s} + \epsilon_* \ ,$$

where x_* is a known vector of covariates (not necessarily present with the data), α_{*r}s are random effects, and ϵ_* is an error, such that $\alpha_{*i} \sim F_{ir}$, $\leq i \leq s$, $\epsilon_* \sim F_0$, where the Fs are unknown distributions, and $\alpha_{*1}, \ldots, \alpha_{*s}, \epsilon_*$

are independent. According to earlier discussion, y_* is independent of the data, $y = (y_i)_{1 \le i \le n}$. It follows that the best (point) predictor of y_*, when β is known, is $\mathrm{E}(y_*|y) = \mathrm{E}(y_*) = x_*'\beta$. Because β is unknown, it is replaced by a consistent estimator, $\hat{\beta}$, which may be the OLS estimator or EBLUE. This results in an empirical best predictor:

$$\hat{y}_* = x_*'\hat{\beta} \,. \tag{7.40}$$

Let $\hat{\delta}_i = y_i - x_i'\hat{\beta}$. Define

$$\hat{F}(x) = \frac{\#\{1 \le i \le n : \hat{\delta}_i \le x\}}{n} = \frac{1}{n}\sum_{i=1}^{n} 1_{(\hat{\delta}_i \le x)} \,. \tag{7.41}$$

Note that, although (7.41) resembles the empirical distribution [e.g., Jiang (2010), sec. 7.1], it is not one in the classic sense, because the $\hat{\delta}_i$s are not independent (the y_is are dependent, and $\hat{\beta}$ depends on all the data). Let $\hat{a} < \hat{b}$ be any numbers satisfying $\hat{F}(\hat{b}) - \hat{F}(\hat{a}) = 1 - \rho$ $(0 < \rho < 1)$. Then, a prediction interval for y_* with asymptotic coverage probability $1 - \rho$ is given by (7.39).

Note that a typical choice of $\hat{a}$, $\hat{b}$ has $\hat{F}(\hat{a}) = \rho/2$ and $\hat{F}(\hat{b}) = 1 - \rho/2$. Another choice would be to select $\hat{a}$ and $\hat{b}$ to minimize $\hat{b} - \hat{a}$, the length of the prediction interval. Usually, $\hat{a}$, $\hat{b}$ are selected such that the former is negative and the latter positive, so that $\hat{y}_*$ is contained in the interval. Also note that, if one considers linear regression as a special case of the LMM, in which the random effects are zero, $\hat{\delta}_i$ is the same as $\hat{\epsilon}_i$, the residual, if $\hat{\beta}$ is the least squares estimator. In this case, $\hat{F}$ is the empirical distribution of the residuals, and the prediction interval (7.39) corresponds to that obtained by the bootstrap method [Efron (1979)]. However, there is one difference. The difference is that our prediction interval is obtained in closed form, rather than by a Monte Carlo method. For more discussion on bootstrap prediction intervals, see, for example, Section 7.3 of Shao and Tu (1995).

Now let us consider a nonstandard LMM (i.e., LMM that is not standard). First, the method developed for standard models may be applied to some of the nonstandard cases. We illustrate with an example.

Example 7.7. Suppose that the data are divided into two parts. The first part satisfies $y_{ij} = x_{ij}'\beta + \alpha_i + \epsilon_{ij}$, $i = 1, \dots, m$, $j = 1, \dots, n_i$, where $\alpha_1, \dots, \alpha_m$ are i.i.d. random effects with mean 0 and distribution F_1; ϵ_{ij}s are i.i.d. errors with mean 0 and distribution F_0, and the αs and ϵs are independent. The second part satisfies $y_k = x_k'\beta + \epsilon_k$, $k = N+1, \dots, N+K$, where $N = \sum_{i=1}^{m} n_i$, and the ϵ_ks are i.i.d. errors with mean 0 and distribution F_0. Note that the random effects only appear in the first part (hence there is no need to use a double index for the second part).

For the first part, let the distribution of $\delta_{ij} = y_{ij} - x_{ij}'\beta$ be F $(= F_0 * F_1)$. For the second part, let $\delta_k = y_k - x_k'\beta$. If β were known, the δ_{ij}s (δ_ks) would be sufficient statistics for F (F_0). Therefore it suffices to consider an estimator of F (F_0) based on the δ_{ij}s (δ_ks). Note that the prediction interval for any future observation is determined either by F or by F_0, depending on to which part the observation corresponds. Now, because β is unknown, it is customary to replace it by $\hat{\beta}$. Thus, a prediction interval for y_*, a future observation corresponding to the first part, is (7.39), where $\hat{y}_* = x_*'\hat{\beta}$, $\hat{a}$, $\hat{b}$ are determined by $\hat{F}(\hat{b}) - \hat{F}(\hat{a}) = 1 - \rho$ with

$$\hat{F}(x) = \frac{1}{N}\#\{(i,j) : 1 \leq i \leq m, 1 \leq j \leq n_i, \hat{\delta}_{ij} \leq x\}$$

and $\hat{\delta}_{ij} = y_{ij} - x_{ij}'\hat{\beta}$.

Similarly, a prediction interval for y_*, a future observation corresponding to the second part, is (7.39), where $\hat{y}_* = x_*'\hat{\beta}$, $\hat{a}$, $\hat{b}$ are determined similarly with $\hat{F}$ replaced by

$$\hat{F}_0(x) = \frac{1}{K}\#\{k : N+1 \leq k \leq N+K, \hat{\delta}_k \leq x\}$$

and $\hat{\delta}_k = y_k - x_k'\hat{\beta}$. The prediction interval has asymptotic coverage probability $1 - \rho$ [see Jiang and Zhang (2002)].

Jiang (1998b) considered estimation of the distributions of the random effects and errors in non-Gaussian LMM. His approach is the following. Consider the EBLUP of the random effects [see (5.5)]:

$$\hat{\alpha}_i = \hat{\sigma}_i^2 Z_i'\hat{V}^{-1}(y - X\hat{\beta}), \quad 1 \leq i \leq s,$$

where $\hat{\beta}$ is the EBLUE of β. The "EBLUP" of the errors can be defined as

$$\hat{\epsilon} = y - X\hat{\beta} - \sum_{i=1}^{s} Z_i\hat{\alpha}_i.$$

It was shown that, if the REML or ML estimators of the variance components are used, then, under suitable conditions,

$$\hat{F}_i(x) = \frac{1}{m_i}\sum_{u=1}^{m_i} 1_{(\hat{\alpha}_{i,u} \leq x)} \xrightarrow{P} F_i(x), \quad x \in C(F_i),$$

where $\hat{\alpha}_{i,u}$ is the uth component of $\hat{\alpha}_i$, $1 \leq i \leq s$, and

$$\hat{F}_0(x) = \frac{1}{n}\sum_{u=1}^{n} 1_{(\hat{\epsilon}_u \leq x)} \xrightarrow{P} F_0(x), \quad x \in C(F_0),$$

where $\hat{\epsilon}_u$ is the uth component of $\hat{\epsilon}$. Here $C(F_i)$ represents the set of all continuity points of F_i, $0 \leq i \leq s$.

For simplicity, assume that all of the distributions, $F_0, \dots, F_s$ are continuous. Let y_* be a future observation we would like to predict. As before, we assume that y_* is independent of y and satisfies a LMM, which can be expressed componentwise as

$$y_i = x_i'\beta + z_{i1}'\alpha_1 + \cdots + z_{is}'\alpha_s + \epsilon_i, \qquad i = 1, \dots, n.$$

This means that y_* can be expressed as

$$y_* = x_*'\beta + \sum_{j=1}^{l} w_j\gamma_j + \epsilon_* ,$$

where x_* is a known vector of covariates (not necessarily present with the data), w_js are known nonzero constants, γ_js are unobservable random effects, and ϵ_* is an error. In addition, there is a partition of the indices $\{1, \dots, l\} = \cup_{k=1}^{q} I_k$, such that $\gamma_j \sim F_{r(k)}$ if $j \in I_k$, where $r(1), \dots, r(q)$ are distinct integers between 1 and s (so $q \le s$); $\epsilon_* \sim F_0$; $\gamma_1, \dots, \gamma_l, \epsilon_*$ are independent. Define

$$\hat{F}^{(j)}(x) = m_{r(k)}^{-1} \sum_{u=1}^{m_{r(k)}} 1_{(w_j\hat{\alpha}_{r(k),u} \le x)}, \quad \text{if } j \in I_k$$

for $1 \le k \le q$. Let

$$\begin{aligned}\hat{F}(x) &= (\hat{F}^{(1)} * \cdots * \hat{F}^{(l)} * \hat{F}_0)(x) \\ &= \left[\left\{\prod_{k=1}^{q} m_{r(k)}^{|I_k|}\right\} n\right]^{-1} \\ &\quad \times \#\{(u_1, \dots, u_l, u) : \sum_{k=1}^{q}\sum_{j \in I_k} w_j\hat{\alpha}_{r(k),u_j} + \hat{\epsilon}_u \le x\}, \qquad (7.42)\end{aligned}$$

where $*$ represents convolution [e.g., Jiang (2007), Appendix C], and $1 \le u_j \le m_{r(k)}$ if $j \in I_k$, $1 \le k \le q$; $1 \le u \le n$. It can be shown that

$$\sup_x |\hat{F}(x) - F(x)| \xrightarrow{P} 0 ,$$

where $F = F^{(1)} * \cdots * F^{(l)} * F_0$, and $F^{(j)}$ is the distribution of $w_j\gamma_j$, $1 \le j \le l$. Note that F is the distribution of $y_* - x_*'\beta$. Let $\hat{y}_*$ be defined by (7.40) with $\hat{\beta}$ being a consistent estimator, and $\hat{a}$, $\hat{b}$ defined by

$$\hat{F}(\hat{b}) - \hat{F}(\hat{a}) = 1 - \rho,$$

where $\hat{F}$ is given by (7.43). Then, the prediction interval (7.39) has asymptotic coverage probability $1 - \rho$.

Jiang and Zhang (2002) carried out a simulation study to compare empirical performance of the prediction interval (7.39) based on OLS estimator, that based on EBLUE, and the standard regression prediction interval, which ignores the presence of random effects and treats the data as independent. They found that prediction intervals (7.39) based on either OLS or EBLUE performed better than the standard regression prediction interval. As for the comparison between OLS-based and EBLUE-based prediction intervals (7.39), the two methods performed similarly in the this study.

7.6.2 *Classified mixed model prediction*

Mixed model prediction (MMP) has been a topic throughout this monograph. See, for example, Chapter 5. The traditional fields of applications include genetics, agriculture, education, and surveys [e.g., Robinson (1991), Rao and Molina (2015)]. This is a field where frequentist and Bayesian approaches found common grounds. Nowadays, new and challenging problems have emerged from such fields as business and health sciences, in addition to the traditional fields, to which methods of MMP are potentially applicable, but not without further methodology and computational developments. Some of these problems occur when interest is at subject level (e.g., personalized medicine), or (small) sub-population level (e.g., county), rather than at large population level (e.g., epidemiology). In such cases, it is possible to make substantial gains in prediction accuracy by identifying a class that a new subject belongs to. This idea was recently implemented by Jiang *et al.* (2018), who proposed an innovative, and interesting method called classified mixed model prediction (CMMP).

As noted in the earlier subsection, there are two types of prediction problems associated with the mixed effects models. The first type, which is encountered more often in practice, is prediction of mixed effects; the second type is prediction of future observation. Let us first consider CMMP for mixed effects; we can then use the method to tackle prediction of future observation.

1. Prediction of mixed effects. Suppose that we have a set of training data, $y_{ij}, i = 1, \dots, m, j = 1, \dots, n_i$ in the sense that their classifications are known, that is, one knows which group, i, that y_{ij} belongs to. The assumed LMM for the training data is

$$y_i = X_i\beta + Z_i\alpha_i + \epsilon_i, \tag{7.43}$$

where $y_i = (y_{ij})_{1\le j\le n_i}$, $X_i = (x_{ij}')_{1\le j\le n_i}$ is a matrix of known covariates,

β is a vector of unknown regression coefficients (the fixed effects), Z_i is a known $n_i \times q$ matrix, α_i is a $q \times 1$ vector of group-specific random effects, and ϵ_i is an $n_i \times 1$ vector of errors. It is assumed that the α_i's and ϵ_i's are independent, with $\alpha_i \sim N(0, G)$ and $\epsilon_i \sim N(0, R_i)$, where the covariance matrices G and R_i depend on a vector ψ of variance components.

The goal is to make a classified prediction for a mixed effect associated with a set of new observations, $y_{\mathrm{n},j}, 1 \le j \le n_{\mathrm{new}}$ (the subscript n refers to "new"). Suppose that

$$y_{\mathrm{n},j} = x_{\mathrm{n}}'\beta + z_{\mathrm{n}}'\alpha_{\mathrm{I}} + \epsilon_{\mathrm{n},j}, \quad 1 \le j \le n_{\mathrm{new}}, \tag{7.44}$$

where $x_{\mathrm{n}}, z_{\mathrm{n}}$ are known vectors, $I \in \{1, \dots, m\}$ but one does not know which element i, $1 \le i \le m$, is equal to I. Furthermore, $\epsilon_{\mathrm{n},j}, 1 \le j \le n_{\mathrm{new}}$ are new errors that are independent with $\mathrm{E}(\epsilon_{\mathrm{n},j}) = 0$ and $\mathrm{var}(\epsilon_{\mathrm{n},j}) = R_{\mathrm{new}}$, and are independent with the α_is and ϵ_is. Note that the normality assumption is not always needed for the new errors, unless prediction interval is concerned (see below). Also, the variance R_{new} of the new errors does not have to be the same as the variance of ϵ_{ij}, the jth component of ϵ_i associated with the training data. The mixed effect that we wish to predict is

$$\theta = \mathrm{E}(y_{\mathrm{n},j}|\alpha_I) = x_{\mathrm{n}}'\beta + z_{\mathrm{n}}'\alpha_{\mathrm{I}}. \tag{7.45}$$

From the training data, one can estimate the parameters, β and ψ. For example, one can use the standard mixed model analysis to obtain ML or REML estimators (e.g., §3.2). Alternatively, one may use the OBP method (see Chapter 5), which is more robust to model misspecifications in terms of the predictive performance. Thus, we can assume that estimators $\hat{\beta}, \hat{\psi}$ are available for β, ψ.

To derive CMMP, let us first assume that there is a match between the random effect, α_I, corresponding to the new observations and one of the random effects associated with the training data. This means that $I = i$ for some $1 \le i \le m$. It then follows that the vectors $y_1, \dots, y_{i-1}, (y_i', \theta)', y_{i+1}, \dots, y_m$ are independent. Thus, we have $\mathrm{E}(\theta|y_1, \dots, y_m) = \mathrm{E}(\theta|y_i)$. By the normal theory [e.g., (5.3)], we have

$$\mathrm{E}(\theta|y_i) = x_{\mathrm{n}}'\beta + z_{\mathrm{n}}'GZ_i'(R_i + Z_iGZ_i')^{-1}(y_i - X_i\beta). \tag{7.46}$$

The right side of (7.46) is the BP under the assumed LMM, if the true β and ψ, are known. Because the latter are unknown, we replace them by $\hat{\beta}$ and $\hat{\psi}$, respectively. The result is the EBP, denoted by $\tilde{\theta}_{(i)}$.

In practice, however, I is unknown and treated as a parameter. In order to identify, or estimate, I, we consider the MSPE of θ by the BP when I is classified as i, that is $\text{MSPE}_i = \text{E}\{\tilde{\theta}_{(i)} - \theta\}^2 = \text{E}\{\tilde{\theta}_{(i)}^2\} - 2\text{E}\{\tilde{\theta}_{(i)}\theta\} + \text{E}(\theta^2)$. Using the expression $\theta = \bar{y}_{\text{n}} - \bar{\epsilon}_{\text{n}}$, where $\bar{y}_{\text{n}} = n_{\text{new}}^{-1}\sum_{j=1}^{n_{\text{new}}} y_{\text{n},j}$ and $\bar{\epsilon}_{\text{n}}$ is defined similarly, we have $\text{E}\{\tilde{\theta}_{(i)}\theta\} = \text{E}\{\tilde{\theta}_{(i)}\bar{y}_{\text{n}}\} - \text{E}\{\tilde{\theta}_{(i)}\bar{\epsilon}_{\text{n}}\} = \text{E}\{\tilde{\theta}_{(i)}\bar{y}_{\text{n}}\}$. Thus, we have the expression:

$$\text{MSPE}_i = \text{E}\{\tilde{\theta}_{(i)}^2 - 2\tilde{\theta}_{(i)}\bar{y}_{\text{n}} + \theta^2\}. \tag{7.47}$$

It follows that the observed MSPE corresponding to (7.47) is the expression inside the expectation. Therefore, a natural idea is to identify I as the index i that minimizes the observed MSPE. Because θ^2 does not depend on i, the minimizer is given by

$$\hat{I} = \text{argmin}_i \left\{\tilde{\theta}_{(i)}^2 - 2\tilde{\theta}_{(i)}\bar{y}_{\text{n}}\right\}. \tag{7.48}$$

The classified mixed-model predictor (CMMP) of θ is then given by $\hat{\theta} = \tilde{\theta}_{(\hat{I})}$.

2. Prediction of future observations. Now suppose that the interest is to predict a future observation, y_{f}, that belongs to an unknown group that matches one of the existing groups. First note that, even though the interest is prediction for a future observation, it is impossible, in general, to do better with CMMP, if one does not have any other observations that are known to be from the same group as y_{f}. For example, take a look at the simplest case when there is no covariates, that is, $X_i\beta = (\mu, \dots, \mu)'$, $Z_i\alpha_i = (\alpha_i, \dots, \alpha_i)'$ in (7.43), and $x_{\text{f}}'\beta = \mu, z_{\text{f}}'\alpha_I = \alpha_I$ in (7.44). In this case, one knows nothing about y_{f}, because (7.44) is no different from (7.43) at the single-component level. Thus, without additional information, one, of course, cannot tell to which group the new observation y_{f} belongs. Therefore, we shall assume that one has some observation(s) that are known to be from the same group as y_{f}, in addition to the training data. Note that this additional group is different from the training data, because we do not know the classification number of the additional group with respect to the training data groups.

Let $y_{\text{n},j}, 1 \le j \le n_{\text{new}}$ be the additional observations. For example, the additional observations may be data collected prior to a medical treatment, and the future observation, y_{f}, is the outcome after the medical treatment that one wishes to predict. Suppose that y_{f} satisfies (7.44), that is $y_{\text{f}} = x_{\text{f}}'\beta + z_{\text{f}}'\alpha_I + \epsilon_{\text{f}}$, where ϵ_{f} is the new error that is independent with the training data. It follows that

$$\begin{aligned}\text{E}(y_{\text{f}}|y_1, \dots, y_m) &= \text{E}(\theta|y_1, \dots, y_m) + \text{E}(\epsilon_{\text{f}}|y_1, \dots, y_m) \\ &= \text{E}(\theta|y_1, \dots, y_m)\end{aligned} \tag{7.49}$$

with $\theta = x_f'\beta + z_f'\alpha_I$. (7.49) shows that the BP for y_f is the same as the BP for θ, which is the right side of (7.46), that is, $\theta_{(i)}$ with x_n, z_n replaced by x_f, z_f, respectively, when $I = i$. Suppose that the additional observations satisfy $y_{n,j} = x_{n,j}'\beta + z_{n,j}'\alpha_I + \epsilon_{n,j}, 1 \le j \le n_{new}$. If $x_{n,j}, z_{n,j}$ do not depend on j (which includes the special case of $n_{new} = 1$), we can treat $y_{n,j}, 1 \le j \le n_{new}$ the same way as the new observations, and identify the classification number, $\hat{I}$, by (7.48). The CMMP of y_f is then given by the right side of (4) with $i = \hat{I}$, β, ψ replaced by $\hat{\beta}, \hat{\psi}$, and x_n, z_n replaced by x_f, z_f, respectively.

A case that is slightly more complicated is when $x_{n,j}$ depends on j. For the simplicity of illustration, let $z_{n,j}'\alpha_I = \alpha_I$. By treating each $y_{n,j}$ as the new observation (with $n_{new} = 1$), and using the CMMP developed in Subsection 2.1, we can obtain CMEP of the mixed effect associated with $y_{n,j}$, given by $\hat{\theta}_{n,j} = x_{n,j}'\hat{\beta} + \hat{\alpha}_{\hat{I}(j)}$, where $\hat{I}(j)$ is the $\hat{I}$ of (7.48) corresponding to $y_{n,j}$. We do this for each $j = 1, \dots, n_{new}$, leading to $\hat{\alpha}_{\hat{I}(j)} = \hat{\theta}_{n,j} - x_{n,j}'\hat{\beta}$, $1 \le j \le n_{new}$. We then take the average, $\widehat{\alpha_I} = n_{new}^{-1} \sum_{j=1}^{n_{new}} \hat{\alpha}_{\hat{I}(j)}$. The CMMP of y_f is then given by $\hat{y}_f = x_f'\hat{\beta} + \widehat{\alpha_I}$.

3. Robustness of CMMP. Recall that CMMP is derived under the assumption that there is a match between the random effect associated with the new observations and one of the random effects associated with the training data. A nice, and somewhat surprising feature of CMMP is that, even if this assumption does not hold, that is, the "match" does not exist, one still gains in prediction accuracy by doing CMMP, pretending that there is a match. In this sense, CMMP is robust to failure of the matching assumption. This robust feature makes the CMMP method (much) more applicable because, in practice, an exact match may well not exist.

To illustrate the robustness property of CMMP, Jiang *et al.* (2018) carried out a simulation study. The training data were generated under the following model:

$$y_{ij} = 1 + 2x_{1,ij} + 3x_{2,ij} + \alpha_i + \epsilon_{ij}, \tag{7.50}$$

$i = 1, \dots, m$, $j = 1, \dots, n$, with $n = 5$, $\alpha_i \sim N(0, G)$, $\epsilon_{ij} \sim N(0, 1)$, and α_i's, ϵ_{ij}'s are independent. The $x_{k,ij}, k = 1, 2$ were generated from the $N(0, 1)$ distribution, then fixed throughout the simulation. There are $K = 10$ new observations, generated under two scenarios. Scenario I: The new observations have the same α_i as the first K groups in the training data ($K \le m$), but independent ϵ's; that is, they have "matches". Scenario II: The new observations have independent α's and ϵ's; that is, they are "unmatched". Note that there are K different mixed effects. CMMP

Table 7.4 **Average MSPE for Prediction of Mixed Effects.** %MATCH = % of times that the new observations were matched to some of the groups in the training data.

	Scenario	σ_α^2	0.1	1	2	3
$m = 10$	I	RP	0.157	1.002	1.940	2.878
	I	CMMP	0.206	0.653	0.774	0.836
	I	%MATCH	91.5	94.6	93.6	93.2
	II	RP	0.176	1.189	2.314	3.439
	II	CMMP	0.225	0.765	0.992	1.147
	II	%MATCH	91.2	94.1	92.6	92.5
$m = 50$	I	RP	0.112	1.013	2.014	3.016
	I	CMMP	0.193	0.799	0.897	0.930
	I	%MATCH	98.7	98.5	98.6	98.2
	II	RP	0.113	1.025	2.038	3.050
	II	CMMP	0.195	0.800	0.909	0.954
	II	%MATCH	98.8	98.7	98.4	98.4

was compared with the standard regression prediction (RP) method in predicting each of the K mixed effects. The average simulated MSPE of the prediction, obtained based on $T = 1000$ simulation runs, are reported in Table 7.4. Two cases, $m = 10$ and $m = 50$, were considered. The CMMP method has a certain way of telling whether or not there is a "match"; of course, it would be totally wrong if the actual match does not exist. Nevertheless, the reported %MATCH is the percentage of times, out of the simulation runs, that CMMP thought there was a match, in which case it uses the CMMP, given below (7.48), for prediction; otherwise it uses RP for prediction.

It appears that, regardless of whether the new observations actually have matches or not, CMMP match them anyway. More importantly, the results show that even a "fake" match still helps. At first, this might sound a little surprising, but it actually makes sense, both practically and theoretically. Think about a business situation. Even if one cannot find a perfect match for a customer, but if one can find a group that is kind of similar, one can still gain in terms of prediction accuracy. This is, in fact, how business decisions are often made. In the simulation study, even if there is no match in terms of the individual random effects, there is at least a "match" in terms of the random effects distribution, that is, the new random effect is generated from the same distribution that has generated the (previous) training-data random effects; therefore, it is not surprising that one can find one among the latter that is close to the new random effect.

Comparing RP with CMMP, PR assumes that the mixed effect is $x_i'\beta$ with nothing extra, while CMMP assumes that the mixed effect is $x_i'\beta$ plus something extra. For the new observation, there is, for sure, something extra, so CMMP is right, at least, in that the extra is non-zero; it then selects the best extra from a number of choices, some of which are better than the zero extra that PR is using. Therefore, it is not surprising that CMMP is doing better, regardless of the actual match (which may or may not exist). In fact, the empirical results reported in Table 7.4 is consistent with the theoretical findings, as shown by the following theorem [see Jiang *et al.* (2018) for detail].

Theorem 7.2. Under regularity conditions, as $m \to \infty$, $(\log m)^{2\nu}/n_{\min} \to 0$, where $n_{\min} = \min_{1 \le i \le m} n_i$, and $n_{\text{new}} \to \infty$, we have

$$\mathrm{E}\{(\hat{\theta}_{\mathrm{n}} - \theta_{\mathrm{n}})^2\} \to 0 \quad \text{and} \quad \liminf \left\{\mathrm{E}(\hat{\theta}_{\mathrm{n,r}} - \theta_{\mathrm{n}})^2\right\} \ge \delta$$

for some constant $\delta > 0$, where θ_{n} is the mixed effect associated with the new observations, $\hat{\theta}_{\mathrm{n}}$ is the CMMP of θ_{n}, and $\hat{\theta}_{\mathrm{n,r}}$ is the RP of θ_{n}.

Note that $m \to \infty$ and $\min_{1 \le i \le m} n_i \to \infty$ means that the information contained in the training data is expanding; while $n_{\text{new}} \to \infty$ suggests that the additional information about the mixed effect associated with the new observations is also growing. Intuitively, with these two sources of sufficient information, one can make accurate prediction, and Theorem 7.2 has just confirmed that. Furthermore, it shows that, asymptotically, CMMP is doing better than RP in terms of MSPE.

3. Prediction intervals. Prediction intervals are of substantial practical interest [e.g., Chatterjee *et al.* (2008)]. Following the model (7.43), (7.44), we assume, in addition, that $\epsilon_{\text{new},j}$ is distributed as $N(0, R)$, where R is the same variance as that of ϵ_{ij} in (7.43). Also write α_I as α_{new}. Still, it is not necessary to assume that α_{new} has the same distribution, or even the same variance, as the α_i in (7.43). Also, α_{new} is understood as either (a) a new random effect or (b) identical to one of the $\alpha_i, 1 \le i \le m$. Consider the following prediction interval for $\theta = x_{\mathrm{n}}'\beta + \alpha_{\text{new}}$:

$$\left[\hat{\theta} - z_{a/2}\sqrt{\frac{\hat{R}}{n_{\text{new}}}},\ \hat{\theta} + z_{a/2}\sqrt{\frac{\hat{R}}{n_{\text{new}}}}\right], \tag{7.51}$$

where $\hat{\theta}$ is the CMEP of θ, $\hat{R}$ is the REML estimator of R, and z_a is the critical value so that $\mathrm{P}(Z > z_a) = a$ for $Z \sim N(0, 1)$. For a future observation, y_{f}, we assume that it shares the same mixed effects as the observed new observations $y_{\mathrm{n},j}$ in (7.44), that is,

$$y_{\mathrm{f}} = \theta + \epsilon_{\mathrm{f}}, \tag{7.52}$$

where $\epsilon_{\rm f}$ is a new error that is distributed as $N(0, R)$, and independent with all of the α's and other ϵ's. We consider the following prediction interval for $y_{\rm f}$:

$$\left[\hat{\theta} - z_{a/2}\sqrt{(1 + n_{\rm new}^{-1})\hat{R}},\ \ \hat{\theta} + z_{a/2}\sqrt{(1 + n_{\rm new}^{-1})\hat{R}}\right], \tag{7.53}$$

where $\hat{\theta}, \hat{R}$ are the same as in (7.51). Jiang *et al.* (2018) proved that, under regularity conditions, both (7.51) and (7.53) have asymptotic coverage probability of $1 - a$ for θ and $y_{\rm f}$, respectively.

4. Classified mixed logistic model prediction. Sun *et al.* (2018) has extended the CMMP method to binary observations under a mixed logistic model. The model assumes that, given the subject-specific random effects, $\alpha_1, \dots, \alpha_m$, binary responses $y_{ij}, i = 1, \dots, m, j = 1, \dots, n_i$ are conditionally independent with the conditional probability satisfying

$$\mathrm{P}(y_{ij} = 1|\alpha) = p_{ij} \ \text{ with } \ \mathrm{logit}(p_{ij}) = x_{ij}'\beta + \alpha_i, \tag{7.54}$$

where $\mathrm{logit}(p) = \log\{p/(1-p)\}$. Here i is the index for subject (e.g., patient, or group of patients), j is the index for observation within the subject (e.g., observation collected at the jth time point, or observation collected from the j patient in the subject group); x_{ij} is a vector of observed covariates, and β is a vector of unknown fixed effects. Furthermore, $\alpha_i, 1 \leq i \leq m$ are random effects, assumed to be independent and distributed as $N(0, \sigma^2)$, where σ^2 is an unknown variance. For any known vector x and function $g(\cdot)$, the BP of the mixed effect, $\theta = g(x'\beta + \alpha_i)$, is given by

$$\begin{aligned} \tilde{\theta} &= \mathrm{E}(\theta|y) \\ &= \frac{\mathrm{E}[g(x'\beta + \sigma\xi)\exp\{y_{i\cdot}\sigma\xi - \sum_{j=1}^{n_i}\log(1 + e^{x_{ij}'\beta + \sigma\xi})\}]}{\mathrm{E}[\exp\{y_{i\cdot}\sigma\xi - \sum_{j=1}^{n_i}\log(1 + e^{x_{ij}'\beta + \sigma\xi})\}]}, \end{aligned} \tag{7.55}$$

where $y_{i\cdot} = \sum_{j=1}^{n_i} y_{ij}$, and the expectations are taken with respect to $\xi \sim N(0, 1)$. Two special cases of (7.55) are the following:

(i) If the covariates are at the cluster level, that is, $x_{ij} = x_i$, and $g(u) = \mathrm{logit}^{-1}(u) = e^u/(1 + e^u)$, then, (7.55) reduces to

$$\tilde{p} = \frac{\mathrm{E}[\mathrm{logit}^{-1}(x'\beta + \sigma\xi)\exp\{y_{i\cdot}\sigma\xi - n_i\log(1 + e^{x_i'\beta + \sigma\xi})\}]}{\mathrm{E}[\exp\{y_{i\cdot}\sigma\xi - n_i\log(1 + e^{x_i'\beta + \sigma\xi})\}]}, \tag{7.56}$$

which is the BP of $p = \mathrm{logit}^{-1}(x'\beta + \alpha_i)$. Note that, in this case, the mixed effect is a subject-specific (conditional) probability, such as the probability of hemorrhage complication of the AT treatment in the ECMO problem discussed in the sequel, for a specific patient.

(ii) If $x = 0$, and $g(u) = u$, (7.55) reduces to

$$\tilde{\alpha}_i = \sigma \frac{\mathrm{E}[\xi \exp\{y_{i\cdot}\sigma\xi - \sum_{j=1}^{n_i} \log(1 + e^{x'_{ij}\beta + \sigma\xi})\}]}{\mathrm{E}[\exp\{y_{i\cdot}\sigma\xi - \sum_{j=1}^{n_i} \log(1 + e^{x'_{ij}\beta + \sigma\xi})\}]}, \tag{7.57}$$

which is the BP of α_i, the subject-specific (e.g., hospital) random effect.

In (7.55)–(7.57), β and σ are understood as the true parameters, which are typically unknown in practice. It is then customary to replace β, σ by their consistent estimators. The results are called empirical BP, or EBP. It is assumed that the sample size for the training data is sufficiently large that the EBP is approximately equal to the BP [Jiang and Lahiri (2001)]. Here, by training data we refer to the data $y_{ij}, 1 \leq i \leq m, 1 \leq j \leq n_i$ described above that satisfy the assumed mixed logistic model.

The main interest is to predict a mixed effect that is associated with a set of new observations. More specifically, let the new, binary observations be $y_{\mathrm{n},k}, k = 1, \dots, n_{\mathrm{new}}$, and the corresponding covariates be $x_{\mathrm{n},k}, k = 1, \dots, n_{\mathrm{new}}$ such that, conditional on a random effect α_I that has the same $N(0, \sigma^2)$ distribution, $y_{\mathrm{n},k}, 1 \leq k \leq n_{\mathrm{new}}$ are independent with

$$\mathrm{P}(y_{\mathrm{n},k} = 1|\alpha_I) = p_{\mathrm{n},k} \text{ and } \mathrm{logit}(p_{\mathrm{n},k}) = x'_{\mathrm{n},k}\beta + \alpha_I, \tag{7.58}$$

where β is the same as in (7.54). Typically, the sample size, n_{new}, for the new observations is limited. If one relies only on the new observations to estimate the mixed effect, say, $p_{\mathrm{n},k}$ for a given k, the available information is limited. Luckily, one has much more than just the new observations. It would be beneficial if one could "borrow strength" from the training data, which are much larger in size. For example, if one knows that $I = i$, then, there is a much larger cluster in the training data, namely, $y_{ij}, j = 1, \dots, n_i$, corresponding to the same cluster-specific random effect, α_I. This cluster in the training data is much larger because, quite often, n_i is much larger than n_{new}. One can also utilize the training data to estimate the unknown parameters, β and σ, which would be much more accurate than using only the new observations. As noted, with accurate estimation of the parameters, the EBP will closely approximate the BP [Jiang and Lahiri (2001)]. Thus, potentially one has a lot more information that can be used to estimate the mixed effect of interest associated with α_I. The difficulty is, however, that I is unknown. In fact, at this point, one does not know the answer to any of the following questions: (I) is there a "match" between I and one of the $1 \leq i \leq m$ corresponding to the training data clusters? and (II) if there is, which one?

It turns out that, as in the development of CMMP, the answer to (I) does not really matter, so far as prediction of the mixed effect is concerned. In other words, even if the actual match does not exist, a CMMP procedure based on the false match still helps in improving prediction accuracy of the mixed effects. Thus, we can simply focus on (II). To illustrate the method, which is called classified mixed logistic model prediction (CMLMP), as an extension of CMMP, let us consider a special case where the covariates are at the cluster level, that is, $x_{ij} = x_i$ for all i, j. Similarly, the covariates for the new observations are also at the cluster level, that is, $x_{\mathrm{n},k} = x_{\mathrm{n}}$.

First assume that there is a match between I, the index for the random effect associated with the new observations, and one of the indexes, $1 \leq i \leq m$, associated with the training-data random effects. However, this match is unknown to us. Thus, as a first step, we need to identify the match, that is, an index $\hat{I} \in \{1, \dots, m\}$ computed from the data, which may be viewed as an estimator of I.

Suppose that $I = i$. Then, by (7.56), the BP of $p_{\mathrm{n}} = \mathrm{P}(y_{\mathrm{n},k} = 1|\alpha_I) = \mathrm{logit}^{-1}(x_{\mathrm{n}}'\beta + \alpha_I) = \mathrm{logit}^{-1}(x_{\mathrm{n}}'\beta + \alpha_i)$ is

$$\tilde{p}_{\mathrm{n},i} = \frac{\mathrm{E}[\mathrm{logit}^{-1}(x_{\mathrm{n}}'\beta + \sigma\xi)\exp\{y_{i\cdot}\sigma\xi - n_i\log(1 + e^{x_i'\beta+\sigma\xi})\}]}{\mathrm{E}[\exp\{y_{i\cdot}\sigma\xi - n_i\log(1 + e^{x_i'\beta+\sigma\xi})\}]}. \quad (7.59)$$

In (7.59), the parameters β, σ are understood as the true parameters, which are unknown in practice. If we replace these parameters by their consistent estimators, such as the ML or GEE estimators [e.g., Jiang (2007), sec. 4.2; also see Chapter 2] based on the training data, we obtain the EBP of p_{n}, denoted by $\hat{p}_{\mathrm{n},i}$.

On the other hand, an "observed" p_{n} is the sample proportion, $\bar{y}_{\mathrm{n}} = n_{\mathrm{new}}^{-1}\sum_{k=1}^{n_{\mathrm{new}}} y_{\mathrm{n},k}$. Our idea is to identify I as the index $1 \leq i \leq m$ that minimizes the distance between $\hat{p}_{\mathrm{n},i}$ and $\bar{y}_{\mathrm{n}}$, that is,

$$\hat{I} = \mathrm{argmin}_{1\leq i\leq m}|\hat{p}_{\mathrm{n},i} - \bar{y}_{\mathrm{n}}|. \quad (7.60)$$

The classified mixed logistic model predictor (CMLMP) of p_{n} is then $\hat{p}_{\mathrm{n},\hat{I}}$.

Although the above development is based on the assumption that a match exists between the random effect corresponding to the new observations and one of the random effects associated with the training data, it was shown [Sun *et al.* (2018)], both theoretically and empirically, that CMLMP enjoys a similar nice behavior as CMMP, that is, even if the actual match does not exist, CMLMP still gains in prediction accuracy compared to the standard logistic regression prediction (SLRP). Furthermore, CMLMP is consistent in predicting the mixed effect associated with the new observations as the size of training data grows and so does the additional information from the new observations. Sun *et al.* (2018) also developed a method of estimating the MSPE of CMLMP as a measure of uncertainty.

5. An application. We conclude this section with an example of real-data application. Thromboembolic or hemorrhagic complications [e.g., Glass *et al.* (1997)] occur in as many as 60% of patients who underwent extracorporeal membrane oxygenation (ECMO), an invasive technology used to support children during periods of reversible heart or lung failure [e.g., Muntean (2002)]. Over half of pediatric patients on ECMO are currently receiving antithrombin (AT) to maximize heparin sensitivity. In a retrospective, multi-center, cohort study of children ($\leq$ 18 years of age) who underwent ECMO between 2003 and 2012, 8,601 subjects participated in 42 free-standing children's hospitals across 27 U.S. states and the District of Columbia known as Pediatric Health Information System (PHIS). Data were de-identified prior to inclusion in the study dataset; however, encrypted medical record numbers allowed for tracking of individuals across multiple hospitalizations. Many of the outcome variables were binary, such as the bleed_binary variable, which is a main outcome variable indicating hemorrhage complication of the treatment; and the DischargeMortalit1Flag variable, which is associated with mortality. Here the treatment refers to AT. Prediction of characteristics of interest associated with the binary outcomes, such as probabilities of hemorrhage complication or those of mortality for specific patients are of considerable interest. Note that the data are also potentially clustered, with the clusters corresponding to the children's hospitals. In addition to the treatment indicator, there were 20 other covariate variables, for which information were available. See Table 7.5 for a list of covariate variables.

We focus on the two outcomes of interest, bleed_binary variable and DischargeMortalit1Flag variable, that were mentioned in the above. The data includes 8601 patients data from 42 hospitals. The numbers of patients in different hospitals range from 3 to 487. We first use a forward-backward (F-B) BIC procedure [e.g., Broman and Speed (2002)] to build a mixed logistic model. Namely, we use a forward selection based on logistic regression to add covariate variables, one by one, until 50% of the variables have been added; we then carry out a backward elimination to drop the variables that have been added, one by one, until all of the variables are dropped. This F-B process generates a sequence of (nested) models, to which the BIC procedure [Schwarz (1978)] is applied to select the model.

The F-B BIC procedure leads to a subset of 12 patient-level covari-

Table 7.5 **ECMO Data Variable Description**

Variable Name	Description
bleed_binary	hemorrhagic (Yes/No)
DischargeMortalit1Flag	mortality at discharge (Yes/No)
LengthOfStay	number of days during hospitalization
MajSurgduringHosp_Count	number of surgeries during hospitalization
MajSurgduringHosp_binary	major surgery during hospitalization (Yes/No)
ecmovol_ind1	ECMO volume: High vs. Low
ecmovol_ind2	ECMO volume: Medium vs. Low
age_ind1	1: $\leq$ 30 days vs. 5: $\geq$ 10 yrs
age_ind2	2: 31 – 364 days vs. 5: $\geq$ 10 yrs
age_ind3	3: 1 – 2.9 yrs vs. 5: $\geq$ 10 yrs
age_ind4	4: 3 – 9.9 yrs vs. 5: $\geq$ 10 yrs
Top5PrincDx	top five common ICD-9 stems [Yes: a patient had at least one of 747, 746, 745, 770, or 756 (see below); No: None]
flag_renal	flag Renal track (Yes/No)
flag_CV	flag Cardiovascular (Yes/No)
flag_GI	flag Gastrointestinal (Yes/No)
flag_hemimm	flag Hematologic/Immunology (Yes/No)
flag_onc	flag Malignancy (Yes/No)
flag_metab	flag Metabolic (Yes/No)
flag_neuromusc	flag Neuromuscular (Yes/No)
flag_congengen	flag Other congenital/Genetic (Yes/No)
flag_resp	respiratory (Yes/No)
ALLecmodays	number of days under ECMO during hospitalization
gender	Sex
AT	antithrombin treatment (Yes: at least one does of antithrombin during hospitalization; No: None)

ates out of a total of more than 20 covariates. The same 12 covariates were selected for both outcome variables. Specifically, in the selected model, the probability of hemorrhage complication (or mortality) is associated with number of days during hospitalization (LengthOfStay), major surgery during hospitalization (MajSurgduringHosp_binary; Yes/No), whether the patient is no more than 30 days old (age_ind1), whether the patient has had at least one of the following: 747–Other congenital anomalies of circulatory system; 746–Other congenital anomalies of heart, excluding endocardial fibroelastosis; 745–Bulbus cordis anomalies and anomalies of cardiac septal closure; 770–Other respiratory conditions of fetus and newborn; 756–Other congenital musculoskeletal anomalies, excluding congenital myotonic chondrodystrophy (Top5PrincDx), whether the patient is flagged for cardiovascular (flag_CV; Yes/No), hematologic/immunology (flag_hemimm; Yes/No), metabolic (flag_metab; Yes/No), neuromuscular (flag_neuromusc; Yes/No), other congenital/genetic (flag_congengen; Yes/No), or respiratory (flag_resp; Yes/No), number of days under ECMO during hospitalization (ALLecmodays), and whether the patient has received the AT treatment (AT; Yes/No).

Out of the 12 patient-level covariates, two are continuous corresponding to number of days during hospitalization and the number of days under ECMO during hospitalization; the rest are binary. In addition to the patient-level covariates, there are two hospital-level covariates, namely, the total number of patients during the 10 year study who did receive AT (*yesat*) and total number of patients that were were included in the 10-year study (*total*). Both hospital-level covariates are continuous. It should be noted that the four continuous covariates need to be standardized before carrying out the CMLMP analysis.

The proposed mixed logistic model includes the above 12 patient-level covariates as well as the 2 hospital-level covariates, plus a hospital-specific random effect that captures the "uncaptured" as well as between-hospital variation. The mixed effects of interest are probabilities of hemorrhage complication corresponding to *bleed_binary*, and mortality probabilities associated with *DischargeMortalit1Flag*, for new observations. Note that, because most of the covariates are at the patient-level, these probabilities are patient-specific. However, the responses are clustered with the clusters corresponding to the hospitals, and there are 42 random effects associated with the hospitals under the mixed logistic model.

In order to test the CMLMP method, we randomly select 5 patients from a given hospital and treat these as the new observations. The rest of the hospitals, and rest of the patients from the same hospital (if any), correspond to the training data. We then use the matching strategy described above, with *yesat* and *total* as the cluster-level covariates that are used to identify the group for the new observations, then compute the CMLMP for each of the 5 selected patients. In addition, MSPE estimate of Sun *et al.* (2018) were computed, whose square root, multiplied by 2, is used as margin of error. This analysis applies to all but one hospital (Hospital #2033), for which only three patients are available. For this hospital all three patients are selected for the new observations, and the CMLMP and margin of error are obtained for all 3 patients. Therefore, for 41 out 42 of these analyses, there is a match between the new observations' group and one of the training data groups; and for 1 analysis there is no such a match.

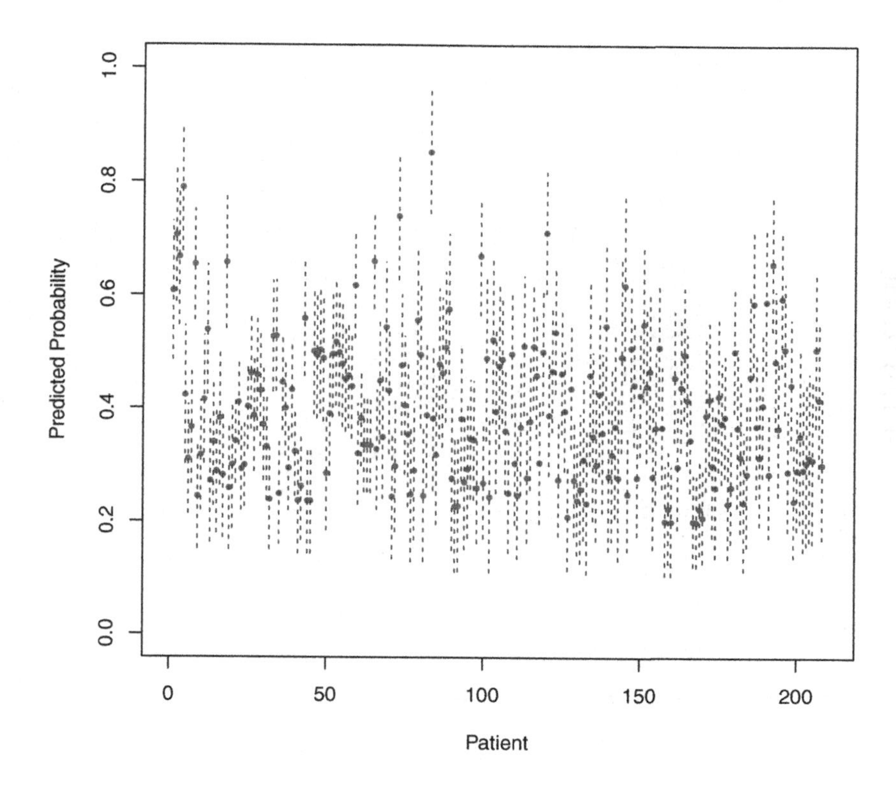

Fig. 7.5 *Predicted Probabilities of Hemorrhage Complication (bleed_ binary) with Margins of Errors: Dash Lines Indicate Margins of Errors*

Overall, the analysis yield a total of 208 predicted probabilities with the corresponding margins of errors.

The results are presented in Figure 7.5 (bleed_binary) and Figure 7.6 (DischargeMortalit1Flag). Note that, for DischargeMortalit1Flag, some of the predicted are close to zero; as a result, the lower margin is negative, and therefore truncated at 0. On the other hand, there is no need for truncation of the lower (or upper) margin for bleed_binary.

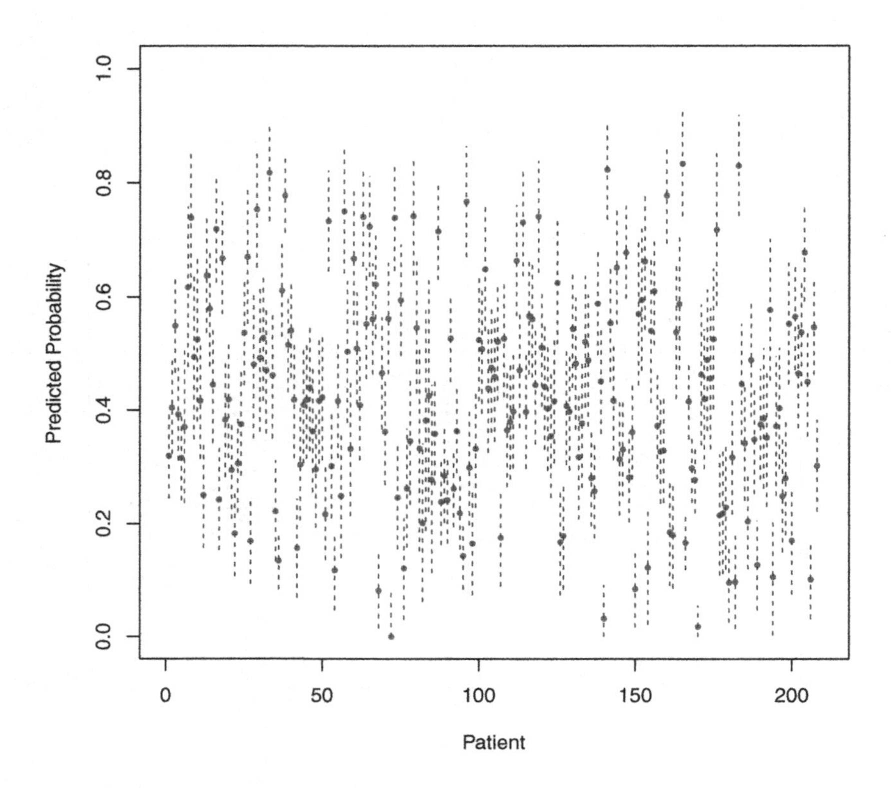

Fig. 7.6 *Predicted Mortality Probabilities (DischargeMortalit1Flag) with Margins of Errors: Dash Lines Indicate Margins of Errors*

7.7 Exercises

7.1. In Example 7.1 of Subsection 7.1, let the true parameters be $\mu = -0.5$, $\sigma^2 = 2.0$, and $\tau^2 = 1.0$. Also, let $m = 100$ and $k_i = 5$, $1 \leq i \leq m$. In the following, the errors are always generated from a normal distribution.

a. Generate the random effects from a normal distribution. Make a Q–Q plot to assess normality of the random effects, using REML estimators of the parameters.

b. Generate the random effects from a double-exponential distribution (with the same variance). Make a Q–Q plot to assess normality of the random effects, again using REML estimators of the parameters.

c. Generate the random effects from a centralized-exponential distribution (with the same variance). Here a centralized-exponential distribution is the distribution of $\xi - \mathrm{E}(\xi)$, where ξ has an exponential distribution. Make a Q–Q plot to assess normality of the random effects, using REML estimators of the parameters.

d. Compare the plots in **a, b,** and **c**. What do you conclude?

7.2. Show that the minimizer of (7.3) is the same as the best linear unbiased estimator (BLUE) for β and the best linear unbiased predictor (BLUP) for γ in the linear mixed model $y = X\beta + Z\gamma + \epsilon$, where $\gamma \sim N(0, \sigma^2 I_q)$, $\epsilon \sim N(0, \tau^2 I_n)$, and γ and ϵ are independent, provided that λ is identical to the ratio τ^2/σ^2. For the definition of BLUE and BLUP, see Section 5.1.

7.3. Prove Lemma 7.1.

7.4. Consider the quadratic spline given by (7.10). Show that the shape of the spline is half circle between 0 and 1 facing up, half circle between 1 and 2 facing down, and half circle between 2 and 3 facing up. Also show that the function is smooth in that it has a continuous derivative.

7.5. Consider the three-point distribution in the paragraph containing (7.12). Verify the moment properties mentioned below (7.12).

7.6. Show that, under (7.16), (7.17), one has $\mathrm{E}(y_i|\eta_i) = b'(\eta_i) \equiv \mu(\theta_i)$ and $\mathrm{var}(y_i|\eta_i) = \phi_i b''(\eta_i) \equiv V(\theta_i)$.

7.7. This exercise is related to Example 7.4.

a. Show that, if the distribution of y is the same as that of $\tilde{y}|y$, the p-value (7.20) has a Uniform$[0,1]$ distribution.

b. Show that the two models described in the example agree in terms of the mean and variance of y_{ij} but not in terms of the covariance between y_{ij} and y_{ik} for $j \neq k$.

7.8. Consider the $\hat{\theta}_{i,\lambda}$ defined by (7.28), where $\hat{\theta}_i$ is the BLUP (see Subsection 5.1), assuming Fay-Herriot model (see Example 5.1) with A being known. Show that the $\lambda_i, 1 \leq i \leq m$ that minimize (7.30) are given by (7.31).

7.9. This exercise has to do with the self benchmarking property of predictors derived under an augmented model; see Section 7.5.

a. This is about self benchmarking of EBLUP derived under an augmented model. Show that, under the Fay-Herriot model (see Example 5.1), the EBLUPs of $\theta_i = x_i'\beta + v_i$, $1 \leq i \leq m$, obtained by fitting the augmented model (7.32), are self benchmarking, that is, they satisfy (7.25).

b. This is about self benchmarking of OBP derived under an augmented model. Let $\hat{\theta}_i$ denote the ith component of the OBP defined by (7.37). Show that $\hat{\theta}_i, 1 \leq i \leq m$ are self benchmarking, that is, they satisfy (7.25).

7.10. Verify the conditional expectation (7.55), and the two special cases (7.56) and (7.57).

Bibliography

Akaike, H. (1973), *Information theory as an extension of the maximum likelihood principle*, in *Second International Symposium on Information Theory* (B. N. Petrov and F. Csaki eds.), pp. 267–281 (Akademiai Kiado, Budapest).

Arvesen, J. N. (1969), *Jackknifing U-statistics*, in *Ann. Math. Statist.* 40, pp. 2076–2100.

Arvesen, J. N. and Schmitz, T. H.(1970), *Robust procedures for variance component problems using the jackknife*, in *Biometrics* 26, pp. 677–686.

Azzalini, A. and Capitanio, A. (2014), *The Skew-Normal and Related Families*, (Cambridge University Press, New York).

Bandyopadhyay, R. (2017), *Benchmarking of Observed Best Predicror*, (Ph.D. Dissertation, Dept. of Stat., Univ. of Calif., Davia, CA).

Bartlett, M. S. (1936), *Some notes on insecticide tests in the laboratory and in the field*, in *J. Roy. Statist. Suppl.* 3, pp. 185–194.

Basawa, I. V., and Rao, B. L. S. P. (1980), *Statistical Inference for Stochastic Processes*, (Academic Press, London).

Battese, G. E., Harter, R. M., and Fuller, W. A. (1988), *An error-components model for prediction of county crop areas using survey and satellite data*, in *J. Amer. Statist. Assoc.* 80, pp. 28–36.

Bhatia, R. (1997), *Matrix Analysis*, (Springer, New York).

Booth, J. G. and Hobert, J. P. (1999), *Maximum generalized linear mixed model likelihood with an automated Monte Carlo EM algorithm*, in *J. Roy. Statist. Soc. B* 61, pp. 265–285.

Breslow, N. E. and Clayton, D. G. (1993), *Approximate inference in generalized linear mixed models*, in *J. Amer. Statist. Assoc.* 88, pp. 9–25.

Bondell, H. D., Krishna, A. and Ghosh, S. K. (2010), *Joint variable selection for fixed and random effects in linear mixed-effects models*, in *Biometrics* 66, pp. 1069–1077.

Broman, K. W. and Speed, T. P. (2002), *A model selection approach for the identification of quantitative trait loci in experiemental crosses*, in *J. Roy. Statist. Soc. Ser. B* 64, pp. 641–656.

Calvin, J. A., and Sedransk, J. (1991), *Bayesian and frequentist predictive inference for the patterns of care studies*, in *J. Amer. Statist. Assoc.* 86, pp.

36–48.

Casella, G. and Berger, R. L. (2002), *Statistical Inference*, (2nd ed.), (Duxbury).

Chakraborty, A., Datta, G. K. and Mandal, A. (2018), *Robust hierarchical Bayes small area estimation for nested error regression model*, in *arXiv:*1702.05832v2..

Chambers, R. (1986), *Outlier robust finite population estimation*, in *J. Amer. Statist. Assoc.* 81, pp. 1063–1069.

Chambers, R. and Tzavidis, N. (2006), *M-quantile models for small area estimation*, in *Biometrika* 93, pp. 255–268.

Chatterjee, S. and Lahiri, P., and Li, H. (2008), *Parametric bootstrap approximation to the distribution of EBLUP, and related prediction intervals in linear mixed models*, in *Ann. Statist.* 36, pp. 1221–1245.

Chaudhuri, S. and Ghosh, M. (2011), *Empirical likelihood for small area estimation*, in *Biometrika* 98, pp. 473–480.

Chen, C. F. (1985), *Robustness aspects of score tests for generalized linear and partially linear regression models*, in *Technometrics* 27, pp. 277–283.

Chen, S. (2012), *Predictive modeling for clustered data with applications, Ph. D. Dissertation*, (Dept. of Statist., Univ. of Calif., Davis, CA).

Chen, S., Jiang, J. and Nguyen, T. (2015), *Observed best prediction for small area counts*, in *J. Survey Statist. Methodology* 3, pp. 136–161.

Chernoff, H., and Lehmann, E. L. (1954), *The use of maximum-likelihood estimates in χ^2 tests for goodness of fit*, in *Ann. Math. Statist.* 25, pp. 579–586.

Cisco Systems Inc. (1996), *NetFlow Services and Applications*, White Paper.

Claeskens, G. and Hart, J. D. (2009), *Goodness-of-fit tests in mixed models (with discussion)*, in *TEST* 18, pp. 213–239.

Copas, J. and Eguchi, S. (2005), *Local model uncertainty and incomplete-data bias (with discussion)*, in *J. Roy. Statist. Soc. B* 67, 459–513.

Datta, G. S. and Ghosh, M. (1991), *Bayesian prediction in linear linear models: Application to small area estimation*, in *Ann. Statist.* 19, pp. 1748–1770.

Datta, G. S. and Lahiri, P. (2000), *A unified measure of uncertainty of estimated best linear unbiased predictors in small area estimation problems*, in *Statist. Sinica* 10, pp. 613–627.

Datta, G. S., Rao, J. N. K. and Smith, D. D. (2005), *On measuring the variability of small area estimators under a basic area level model*, in *Biometrika* 92, pp. 183–196.

Datta, G. S., Kubokawa, T., Rao, J. N. K., and Molina, I. (2011), *Estimation of mean squared error of model-based small area estimators*, in *TEST* 20, pp. 367–388.

de Leeuw, J. (1992), *Introduction to Akaike (1973) information theory and an extension of the maximum likelihood principle*, in *Breakthroughs in Statistics* (S. Kotz and N. L. Johnson eds.), Vol. 1, pp. 599–609 (Springer, London).

Demidenko, E. (2013), *Mixed Models: Theory and Application with R*, (2nd ed.), (Wiley, New York).

Dempster, A., Laird, N., and Rubin, D. (1977), *Maximum likelihood from incomplete data via the EM algorithm (with discussion)*, in *J. Roy. Statist. Soc. B* 39, pp. 1–38.

Dempster, A. P. and Ryan, L. M. (1985), *Weighted normal plots*, in *J. Amer. Ststist. Assoc.* 80, pp. 845–850.

Diggle, P. J., Liang, K. Y., and Zeger, S. L. (1994), *Analysis of Longitudinal Data*, (Oxford Univ. Press).

Diggle, P. J., Heagerty, P., Liang, K. Y., and Zeger, S. L. (2002), *Analysis of Longitudinal Data* (2nd ed.), (Oxford Univ. Press).

Dzhaparidze, K. (1986), *Parameter Estimation and Hypothesis Testing in Spectral Analysis of Stationary Time Series*, (Springer, New York).

Efron, B. (1979), *Bootstrap method: Another look at the jackknife*, in *Ann. Statist.* 7, pp. 1–26.

Efron, B. and Hinkley, D. V. (1978), *Assessing the accuracy of the maximum likelihood estimator: observed versus expected Fisher information*, in *Biometrika* 65, pp. 457–487.

Efron, B. and Tibshirani, R. J. (1993), *An Introduction to the Bootstrap*, (Chapman & Hall/CRC).

Efron, B. and Tibshirani, R. (2007), *On testing the significance of sets of genes*, in *Ann. Appl. Statist.* 1, pp. 107–129.

Fabrizi, E. and Lahiri, P. (2013), *A design-based approximation to the Bayes information criterion in finite population sampling*, in *Statistica* 73, pp. 289–301.

Fan, J. and Yao, Q. (2003), *Nonlinear Time Series: Nonparametric and Parametric Methods*, (Springer, New York).

Fay, R. E. and Herriot, R. A. (1979), *Estimates of income for small places: an application of James-Stein procedures to census data*, in *J. Amer. Statist. Assoc.* 74, pp. 269–277.

Ferrante, M. R. and Trivisano, C. (2010), *Small area estimation of the number of firms' recruits by using multivariate models for count data*, in *Survey Methodology* 36, pp. 171–180.

Fisher, R. A. (1922), *On the interpretation of chi-square from contingency tables, and the calculation of P*, in *J. Roy. Statist. Soc.* 85, 87–94.

Foutz, R. V., and Srivastava, R. C. (1977), *The performance of the likelihood ratio test when the model is incorrect*, in *Ann. Statist.* 5, pp. 1183–1194.

Friedman, J. (1991), *Multivariate adaptive regression splines (with discussion)*, in *Ann. Statist.* 19, pp. 1–67.

Fuller, W. A. (2009), *Sampling Statistics*, (Wiley, Hoboken, NJ).

Ganesh, N. (2009), *Simultaneous credible intervals for small area estimation problems*, in *J. Multivariate Anal.* 100, pp. 1610–1621.

Gershunskaya, J. (2018), *Robust empirical best small area finite population mean estimation using a mixture model*, in *Calcutta Statist. Assoc. Bull.* 69, pp. 183–204.

Ghosh, M., Nangia, N., and Kim, D. (1996), *Estimation of median income of four-person families: A Bayesian time series approach*, in *J. Amer. Statist. Assoc.* 91, 1423–1431.

Ghosh, M., Natarajan, K., Stroud, T. W. F. and Carlin, B. P. (1998), *Generalized linear models for small-area estimation*, in *J. Amer. Statist. Assoc.* 93, pp. 273–282.

Glass P, Bulas, D. I., Wagner, A. E., *et al.* (1997), *Severity of brain injury following neonatal extracorporeal membrane oxygenation and outcome at age 5 years*, in *Dev. Med. Child neurol.* 39, pp. 441–448.

Guo, W. (2002), *Functional mixed effects models*, in *Biometrics* 58, pp. 121–128.

Gourieroux, C., and Monfort, A. (1995), *Statistics and Econometric Models, Vol. 2*, (Cambridge Univ. Press).

Hajarisman, N. (2013), *Two-level hierarchical Bayesian Poisson models for small area estimation of infant mortality rates, Ph. D. Dissertation*, (Dept. of Math. Natural Sci., Bogor Agricultural Univ., Indonesia).

Hampel, F. R., Ronchetti, E. M., Rousseeuw, P. J., and Stahe, W. A. (1986), *Robust Statistics: The Approach Based on Influence Functions*, (Wiley, New York).

Hall, P., and Maiti, T. (2006), *Nonparametric estimation of mean-squared prediction error in nested-error regression models*, in *Ann. Statist.* 34, pp. 1733–1750.

Hand, D. and Crowder, M. (2002), *Practical Longitudinal Data Analysis*, (Chapman and Hall, London).

Hannan, E. J. and Quinn, B. G. (1979), *The determination of the order of an autoregression*, in *J. Roy. Statist. Soc. B* 41, pp. 190–195.

Hansen, L. P. (1982), *Large sample properties of generalized method of moments estimators*, in *Econometrica* 50, pp. 1029–1054.

Hartley, H. O. and Rao, J. N. K., *Maximum likelihood estimation for the mixed analysis of variance model*, in *Biometrica* 54, pp. 93–108.

Hastie, T. J. and Tibshirani, R. J. (1990), *Generalized Additive Models*, (Chapman & Hall, CRC).

Hayes, P. M. *et al.* (1993), *Quantitative trait locus effects and environmental interaction in a sample of North American barley germ plasm*, in *Theor. Appl. Genet.* 87, pp. 392–401.

He, X., Zhu, Z.-Y., and Fung, W. K. (2002), *Estimation in a semiparametric model for longitudinal data with unspecified dependence structure*, in *Biometrika* 89, pp. 579–590.

Hedeker, D., Gibbons, R. D., and Flay, B. R. (1994), *Random-effects regression models for clustered data with an example from smoking prevention research*, in *J. Consulting Clinical Psych.* 62, pp. 757–765.

Henderson, C. R. (1948), *Estimation of general, specific and maternal combining abilities in crosses among inbred lines of swine, Ph. D. Dissertation*, (Iowa State Univ., Ames, IA).

Heritier, S., and Ronchetti, E. (1994), *Robust bounded-influence tests in general parametric models*, in *J. Amer. Statist. Assoc.* 89, pp. 897–904.

Heyde, C. C. (1994), *A quasi-likelihood approach to the REML estimating equations*, in *Statist. Probab. Letters* 21, pp. 381–384.

Heyde, C. C. (1997). *Quasi-likelihood and Its Application*, (Springer, New York).

Hu, K., Choi, J., Sim, A., and Jiang, J. (2013), *Best predictive generalized linear mixed model with predictive lasso for high-speed network data analysis*, in *International J. Statist. Probab.* 4, pp. 132i–148.

Huber, P. J. (1964), *Robust estimation of a location parameter*, in *Ann. Math.*

Statist. 35, pp. 73–101.

Huber, P. J. (1981). *Robust Statistics*, (Wiley, New York).

Ibrahim, J. G., Zhu, H., Carcia, R. I., and Guo, R. (2011), *Fixed and random effects selection in mixed effects models*, in *Biometrics* 67, pp. 495–503.

Ibrahim, J. G., Zhu, H., and Tang, N. (2008), *Model selection criteria for missing-data problems using the EM algorithm*, in *J. Amer. Statist. Assoc.* 103, pp. 1648–1658.

Jiang, J. (1996), *REML estimation: Asymptotic behavior and related topics*, in *Ann. Statist.* 24, pp. 255–286.

Jiang, J. (1997), *Wald consistency and the method of sieves in REML estimation*, in *Ann. Statist.* 25, pp. 1781–1803.

Jiang, J. (1998a), *Consistent estimators in generalized linear mixed models*, in *J. Amer. Statist. Assoc.* 93, pp. 720–729.

Jiang, J. (1998b), *Asymptotic properties of the empirical BLUP and BLUE in mixed linear models*, in *Statistica Sinica* 8, pp. 861–885.

Jiang, J. (2001), *Goodness-of-fit tests for mixed model diagnostics*, in *Ann. Statist.* 29, pp. 1137–1164.

Jiang, J. (2003), *Empirical method of moments and its applications*, in *J. Statist. Plann. Inference* 115, pp. 69–84.

Jiang, J. (2005), *Partially observed information and inference and inference about non-Gaussian mixed linear models*, in *Ann. Statist.* 33, pp. 2695–2731.

Jiang, J. (2007), *Linear and Generalized Linear Mixed Models and Their Applications*, (Springer, New York).

Jiang, J. (2010), *Large Sample Techniques for Statistics*, (Springer, New York).

Jiang, J. (2017), *Large Asymptotic Analysis of Mixed Effects Models: Theory, Applications, and Open Problems*, (Chapman & Hall/CRC, Boca Raton, FL).

Jiang, J. (2012), *On robust versions of classical tests with dependent data*, in *Nonparametric Statistical Methods and Related Topics - A Festschrift in Honor of Professor P. K. Bhattacharya on the Occasion of His 80th Birthday*, J. Jiang, G. G. Roussas, F. J. Samaniego eds., pp. 77–99, (World Scientific, Singapore).

Jiang, J., Jia, H., and Chen, H. (2001), *Maximum posterior estimation of random effects in generalized linear mixed models*, in *Statistica Sinica* 11, pp. 97–120.

Jiang, J. and Lahiri (2001), *Empirical best prediction for small area inference with binary data*, in *Ann. Inst. Statist. Math.* 53, 217–243.

Jiang, J. and Lahiri (2005), *Mixed model prediction and small area estimation (with discussion)*, in *TEST* 15, 1–96.

Jiang, J., Lahiri, P. and Wan, S. (2002), *A unified jackknife theory for empirical best prediction with M-estimation*, in *Ann. Statist.* 30, pp. 1782–1810.

Jiang, J. and Nguyen, T. (2009), Comments on: Goodness-of-fit tests in mixed models by G. Claeskens and J. D. Hart, *TEST* 18, 248–255.

Jiang, J. and Nguyen, T. (2012), *Small area estimation via heteroscedastic nested-error regression*, in *Canadian J. Statist.* 40, pp. 588–603.

Jiang, J., Nguyen, T. and Rao, J. S. (2009), *A simplified adaptive fence procedure*, in *Statist. Probab. Letters* 79, pp. 625–629.

Jiang, J., Nguyen, T. and Rao, J. S. (2010), *Fence method for nonparametric small area estimation*, in *Survey Methodology* 36, pp. 3–11.

Jiang, J., Nguyen, T. and Rao, J. S. (2011a), *Best predictive small area estimation*, in *J. Amer. Statist. Assoc.* 106, pp. 732–745.

Jiang, J., Nguyen, T. and Rao, J. S. (2011b), *Invisible fence method and the identification of differentially expressed gene sets*, in *Statist. Interface* 4, pp. 403–415.

Jiang, J. and Nguyen, T. (2015), *The Fence Methods*, (World Scientific, Singapore).

Jiang, J., Nguyen, T. and Rao, J. S. (2015a), *The E-MS algorithm: Model selection with incomplete data*, in *J. Amer. Statist. Assoc.* 110, pp. 1136–1147.

Jiang, J., Nguyen, T. and Rao, J. S. (2015b), *Observed best prediction via nested-error regression with potentially misspecified mean and variance*, in *Survey Methodology* 41, pp. 37–55.

Jiang, J., Luan, Y. and Wang, Y.-G. (2007), *Iterative estimating equations: Linear convergence and asymptotic properties*, in *Ann. Statist.* 35, pp. 2233–2260.

Jiang, J. and Rao, J. S. (2003), *Consistent procedures for mixed linear model selection*, in *Sankhya* 65A, 23–42.

Jiang, J., Rao, J. S., Fan, J. and Nguyen, T. (2018), *Classified mixed model prediction*, in *J. Amer. Statist. Assoc.* 113, pp. 269–279.

Jiang, J., Rao, J. S., Gu, Z. and Nguyen, T. (2008), *Fence methods for mixed model selection*, in *Ann. Statist.* 36, pp. 1669–1692.

Jiang, J. and Torabi, M. (2018), A unified approach to goodness-of-fit tests with application to small area estimation, Technical Report.

Jiang, J. and Zhang, W. (2001), *Robust estimation in generalized linear mixed models*, in *Biometrika* 88, pp. 753–765.

Jiang, J. and Zhang, W. (2002), *Distribution-free prediction intervals in mixed linear models*, in *Statistica Sinica* 12, pp. 537–553.

Jin, X., Carlin, B. P., and Banerjee, S. (2005), *Generalized hierarchical multivariate CAR models for area data*, in *Biometrics* 61, pp. 950–961.

Jung, S.-H. (1996), *Quasi-likelihood for median regression models*, in *J. Amer. Statist. Assoc.* 91, pp. 251–257.

Karim, M. R. and Zeger, S. L. (1992), *Generalized linear models with random effects: Salamander mating revisited*, in *Biometrics* 48, pp. 631–644.

Kass, R. E. and Wassermann L. (1995), *A reference test for nested hypotheses and its relationship to the Schwartz criterion*, in *J. Amer. Statist. Assoc.* 90, pp. 928–934.

Kauermann, G. (2005), *A note on smoothing parameter selection for penalized spline smoothing*, in *J. Statist. Planning & Inference* 127, pp. 53–69.

Kent, J. T. (1982), *Robustness properties of likelihood ratio tests*, in *Biometrika* 69, pp. 19–27.

Khuri, A. I., Mathew, T., and Sinha, B. K. (1998), *Statistical tests for mixed linear models*, (Wiley, New York).

Kim, H. J., and Cai, L. (1993), *Robustness of the likelihood ratio test for a change in simple linear regression*, in *J. Amer. Statist. Assoc.* 88, pp. 864–871.

Koenker, P. and Bassett, G. (1978), *Regression quantiles*, in *Econometrica* 46, pp. 33–50.

Koopmans, L. H. (1995), *The Spectral Analysis of Time Series*, (Elsevier).

Kott, P. S. (1991), *Robust small domain estimation using random effects modelling*, in *Survey Methodology* 15, pp. 3–12.

Krafty, R. T., Hall, M. and Guo, W. (2011), *Functional mixed effects spectral analysis*, in *Biometrika* 98, pp. 583–598.

Lahiri, P. and Rao, J. N. K. (1995), *Robust estimation of mean squared error of small area estimators*, in *J. Amer. Statist. Assoc.* 90, pp. 758–766.

Lai, P. Y. and Lee, S. (2005), *An overview of asymptopic properties of L_p regression under general classes of error distributions*, in *J. Amer. Statist. Assoc.* 100, pp. 446–458.

Lander, E. S., and Botstein, D. (1989), *Mapping Mendelian factors underlying quantitative traits using RFLP linkage maps*, in *Genetics* 121, pp. 185–199.

Lange, N. and Ryan, L. (1989), *Assessing normality in random effects models*, in *Ann. Statist.* 17, pp. 624–642.

Lee, L. F. (1992), *On the efficiency of methods of simulated moments and maximum simulated likelihood estimation of discrete response models*, in *Econometric Theory* 8, pp. 518–552.

Lehmann, E. L. (1999), *Elements of Large-Sample Theory*, (Springer, New York).

Lehmann, E. L. and Casella, G. (1998), *Theory of Point Estimation, 2nd ed.*, (Springer, New York).

Liang, K. Y. and Zeger, S. L. (1986), *Longitudinal data analysis using generalized linear models*, in *Biometrika* 73, pp. 13–22.

Lin, X. and Breslow, N. E. (1996), *Bias correction in generalized linear mixed models with multiple components of dispersion*, in *J. Amer. Statist. Assoc.* 91, pp. 1007–1016.

Little, R. J. A. and Rubin, D. B. (2002), *Statistical Analysis with Missing Data*, (2nd ed.), (Wiley, New York).

Lombardía, M. J. and Sperlich, S. (2008), *Semiparametric inference in generalized mixed effects models*, in *J. Roy. Statist. Soc. B* 70, pp. 913–930.

Luo, Z. W. *et al.* (2007), *SFP genotyping from affymetrix arrays is robust but largely detects cis-acting expression regulators*, in *Genetics* 176, pp. 789–800.

Malec, D., Sedransk, J., Moriarity, C. L., and LeClere, F. B. (1997), *Small area inference for binary variables in the National Health Interview Survey*, in *J. Amer. Statist. Assoc.* 92, 815–826.

McCullagh, P. and Nelder, J. A. (1989), *Generalized Linear Models* (2nd ed.), (Chapman and Hall, London).

McCulloch, C. E., Searle, S. R., and Neuhaus, J. M. (2008), *Generalized, Linear, and Mixed Models*, (2nd ed.), (Wiley, Hoboken, NJ).

McFadden, D. (1989), *A method of simulated moments for estimation of discrete response models without numerical integration*, in *Econometrika* 57, pp. 995–1026.

Moore, D. S. (1978), Chi-square tests, in *Studies in Statistics* (R. V. Hogg, ed.),

Mathematical Society of America, Providence, RI.
Morris, C. N. and Christiansen, C. L. (1995), *Hierarchical models for ranking and for identifying extremes with applications*, in *Bayes Statistics* 5, (Oxford Univ. Press).
Morton, R. (1987), *A generalized linear model with nested strata of extra-Poisson variation*, in *Biometrika* 74, pp. 247–257.
Mou, J. (2012), *Two-stage fence methods in selecting covariates and covariance for longitudinal data, Ph. D. Dissertation*, (Dept. of Statist., Univ. of Calif., Davis, CA).
Müller, S., Scealy, J. L., and Welsh, A. H. (2013), *Model selection in linear mixed models*, in *Statist. Sci.* 28, pp. 135–167.
Münnich, R., Burgard, J. P., and Vogt, M. (2009), *Small area estimation for population counts in the German Census 2011*, in *Section on Survey Research Methods, JSM 2009, Washington, D.C.*.
Muntean W. (2002), *Fresh frozen plasma in the pediatric age group and in congenital coagulation factor deficiency*, in *Thromb. Res.* 107, S29-S32, pp. 0049–3848.
Newey, W. K. (1985), Generalized method of moments specification testing, *J. Econometrics* 29, 229–256.
Nishii, R. (1984), *Asymptotic properties of criteria for selection of variables in multiple regression*, in *Ann. Statist.* 12, pp. 758–765.
Opsomer, J. D., Breidt, F. J., Claeskens, G., Kauermann, G. & Ranalli, M. G. (2008), *Nonparametric small area estimation using penalized spline regression*, in *J. Roy. Statist. Soc. B* 70, pp. 265–286.
Owen, A. B. (1988), *Empirical likelihood ratio confidence intervals for a single functional*, in *Biometrika* 75, pp. 237–249.
Owen, A. B. (2001), *Empirical Likelihood*, (Chapman & Hall).
Pan, W. (2001), *On the robust variance estimator in generalised estimating equations*, in *Biometrika* 88, pp. 901–906.
Pfeffermann, D. (2013), *New important developments in small area estimation*, in *Statist. Sci.* 28, 40–68.
Pierce, D. (1982), *The asymptotic effect of substituting estimators for parameters in certain types of statistics*, in *Ann. Statist.* 10, pp. 475–478.
Prasad, N. G. N. and Rao, J. N. K. (1990), *The estimation of mean squared errors of small area estimators*, in *J. Amer. Statist. Assoc.* 85, pp. 163–171.
Press, W. H., Teukolsky, S. A., Vetterling, W. T. and Flannery, B. P. (1997), *Numerical Recipes in C—The Arts of Scientific Computing*, (2nd ed.), (Cambridge Univ. Press).
Rady, E. A., Kilany, N. M. and Eliwa, S. A. (2015) *Estimation in mixed-effects functional ANOVA models*, in *J. Multivariate Anal.* 133, pp. 346–355.
Rao, C. R., and Wu, Y. (1989), *A strongly consistent procedure for model selection in a regression problem*, in *Biometrika* 76, pp. 369–374.
Rao, J. N. K. and Molina, I. (2015), *Small Area Estimation* (2nd ed.), (Wiley, New York).
Richardson, A. M. and Welsh, A. H. (1994), *Asymptotic properties of restricted maximum likelihood (REML) estimates for hierarchical mixed linear models*,

in *Austral. J. Statist.* 36, pp. 31–43.

Rice, J. A. (1995), *Mathematical Statistics and Data Analysis*, 2nd ed., Duxbury Press, Belmont, CA.

Richardson, A. M. and Welsh, A. H. (1996), *Covariate screening in mixed linear models*, in *J. Multivariate Anal.* 58, pp. 27–54.

Robinson, G. K. (1991), *That BLUP is a good thing: The estimation of random effects (with discussion)*, in *Statist. Sci.* 6, pp. 15–51.

Schrader, R. M., Hettmansperger, T. P. (1980), *Robust analysis of variance based upon a likelihood ratio criterion*, in *Biometrika* 67, pp. 93–101.

Schwarz, G. (1978), *Estimating the dimension of a model*, in *Ann. Statist.* 6, pp. 461–464.

Sen, A. and Srivastava, M. (1990), *Regression Analysis: Theory, Methods, and Applications*, (Spriner, New York).

Searle, S. R. (1971), *Linear Models*, (Wiley, New York).

Searle, S. R., Casella G. and McCulloch, C. E. (1992), *Variance Components*, (Wiley, New York).

Self, S. G. and Liang, K. Y. (1987), *Asymptotic properties of maximum likelihood estimators and likelihood ratio tests under nonstandard conditions*, in *J. Amer. Statist. Assoc.* 82, pp. 605–610.

Shao, J. (1993), *Linear model selection by cross-validation*, in *J. Amer. Statist. Assoc.* 88, pp. 486–494.

Shao, J. and Tu, D. (1995), *Jackknife and Bootstrap*, (Springer, New York).

Shibata, R. (1984), *Approximate efficiency of a selection procedure for the number of regression variables*, in *Biometrika* 71, pp. 43–49.

Silvapulle, M. J. (1992), *Robust Wald-type tests of one-sided hypotheses in the linear model*, in *J. Amer. Statist. Assoc.* 87, pp. 156–161.

Sinha, S. K. and Rao, J. N. K. (2009), *Robust small area estimation*, in *Canadian J. Statist.* 37, pp. 381–399.

Subramanian, A., Tamayo, P., Mootha, V. K., Mukherjee, S., Ebert, B. L., Gillette, M. A., Paulovich, A., Pomeroy, S. L., Golub, T. P., Lander, E. S. and Mesirov, J. P. (2005), *Gene set enrichment analysis: A knowledge-based approach for interpreting genome-wide expression profiles hypotheses in the linear model*, in *Proc. Natl. Acad. Sci. USA* 102, pp. 15545–15550.

Sun, H., Nguyen, T., Luan, Y., and Jiang, J. (2018), *Classified mixed logistic model prediction*, *J. Multivariate Anal.*, in press.

Thall, P. F. and Vail, S. C. (1990), *Some covariance models for longitudinal count data with overdispersion*, in *Biometrics* 46, pp. 657–671.

Torabi, M. (2012), *Likelihood inference in generalized linear mixed models with two components of dispersion using data cloning*, in *Comput. Statist.Data Anal.* 56, pp. 4259–4265.

Torabi, M. (2014), *Spatial generalized linear mixed models with multivariate CAR models for area data*, in *Spatial Statist.* 10, pp. 12–26.

Vaida, F. and Blanchard, S. (2005), *Conditional Akaike information for mixed-effects models*, in *Biometrika* 92, pp. 351–370.

Verbeke, G., Molenberghs, G., and Beunckens, C. (2008), *Formal and informal model selection with incomplete data*, in *Statist. Sci.* 23, 201–218.

Wahba, G. (1978), *Improper priors, spline smoothing and the problem of guarding against model errors in regression*, in *J. Roy. Statist. Soc. B* 40, pp. 364–372.

Wahba, G. (1983), *Bayesian confidence intervals for the cross-validated smoothing spline*, in *J. Roy. Statist. Soc. B* 45, pp. 133–150.

Wand, M. (2003), *Smoothing and mixed models*, in *Comput. Statist.* 18, pp. 223–249.

Wang, Y.-G., Bai, Z.-D., and Jiang, J. (2015), *M-estimation for analysis of longitudinal data*, *Electronic J. Statist.*, revised.

Wang, J., Fuller, W. A. and Qu, Y. (2008), *Small area estimation under a restriction*, *Survey Methodology* 34, pp. 29–36.

Weiss, L. (1975), *The asymptotic distribution of the likelihood ratio in some nonstandard cases*, in *J. Amer. Statist. Assoc.* 70, pp. 204–208.

Welham, S. J., and Thompson, R. (1997), *Likelihood ratio tests for fixed model terms using residual maximum likelihood*, in *J. Roy. Statist. Soc. B* 59, pp. 701–714.

White, H. (1982), *Maximum likelihood estimation of misspecified models*, in *Econometrika* 50, pp. 1–25.

Yan, G. and Sedransk, J. (2007), *Bayesian diagnostic techniques for detecting hierarchical structure*, in *Bayesian Anal.* 2, pp. 735–760.

Yan, G. and Sedransk, J. (2010), *A note on Bayesian residuals as a hierarchical model diagnostic technique*, in *Statist. Papers* 51, pp. 1–10.

Ye, J. (1998), *On measuring and correcting the effects of data mining and model selection*, in *J. Amer. Statist. Assoc.* 93, pp. 120–131.

You, Y. and Rao, J. N. K. (2002), *A pseudo-empirical best linear unbiased prediction approach to small area estimation using survey weights*, in *Canadian J. Statist.* 30, pp. 431–439.

Zhan, H., Chen, X., and Xu, S. (2011), *A stochastic expectation and maximization algorithm for detecting quantitative trait-associated genes*, in *Bioinformatics* 27, pp. 63–69.

Zheng, X., and Loh, W.-Y. (1995), *Consistent variable selection in linear models*, in *J. Amer. Statist. Assoc.* 90, pp. 151–156.

Zou, H. (2006), *The adaptive Lasso and its oracle properties*, in *J. Amer. Statist. Assoc.* 101, pp. 1418–1429.

Index

Printed in the United States
By Bookmasters